T0279349

Multi-Criteria Decision Analysis

Multi-Criteria Decision Analysis

Methods and Software

Alessio Ishizaka

Reader in Decision Analysis, Portsmouth Business School
University of Portsmouth, UK

Philippe Nemery

Senior Research Scientist, SAP Labs – China, Shanghai, PRC

This edition first published 2013

Registered office
John Wiley & Sons Ltd, The Atrium, Southern Gate, Chichester, West Sussex, PO19 8SQ, United Kingdom

For details of our global editorial offices, for customer services and for information about how to apply for permission to reuse the copyright material in this book please see our website at www.wiley.com.

Library of Congress Cataloging-in-Publication Data

Ishizaka, Alessio.
Multi-criteria decision analysis : methods and software / Alessio Ishizaka, Philippe Nemery.
pages cm
Includes bibliographical references and index.
ISBN 978-1-119-97407-9 (cloth)
1. Multiple criteria decision making. 2. Multiple criteria decision making–Data processing. 3. Decision support systems. I. Nemery, Philippe. II. Title.
T57.95.I84 2013
003′.56–dc23

2013004490

A catalogue record for this book is available from the British Library.

ISBN: 978-1-119-97407-9

Typeset in 10/12pt Times by Aptara Inc., New Delhi, India

1 2013

Contents

Foreword

The growing recognition that decision makers will often try to achieve multiple, and usually conflicting, objectives has led during the last three decades to the development of multi-criteria decision analysis (MCDA). This is now a vast field of research, with its scientific community and its specialized journals, as well as a large and growing number of real-world applications, for supporting both public policy making and decisions by private corporations.

Students and practitioners coming to the field, however, will be surprised by the plethora of alternative methods, overloaded by the array of software available, and puzzled by the diversity of approaches that an analyst needs to choose from. For precisely these reasons, this book is a very welcome event for the field. Alessio Ishizaka and Philippe Nemery have managed to provide an accessible, but rigorous, introduction to the main existing MCDA methods available in the literature.

There are several features of the book that are particularly innovative. First, it provides a balanced assessment of each method, and positions them in terms of the type of evaluation that the decision requires (a single choice among alternatives, the ranking of all alternatives, the sorting of alternatives into categories, or the description of consequences) and the level of preference information that each method requires (from utility functions to no preference information). This taxonomy helps both researchers and practitioners in locating adequate methods for the problems they need to analyze.

Second, the methods are presented with the right level of formulation and axiomatization for an introductory course. This makes the book accessible to anyone with a basic quantitative background. Readers who wish to learn in greater depth about a particular method can enjoy the more advanced content covered 'in the black box' of each chapter.

Third, the book illustrates each method with widely available and free software. This has two major benefits. Readers can easily see how the method works in practice via an example, consolidating the knowledge and the theoretical content. They can also reflect on how the method could be used in practice, to facilitate real-world decision-making processes.

Fourth, instructors using the book, as well as readers, can benefit from the companion website (`www.wiley.com/go/multi_criteria_decision_analysis`) and the availability of software files and answers to exercises.

This book should therefore be useful reading for anyone who wants to learn more about MCDA, or for those MCDA researchers who want to learn more about other MCDA methods and how to use specialized software to support multi-criteria decision making.

Gilberto Montibeller
Department of Management
London School of Economics

Acknowledgements

We are indebted to Kimberley Perry for her patience and constructive feedback while reviewing the manuscript. We would like to thank Ian Stevens and Alfred Quintano, who proofread a chapter.

We wish to express our sincere gratitude to Prof. Roman Słowiński, Poznań University of Technology; Dawid Opydo, BS Consulting Dawid Opydo; Tony Kennedy, Ventana Systems UK; Prof. Boris Yatsalo and Dr Sergey Gritsyuk for their suggestions.

We are grateful to the following organizations which granted us the permission to reproduce screenshots of their software: BS Consulting Dawid Opydo, Creative Decision Foundations, Ventana Systems UK Ltd, Lamsade Université Paris-Dauphine, Poznań University of Technology, BANA Consulting Lda, Smart-Picker, Obninsk State Technical University of Nuclear Power Engineering, Prof. Tim Coelli (The University of Queensland), Prof. Michel Deslierres (Univeristé de Moncton).

Last, but not least, we would like to thank all our students who have provided us with constant feedback and new ideas.

1

General introduction

1.1 Introduction

People face making decisions both in their professional and private lives. A manager in a company, for example, may need to evaluate suppliers and develop partnerships with the best ones. A household may need to choose an energy supplier for their family home. Students cannot ignore university rankings. Often candidates for a job vacancy are 'ranked' based on their experience, performance during the interview, etc.

As well as ranking and choice problems, there are also classification problems that have existed since classical times. In the fourth century BC, the ancient Greek philosopher Epicurus arranged human desires into two classes: vain desires (e.g. the desire for immortality) and natural desires (e.g. the desire for pleasure). These classifications were supposed to help in finding inner peace. Nowadays, classification problems occur naturally in daily life. A doctor, for instance, diagnoses a patient on the basis of their symptoms and assigns them to a pathology class to be able to prescribe the appropriate treatment. In enterprise, projects are often sorted into priority-based categories. Not long ago, a study showed that over 20 million Brazilians have moved from the lower social categories (D and E) to category C, the first tier of the middle class, and are now active consumers due to an increase in legal employment (Observador 2008). Hurricanes or cyclones are sorted into one of the five Saffir–Simpson categories based on their wind speed, superficial pressure and tide height.

All of these examples show that delicate decision problems arise frequently. Decision problems such as ranking, choice and sorting problems are often complex as they usually involve several criteria. People no longer consider only one criterion (e.g. price) when making a decision. To build long-term relationships, make sustainable and environmentally friendly decisions, companies consider multiple criteria in their decision process.

Multi-Criteria Decision Analysis: Methods and Software, First Edition. Alessio Ishizaka and Philippe Nemery.

Table 1.1 Category of decision problems.

Decision	Time perspective	Novelty	Degree of structure	Automation
Strategic	long term	new	low	low
Tactical	medium term	adaptive	semi-structured	middle
Operational	short term	every day	well defined	high

Most of the time, there is no one, perfect option available to suit all the criteria: an 'ideal' option does not usually exist, and therefore a compromise must be found. To address this problem the decision maker can make use of naïve approaches such as a simple weighted sum. The weighted sum, described in Section 4.3.1, is a special case of a more complex method and can only be applied with the right precautions (correct normalization phase, independent criteria, etc.) to enable sensible outputs. In reality, this approach is unrefined as it assumes linearity of preferences which may not reflect the decision maker's preferences. For example, it cannot be assumed that a wage of £4000 is twice as good as one of £2000. Some people would see their utility of preference improved by a factor of 5 with a wage of £4000. This cannot always be modelled with a weighted sum.

Multi-criteria decision analysis (MCDA) methods have been developed to support the decision maker in their unique and personal decision process. MCDA methods provide stepping-stones and techniques for finding a compromise solution. They have the distinction of placing the decision maker at the centre of the process. They are not automatable methods that lead to the same solution for every decision maker, but they incorporate subjective information. Subjective information, also known as preference information, is provided by the decision maker, which leads to the compromise solution.

MCDA is a discipline that encompasses mathematics, management, informatics, psychology, social science and economics. Its application is even wider as it can be used to solve any problem where a significant decision needs to be made. These decisions can be either tactical or strategic, depending on the time perspective of the consequences (Table 1.1).

A large number of methods have been developed to solve multi-criteria problems. This development is ongoing (Wallenius et al. 2008) and the number of academic MCDA-related publications is steadily increasing. This expansion is among others due to both the efficiency of researchers and the development of specific methods for the different types of problem encountered in MCDA. The software available, including spreadsheets containing method computations, *ad hoc* implementations, off-the-shelf, web or smartphone applications, has made MCDA methods more accessible and contributed to the growth in use of MCDA methods amongst researchers and the user community.

The aim of this book is to make MCDA methods even more intelligible to *novice* users such as students, or practitioners, but also to confirmed researchers. This book is ideal for people taking the first step into MCDA or specific MCDA methods. The cases studies and exercises effectively combine the mathematical and practical approach. For each method described in this book, an intuitive explanation and interpretation of the method is set out, followed by a detailed description of the software best suited to the method. Free or free trial version software has been intentionally chosen, as it allows the reader to better understand the main ideas behind the methods by practising with the exercises in this book. Furthermore, the user has access to a Microsoft Excel spreadsheet containing an 'implementation' of each method. Software files and answers to the exercises can be downloaded from the companion website, indicated by the icon in the book. The selected software and exercises allow the user to observe the impact of changes to the data on the results. The use of software enables the decision maker or analyst to communicate and justify decisions in a systematic way.

Each chapter contains a section ('In the black box') where scientific references and further reading are indicated for those interested in a more in-depth description or detailed understanding of the methods. Each chapter concludes with extensions of the methods to *other* decision problems, such as group decision or sorting problems.

This first chapter describes the different type of decision problems to be addressed in this book. This is followed by the introduction of the MCDA method best suited to solving these problems along with the corresponding software implementation. As several methods can solve similar problems, a section devoted to choosing an appropriate method has also been included. The chapter concludes with an outline of the book.

1.2 Decision problems

On any one day people face a plethora of different decisions. However, Roy (1981) has identified four main types of decision:

1. The choice problem. The goal is to select the single best option or reduce the group of options to a subset of equivalent or incomparable 'good' options. For example, a manager selecting the right person for a particular project.
2. The sorting problem. Options are sorted into ordered and predefined groups, called categories. The aim is to then regroup the options with similar behaviours or characteristics for descriptive, organizational or predictive reasons. For instance, employees can be evaluated for classification into different categories such as 'outperforming employees', 'average-performing employees' and 'weak-performing emplyees'. Based on these classifications, necessary measures can be taken. Sorting methods are useful for repetitive or automatic use. They can also be used as an initial screening to reduce the number of options to be considered in a subsequent step.
3. The ranking problem. Options are ordered from best to worst by means of scores or pairwise comparisons, etc. The order can be partial if incomparable options are considered, or complete. A typical example is the ranking of universities according to several criteria, such as teaching quality, research expertise and career opportunities.

4. The description problem. The goal is to describe options and their consequences. This is usually done in the first step to understand the characteristics of the decision problem.

Additional problem types have also been proposed in the MCDA community:

5. Elimination problem. Bana e Costa (1996) proposed the elimination problem, a particular branch of the sorting problem.
6. Design problem. The goal is to identify or create a new action, which will meet the goals and aspirations of the decision maker (Keeney 1992)

To this list of problems the 'elicitation problem' can be added as it aims to elicit the preference parameters (or subjective information) for a specific MCDA method. Moreover, when the problem involves several decision makers, an appropriate group decision method needs to be used.

Many other decision problems exist, often combining several of the problems listed above. However, this book concentrates on the first four decision problems and presents extensions of some of the methods that allow, for example, group, elicitation and description problems also to be addressed.

1.3 MCDA methods

To solve the problems defined in the previous section, *ad hoc* methods have been developed. In this book, the most popular MCDA methods are described along with their variants. Table 1.2 presents these methods and the decision problems they solve. There are many more decision methods than those presented in Table 1.2, but this book confines itself to the most popular methods that have a supporting software package.

Table 1.2 MCDA problems and methods.

Chapter	Choice problems	Ranking problems	Sorting problems	Description problems
2	AHP	AHP	AHPSort	
3	ANP	ANP		
4	MAUT/UTA	MAUT/UTA	UTADIS	
5	MACBETH	MACBETH		
6	PROMETHEE	PROMETHEE	FlowSort	GAIA, FS-Gaia
7	ELECTRE I	ELECTRE III	ELECTRE-Tri	
8	TOPSIS	TOPSIS		
9	Goal Programming			
10	DEA	DEA		
11	Multi-methods platform that supports various MCDA methods			

Table 1.3 MCDA software programs.

Problems	MCDA Methods	Software
Ranking, description, choice	PROMETHEE – GAIA	Decision Lab, D-Sight, **Smart Picker Pro**, Visual Promethee
Ranking, choice	PROMETHEE	DECERNS
	ELECTRE	Electre IS, **Electre III-IV**
	UTA	**Right Choice**, UTA+, DECERNS
	AHP	**MakeItRational**, ExpertChoice, Decision Lens, HIPRE 3+, RightChoiceDSS, Criterium, EasyMind, Questfox, ChoiceResults, 123AHP, DECERNS
	ANP	**Super Decisions**, Decision Lens
	MACBETH	**M-MACBETH**
	TOPSIS	DECERNS
	DEA	**Win4DEAP**, Efficiency Measurement System, DEA Solver Online, DEAFrontier, DEA-Solver PRO, Frontier Analyst
Choice	Goal Programming	-
Sorting, description	FlowSort - FS-GAIA	**Smart Picker Pro**
Sorting	ELECTRE-Tri	**Electre Tri**, IRIS
	UTADIS	-
	AHPSort	-

1.4 MCDA software

Researchers and commercial companies have developed various software programs over the last decade to help users structure and solve their decision problems. The aim of this book is not to describe all existing software, but to narrow the list down to the packages that apply to the methods described. A non-exhaustive list of the programs available is given in Table 1.3. The software packages represented in this book are in bold. Let us remark that the user has access to all the Microsoft Excel spreadsheets on the companion website.

1.5 Selection of MCDA methods

Considering the number of MCDA methods available, the decision maker is faced with the arduous task of selecting an appropriate decision support tool, and often the choice can be difficult to justify. None of the methods are perfect nor can they be applied to all problems. Each method has its own limitations, particularities, hypotheses, premises and perspectives. Roy and Bouyssou (1993) say that 'although the great diversity of MCDA procedures may be seen as a strong point, it can also be a weakness. Up to now, there has been no possibility of deciding whether one method makes more sense than another in a specific problem situation. A systematic axiomatic analysis of decision procedures and algorithms is yet to be carried out.'

Guitouni et al. (1999) propose an initial investigative framework for choosing an appropriate multi-criteria procedure; however, this approach is intended for experienced researchers. The next paragraphs give some guidance on selecting an appropriate method according to the decision problem, which will avoid an arbitrary adoption process.

There are different ways of choosing appropriate MCDA methods to solve specific problems. One way is to look at the required input information, that is, the data and parameters of the method and consequently the modelling effort, as well as looking at the outcomes and their granularity (Tables 1.4 and 1.5). This approach is supported by Guitouni et al. (1999).

If the 'utility function' for each criterion (a representation of the perceived utility given the performance of the option on a specific criterion) is known, then MAUT (Chapter 4) is recommended. However, the construction of the utility function requires a lot of effort, but if it is too difficult there are alternatives. Another way is by using pairwise comparisons between criteria and options. AHP (Chapter 2) and MACBETH (Chapter 5) support this approach. The difference is that comparisons are evaluated on a ratio scale for AHP and on an interval scale for MACBETH. The decision maker needs to know which scale is better suited to yield their preferences. The drawback is that a large quantity of information is needed.

Another alternative way is to define key parameters. For example, PROMETHEE (Chapter 6) only requires indifference and preference thresholds, whilst ELECTRE (Chapter 7) requires indifference, preference and veto thresholds. There exist so-called elicitation methods to help defining these parameters, but if the user wants to avoid those methods or parameters, TOPSIS (Chapter 8) can be used because only ideal and anti-ideal options are required. If criteria are dependent, ANP (Chapter 3) or the Choquet integral[1] can be used.

The modelling effort generally defines the richness of the output. One advantage to defining utility functions is that the options of the decision problem have a global score. Based on this score, it is possible to compare all options and rank them from best to worst, with equal rankings permitted. This is defined as a complete ranking. This approach is referred to as the full aggregation approach where a bad score on one criterion can be compensated by a good score on another criterion.

Outranking methods are based on pairwise comparisons. This means that the options are compared two-by-two by means of an outranking or preference degree. The preference or outranking degree reflects how much better one option is than

[1] This method has not been described in this book because it is not supported by a software package.

Table 1.4 Required inputs for MCDA ranking or choice method.

	Inputs	Effort input	MCDA method	Output
Ranking/choice problem	utility function	Very HIGH ↑	**MAUT**	Complete ranking with scores
	pairwise comparisons on a ratio scale and interdependencies		**ANP**	Complete ranking with scores
	pairwise comparisons on an interval scale		**MACBETH**	Complete ranking with scores
	pairwise comparisons on a ratio scale		**AHP**	Complete ranking with scores
	indifference, preference and veto thresholds		**ELECTRE**	Partial and complete ranking (pairwise outranking degrees)
	indifference and preference thresholds		**PROMETHEE**	Partial and complete ranking (pairwise preference degrees and scores)
	ideal option and constraints		**Goal programming**	Feasible solution with deviation score
	ideal and anti-ideal option		**TOPSIS**	Complete ranking with closeness score
	no subjective inputs required	↓ Very LOW	**DEA**	Partial ranking with effectiveness score

Table 1.5 Required inputs for MCDA sorting methods.

	Inputs	Effort Input	MCDA method	Output
Sorting method	utility function	HIGH	UTADIS	Classification with scoring
	pairwise comparisons on a ratio scale	↑	AHPSort	Classification with scoring
	indifference, preference and veto thresholds	↓	ELECTRE-TRI	Classification with pairwise outranking degrees
	indifference and preference thresholds	LOW	FLOWSORT	Classification with pairwise outranking degrees and scores

another. It is possible for some options to be incomparable. The comparison between two options is difficult as they have different profiles: one option may be better based one set of criteria and the other better based on another set of criteria. These incomparabilities mean that a complete ranking is not always possible, which is referred to as a partial ranking. The incomparability is a consequence of the non-compensatory aspect of those methods. When facing a decision problem, it is important to define the type of output required from the beginning (presented in Tables 1.4 and 1.5).

Goal programming and data envelopment analysis (DEA) are also part of the MCDA family but are used in special cases. In goal programming, an ideal goal can be defined subject to feasibility constraints. DEA is mostly used for performance evaluation or benchmarking, where no subjective inputs are required.

1.6 Outline of the book

Following this introduction, in which general concepts of MCDA are explained, nine chapters describe the major MCDA methods. Each chapter can be read independently, and they are grouped into three sections, according to their approach:

- *Full aggregation approach* (or American school). A score is evaluated for each criterion and these are then synthesized into a global score. This approach assumes compensable scores, i.e. a bad score for one criterion is compensated for by a good score on another.
- *Outranking approach* (or French school). A bad score may not be compensated for by a better score. The order of the options may be partial because the notion of incomparability is allowed. Two options may have the same score, but their behaviour may be different and therefore incomparable.

- *Goal, aspiration or reference level approach.* This approach defines a goal on each criterion, and then identifies the closest options to the ideal goal or reference level.

Most chapters are divided into four sections, with the exception of specific MCDA methods, as extensions do not exist. Specific objectives are as follows:

- *Essential concepts.* The reader will be able to describe the essentials of the MCDA method.

- *Software.* The reader will be able to solve MCDA problems using the corresponding software.

- *In the black box.* The reader will understand the calculations behind the method. An exercise in Microsoft Excel facilitates this objective.

- *Extensions.* The reader will be able to describe the extensions of the MCDA methods to other decision problems, such as sorting or group decisions.

The book concludes with a description of the integrated software DECERNS, which incorporates six MCDA methods and a Geographical Information System. Linear programming, the underlying method for MACBETH and goal programming, is explained in the Appendix.

References

Bana e Costa, C. (1996). Les problématiques de l'aide à la décision: Vers l'enrichissement de la trilogie choix–tri–rangement. *RAIRO – Operations Research*, *30*(2), 191–216.

Guitouni, A., Martel, J., and Vincke, P. (1999). A framework to choose a discrete multicriterion aggregation procedure. *Technical Report.*

Keeney, R. (1992). *Value-Focused Thinking: A Path to Creative Decision Making.* Cambridge, MA: Harvard University Press.

Observador (2008). The growth of class 'C' and its electoral importance. *Observador*, 31 March, p. 685.

Roy, B. (1981). The optimisation problem formulation: Criticism and overstepping. *Journal of the Operational Research Sociey*, *32*(6), 427–436.

Roy, B., and Bouyssou, D. (1993). *Aide multicritère à la décision: Méthodes et cas.* Paris: Economica.

Wallenius, J. D., Dyer, J.S., Fishburn, P.C., Steuer, R.E., Zionts, S., and Deb, K. (2008). Multiple criteria decision making, multiattribute utility theory: Recent accomplishments and what lies ahead. *Management Science*, *54*(7), 1336–1349.

Part I

FULL AGGREGATION APPROACH

2

Analytic hierarchy process

2.1 Introduction

This chapter explains the theory behind and practical uses of the analytic hierarchy process (AHP) method as well as its extensions. *MakeItRational*, a software package that helps to structure problems and calculate priorities using AHP, is described. Section 2.3 is designed for readers interested in the methodological background of AHP. Section 2.4 covers the extensions of AHP in group decision, sorting, scenarios with incomparability and large size problems.

The companion website provides illustrative examples with *Microsoft Excel*, and case studies and examples with *MakeItRational*.

2.2 Essential concepts of AHP

AHP was developed by Saaty (1977, 1980). It is a particularly useful method when the decision maker is unable to construct a utility function, otherwise MAUT is recommended (Chapter 4). To use AHP the user needs to complete four steps to obtain the ranking of the alternatives. As with any other MCDA method, the problem first has to be structured (Section 2.2.1). Following this, scores – or priorities, as they are known in AHP – are calculated based on the pairwise comparisons provided by the user (Section 2.2.2). The decision maker does not need to provide a numerical judgement; instead a relative verbal appreciation, more familiar to our daily live, is sufficient. There are two additional steps that can be carried out: a consistency check (Section 2.2.3) and a sensitivity analysis (Section 2.2.4). Both steps are optional but recommended as confirmation of the robustness of the results. The consistency check is common in all methods based on pairwise comparisons like AHP. The supporting software of *MakeItRational* facilitates the sensitivity analysis.

Multi-Criteria Decision Analysis: Methods and Software, First Edition. Alessio Ishizaka and Philippe Nemery.

2.2.1 Problem structuring

AHP is based on the motto *divide and conquer*. Problems that require MCDA techniques are complex and, as a result, it is advantageous to break them down and solve one 'sub-problem' at a time. This breakdown is done in two phases of the decision process during:

- the problem structuring and
- the elicitation of priorities through pairwise comparisons.

The problem is structured according to a hierarchy (e.g. Figure 2.2) where the top element is the goal of the decision. The second level of the hierarchy represents the criteria, and the lowest level represents the alternatives. In more complex hierarchies, more levels can be added. These additional levels represent the sub-criteria. In any case, there are a minimum of three levels in the hierarchy.

Throughout this chapter, a shop location problem (Case Study 2.1) will be considered to illustrate the different steps of the AHP process.

Case Study 2.1

A businessman wants to open a new sports shop in one of three different locations:

(a) **A shopping centre.** The shopping centre has a high concentration of a variety of shops and restaurants. It is a busy area, with a mix of customers and people walking around. Shops regularly use large displays and promotions to attract potential customers. As demand for these retail units is low, the rental costs are reasonable.

(b) **The city centre.** The city centre is a busy area, and a meeting point for both young people and tourists. Attractions such as dance shows, clowns and market stalls are often organized, which attract a variety of visitors. The city centre has several small shops located at ground level in historical buildings, which suggests high rental costs. These shops have a high number of customers and are often in competition.

(c) **A new industrial area.** The new industrial estate is in the suburbs of the city, where several businesses have recently been set up. Some buildings have been earmarked for small shops, but on the whole it has been difficult to attract tenants, which means that rental costs are currently low. Customers of the existing shops mainly work in the area and only a few customers come from the surrounding towns or cities to shop here.

Given the description of the problem, four criteria will be considered in making the final decision (Table 2.1).

Table 2.1 Criteria for shop location decision.

Criterion	Explanation
Visibility	Probability that a random passer-by notices the shop
Competition	Level of competition in the area
Frequency	Average number of customers in similar shops in the area
Rental cost	Average rental cost by square metre

Figure 2.1 represents the hierarchy of Case Study 2.1. It has three levels, the minimum required to solve a problem with AHP. Other sub-criteria could be considered, for example, the competition criterion could be broken down into two sub-criteria: direct and indirect competition. Direct competition would be the number of other sports shops. Indirect competition would represent other types of shop, which could distract potential customers. To keep the example simple, additional levels will not be considered at this stage.

Each lower level is prioritized according to its immediate upper level. The appropriate question to ask with regard to prioritization depends on the context and sometimes on the decision maker. For example, in order to prioritize the criteria of level 2 with regard to the goal 'location of a sports shop', an appropriate question would be: 'Which criterion is most important for choosing the location of the sports shop and to what extent?' On the other hand, the alternatives in level 3 must be prioritized with regard to each criterion in level 2. In this case, an appropriate question would

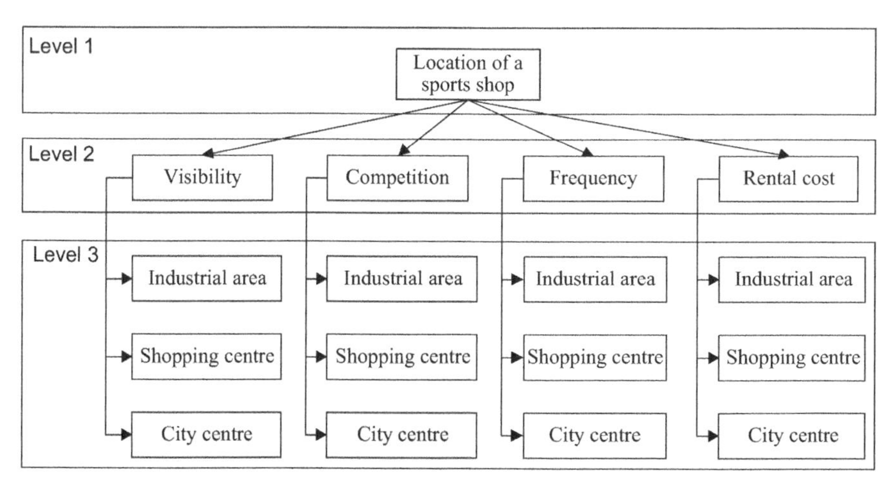

Figure 2.1 Hierarchy of decision levels for Case Study 2.1.

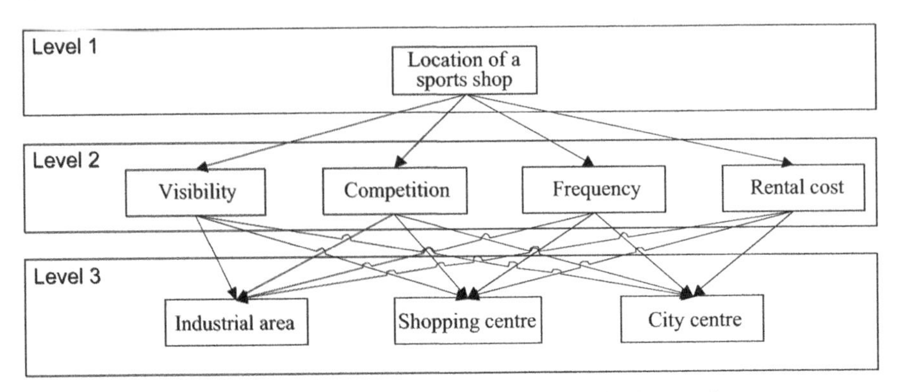

Figure 2.2 Traditional representation of the hierarchy.

be: 'Which alternative is preferable to fulfil the given criterion and to what extent?' In Case Study 2.1, five different prioritizations are required:

- four local prioritizations of alternatives with regard to each criterion and
- one criteria prioritization.

The aggregation of the local and criteria prioritizations leads to global prioritizations.

As Figure 2.1 contains redundant information at the lowest level, the alternatives in the hierarchy are often not repeated or are connected as in Figure 2.2.

2.2.2 Priority calculation

A priority is a score that ranks the importance of the alternative or criterion in the decision. Following the problem-structuring phase (see Section 2.2.1), three types of priorities have to be calculated:

- **Criteria priorities**. Importance of each criterion (with respect to the top goal).
- **Local alternative priorities**. Importance of an alternative with respect to one specific criterion.
- **Global alternative priorities**. Priority criteria and local alternative priorities are intermediate results used to calculate the global alternative priorities. The global alternative priorities rank alternatives with respect to all criteria and consequently the overall goal.

The criteria and local alternatives priorities are calculated using the same technique. Instead of directly allocating performances to alternatives (or criteria) as in the other techniques from the American school (see MAUT, Chapter 4), AHP uses

Table 2.2 The 1–9 fundamental scale.

Degree of importance	Definition
1	**Equal importance**
2	Weak
3	**Moderate importance**
4	Moderate plus
5	**Strong importance**
6	Strong plus
7	**Very strong or demon-strated importance**
8	Very, very strong
9	**Extreme importance**

pairwise comparisons. Psychologists often use this technique (Yokoyama 1921; Thurstone 1927), for example, to evaluate the food preference of a cat by presenting two dishes at a time. The cat indicates its preference by eating one dish. The psychologists argue that it is easier and more accurate to express a preference between only two alternatives than simultaneously among all the alternatives. The use of pairwise comparisons (called paired comparisons by psychologists) is generally evaluated on the fundamental 1–9 scale. The conversion from verbal to numerical scale is given in Table 2.2. Psychologists suggest that a smaller scale, say 1–5, would not give the same level of detail in a data set, and that the decision maker would be lost in a larger scale: for example, on a scale of 1–100, it is difficult for the decision maker to distinguish between a score of 62 and 63. In practice, there is no fixed rule and other scales have been proposed (Section 2.4.2).

The comparisons are collected in a matrix (Example 2.1).

Example 2.1 The comparison matrix in Figure 2.3 gathers the pairwise comparisons between the criteria. All comparisons are positive. The comparisons on the main diagonal are 1 because a criterion is compared with itself. The matrix is reciprocal because the upper triangle is the reverse of the lower triangle, for example visibility is 1/4 as important as competition and competition is 4 times as important as visibility.

The advantage of precision requires more effort, especially when there are a large number of criteria or alternatives. The number of necessary comparisons for each comparison matrix is

$$\frac{n^2 - n}{2} \tag{2.1}$$

	Visibility	Competition	Frequency	Rental costs
Visibility	**1**	1/4	1/5	2
Competition	4	**1**	1/2	1
Frequency	5	2	**1**	4
Rental costs	1/2	1	1/4	**1**

Figure 2.3 Comparison matrix.

where n is the number of alternatives/criteria. This formula can be explained as follows:

- n^2 is the total number of comparisons that can be written in a matrix.
- n of these represent the comparison of the alternative with itself (on the main diagonal). The evaluation is 1 and therefore not required (shown in bold in Figure 2.3).
- As the matrix is reciprocal, only half of the comparisons are required. The other half are automatically calculated from the first half.

For example, in Figure 2.3 we have $n = 4$, therefore the number of comparisons to provide is $(4^2 - 4)/2 = 6$.

Even though the squared number is reduced by n and divided by 2, the required number of comparisons can be very high. For example, 10 alternatives lead to 45 questions for each criterion. The effort required to complete the matrix is time-consuming and can be discouraging. In Section 2.5.3, ways to deal with this quadratic increase in the number of comparisons will be discussed.

From these comparison matrices, the software will calculate the local and criteria priorities; see Section 2.4.4, where the calculation of these priorities is explained. Finally, it aggregates these two priorities to establish the global priority. Priorities only make sense if they are derived from consistent or near-consistent matrices, and as a result a consistency check must be performed, to which we now turn.

2.2.3 Consistency check

When the matrix is complete, a consistency check may be performed to detect possible contradictions in the entries. When several successive pairwise comparisons are presented, they may contradict each other. The reasons for these contradictions could be, for example, vaguely defined problems, a lack of sufficient information (known as bounded rationality), uncertain information or lack of concentration. Suppose that the decision maker, as an example, gives the following pairwise comparisons:

- The shopping centre is **two times** more visible than the city centre.
- The city centre is **three times** more visible than the industrial area.
- The industrial area is **four times** more visible than the shopping centre.

The third assertion is inconsistent as determined from the two first assertions; the industrial area is six times more visible than the shopping centre (2×3). Human nature is often inconsistent, for example, in football it is possible for the team at the top of the table to lose against the team at the bottom of the table. To allow this inconsistent reality, AHP allows up to a 10% inconsistency compared to the average inconsistency of 500 randomly filled matrices. A calculation is done by the supporting software and indicates if a matrix needs to be reconsidered due to its high inconsistency (a detailed description is available in the black box Section 2.4.3).

2.2.4 Sensitivity analysis

The last step of the decision process is the sensitivity analysis, where the input data is slightly modified to observe the impact on the results. As complex decision models are often inherently ill defined, the sensitivity analysis allows different scenarios to be generated. These different scenarios may result in other rankings, and further discussion may be needed to reach a consensus. If the ranking does not change, the results are said to be *robust* – otherwise they are sensitive. The sensitivity analysis in *MakeItRational* is performed by varying the weight of the criteria and observing the impact on the global alternative priority.

Exercise 2.1

The following multiple-choice questions test your knowledge of the basics of AHP. Only one answer is correct. Answers can be found on the companion website.

1. What does AHP stand for?
   a) Analytic Hierarchy Program
   b) Analytic Hierarchy Process
   c) Analytic Hierarchical Programming
   d) Analytical Hierarchy Partitioning
2. What is the typical Saaty scale?
   a) A 1–5 scale
   b) A 1–9 scale
   c) A 1–10 scale
   d) A 1–100 scale
3. What is the main purpose of AHP?
   a) AHP prioritizes alternatives based on criteria and constraints
   b) AHP assigns goals to alternatives

c) AHP prioritizes alternatives based on criteria

d) AHP assigns criteria to alternatives

4. Pairwise comparisons in AHP are based on which scale?

a) Ratio scale

b) Interval scale

c) Ordinal scale

d) Nominal scale

5. How many pairwise comparisons are required to rank five criteria?

a) 25

b) 20

c) 15

d) 10

2.3 AHP software: MakeItRational

The available AHP software has greatly contributed to the success of the AHP method. The software incorporates intuitive graphical user interfaces, automatic calculation of priorities and inconsistencies, and provides several ways of processing a sensitivity analysis (Ishizaka and Labib 2009). At the time of writing there are several software packages: Expert Choice, Decision Lens, HIPRE 3+, RightChoiceDSS, Criterium, EasyMind, Questfox, ChoiceResults and 123AHP, as well as the option of adapting a template in *Microsoft Excel* (e.g. see Exercise 2.3).

This section describes the AHP web software *MakeItRational*, available from http://makeitrational.com. This software was chosen because of its simplicity and the free trial version available (Kaspar and Ossadnik 2013). The disadvantage of the free version is that models cannot be saved, but as *MakeItRational* is an online software package, it is automatically updated. Data is stored on the web server, although a server edition can be purchased, which allows the data to be saved on computer. When using *MakeItRational*, it is not necessary to know *how* priorities are calculated, only *what* should be ranked. This section describes the graphical user interface. The four steps introduced in Section 2.2 will be followed by navigating between the top tabs (Figure 2.5) of the software.

2.3.1 Problem structuring

For problem structuring, three tabs are necessary.

- Project tab: Name the project (this is needed to save it) and enter a description (optional).

- Alternatives tab: Enter a minimum of two alternatives.
- Criteria tab: Enter a minimum of two criteria.

2.3.2 Preferences and priority calculation

The Evaluation tab in Figure 2.4 displays the pairwise comparisons needed to calculate the priorities. The user first has to select the Goal in the left panel, and the right panel will ask for pairwise evaluations of the criteria. For example, in Figure 2.4, *Competition* has been evaluated as twice as important as *Frequency*. When this step is complete, the user will need to select the first criterion from the left panel, where again the right panel will ask for pairwise comparisons. In Figure 2.5, the *City centre* has been evaluated as 5 times as important as *the Industrial area* with regard to *Competition*. This process is repeated for each criterion.

MakeItRational allows a direct rating of the alternatives/criteria if they are already known. For example, in Figure 2.6, the exact frequency of people per hour for each alternative is known; therefore the precise amount can be entered. Note that the criterion needs to be maximized. For criteria to minimize, the score needs to be inverted, for example x becomes $1/x$. If all evaluation preferences are rated directly, then the weighted sum is used. In this case, the decision maker needs to know the utility function either implicitly or explicitly. *MakeItRational* is able to support both methods because they share several common features.

Priorities will be automatically calculated by *MakeItRational* after a consistency check.

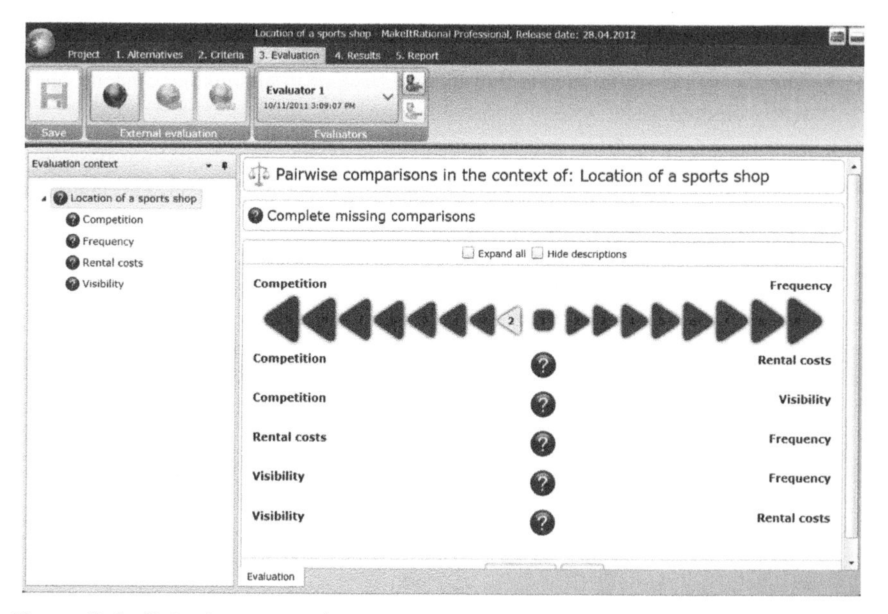

Figure 2.4 Pairwise comparisons of criteria in MakeItRational. *Reproduced by permission of BS Consulting Dawid Opydo.*

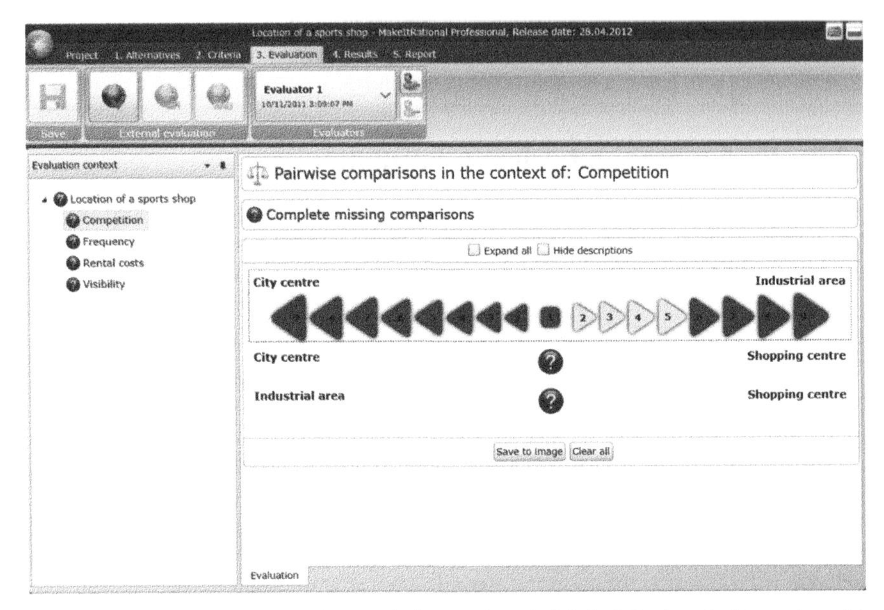

Figure 2.5 Pairwise comparisons of alternatives in MakeItRational. *Reproduced by permission of BS Consulting Dawid Opydo.*

2.3.3 Consistency check

MakeItRational has various consistency checks represented by the icons on the left pane of the tab (Figure 2.6). Table 2.3 explains the status.

- The *Complete* status means that all pairwise comparisons have been consistently entered.
- In the *Enough* status, not all pairwise comparisons are entered but those provided can be used to estimate the missing ones (Section 2.4.4.1). This status can be used when a large number of alternatives are evaluated in order to decrease the number of required pairwise comparisons. Therefore, comfortable comparisons should be entered first.

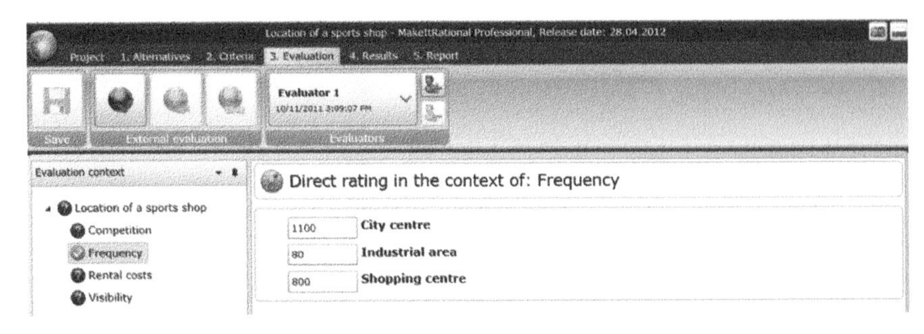

Figure 2.6 Direct rating of alternatives in MakeItRational. *Reproduced by permission of BS Consulting Dawid Opydo.*

Table 2.3 Preference status. Reproduced by permission of BS Consulting Dawid Opydo.

Icon	Status	Description
	Complete	All judgements in the context of this criterion have been provided. The entered pairwise comparisons are consistent (CR $< 10\%$).
	Enough	There are some empty judgements in the context of this criterion but weights/scores can be calculated.
	Inconsistency	The entered pairwise comparisons are inconsistent (CR $> 10\%$).
	Contradictory	Contradictory pairwise comparisons in the context of this criterion.
	Missing	There is not enough data to calculate weights/scores.
	Error	The decision problem contains only one criterion or one alternative.

- The *Inconsistency* status recommends the revision of pairwise comparisons in order to decrease inconsistency. The consistency ratio (CR) should be lower than 10% to be considered acceptable. *MakeItRational* will recommend which comparison to modify. For example, in Figure 2.7, the most inconsistent comparison is between visibility and rental costs. *MakeItRational* recommends modifying the comparison to 4.

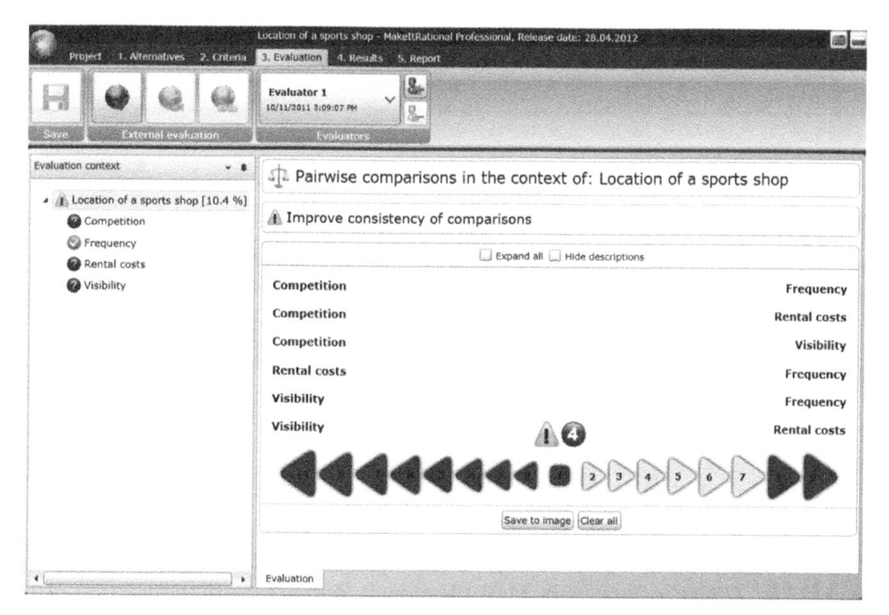

Figure 2.7 Inconsistency and recommended comparisons. Reproduced by permission of BS Consulting Dawid Opydo.

- The *Contradictory* status indicates logically impossible cardinal preferences. For example, I prefer the *shopping centre* to the *city centre*, I prefer the *city centre* to the *industrial area*, and I prefer the *industrial area* to the *shopping centre*, which induces an impossible preference cycle:

$$\textit{shopping centre} > \textit{city centre} > \textit{industrial area} > \textit{shopping centre}.$$

- The *Missing* status indicates that not enough data has been provided to calculate priorities.
- The *Error* status indicates an error in the problem structuring: a criterion contains only one sub-criterion or the problem contains only one alternative.

Priorities will be calculated for the first four matrix statuses, but it is strongly recommended to revise pairwise comparisons for the *Inconsistency* and *Contradictory* status.

2.3.4 Results

Figure 2.8 shows the global priorities of the alternatives with regard to the goal 'Location selection for a sports shop'. The results are displayed with scores and stacked bar diagrams for better visualization. It can be seen that the city centre is the preferred alternative, especially because of its high frequency. In the chart data

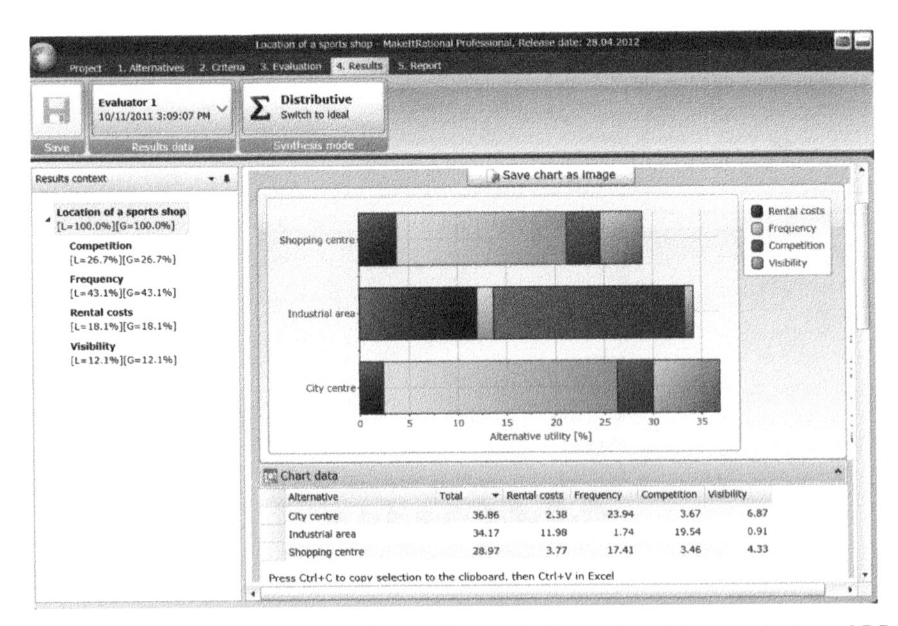

Figure 2.8 Global priorities in MakeItRational. *Reproduced by permission of BS Consulting Dawid Opydo.*

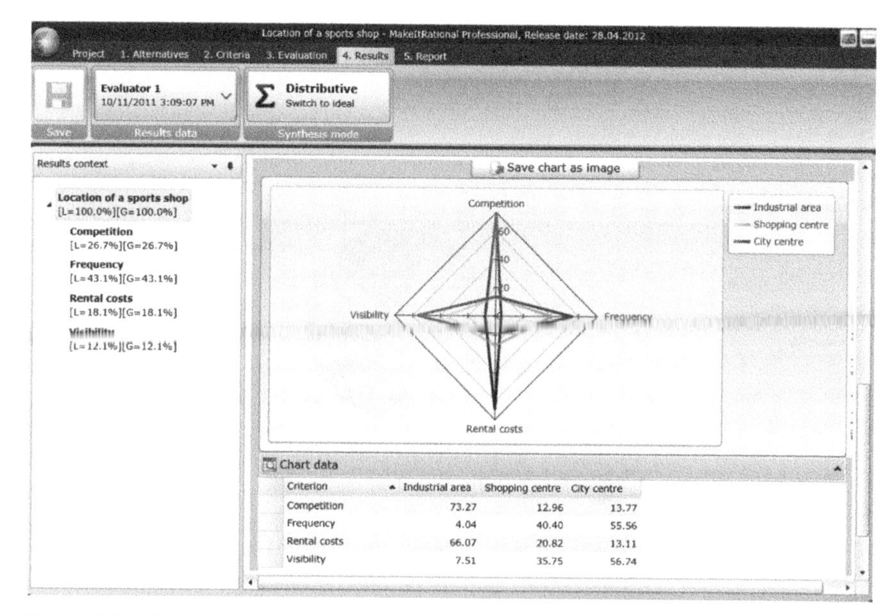

Figure 2.9 Local priority in MakeItRational. *Reproduced by permission of BS Consulting Dawid Opydo.*

of Figure 2.8, it can be seen that frequency contributes 23.94 towards the total score of 36.86.

MakeItRational allows the local (Figure 2.9) and criteria priorities (Figure 2.10) to be seen.

Figure 2.9 displays the unweighted local priorities in a spider diagram and the scores in the table immediately below. This representation allows a visualization of the strengths and weaknesses of each alternative. In this case, the industrial area is very strong on the competition criterion but very weak on the visibility and frequency criteria. The shopping centre scores very high on the frequency and visibility criteria. Figure 2.10 displays the criteria priorities in a pie chart and the scores in the table beneath it.

2.3.5 Sensitivity analysis

On the same Results tab, a sensitivity analysis in *MakeItRational* allows the impact of the changes of one criterion weight over the global priority to be seen. For example, in Figure 2.11, if the current rental costs weight of 18.1% is increased to over 22.35%, then the preferred alternative is no longer the city centre but the industrial area.

Finally, the results can be collected in a report and downloaded in different formats. An example of a report can be downloaded from the companion website (Report_MakeItRational.pdf).

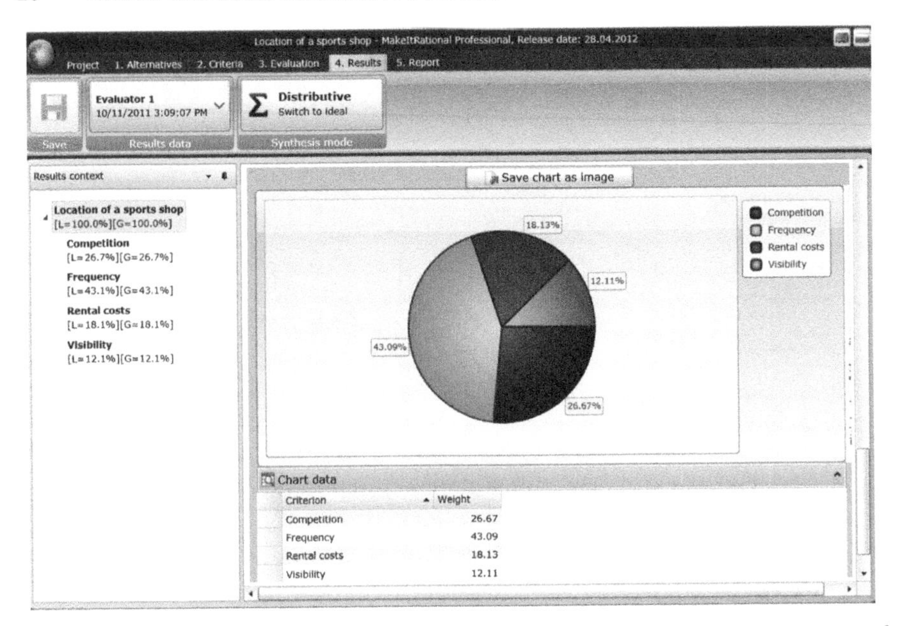

Figure 2.10 Criterion priorities in MakeItRational. *Reproduced by permission of BS Consulting Dawid Opydo.*

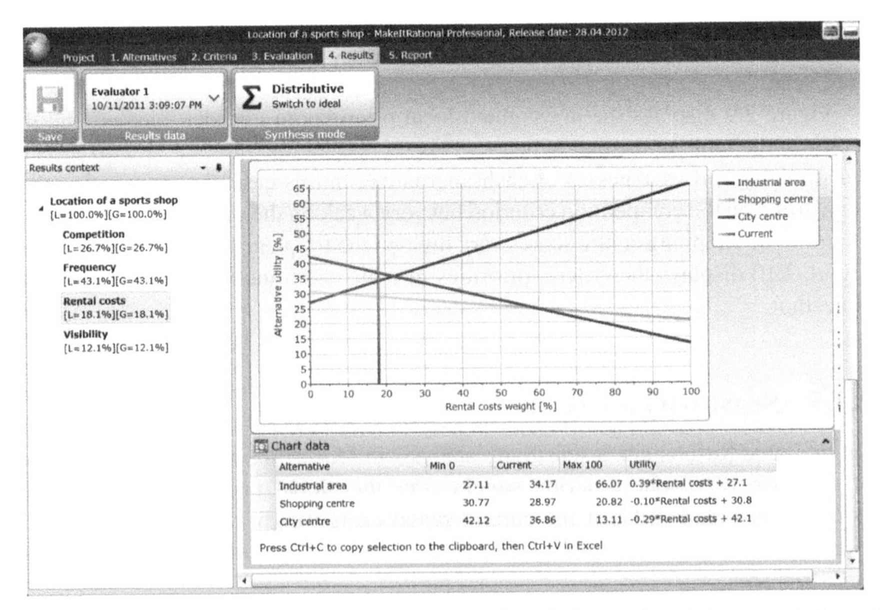

Figure 2.11 Sensitivity analysis in MakeItRational. *Reproduced by permission of BS Consulting Dawid Opydo.*

Exercise 2.2

In this exercise, the sports shop problem in Case Study 2.1 will be solved with the *MakeItRational* software.

Learning Outcomes

- Structure a problem in *MakeItRational*
- Enter pairwise comparisons
- Understand the results
- Conduct a sensitivity analysis

Tasks

a) Open the webpage http://makeitrational.com/demo. The free version has the full functionalities but the problem cannot be saved.

b) Read the description of Case Study 2.1, on page 14.

c) Give your decision project a name (*Project* tab).

d) Enter the alternatives (*Alternatives* tab).

e) Enter the criteria (*Criteria* tab).

f) Enter the pairwise comparisons (*Evaluation* tab). Are they consistent?

g) Read your global ranking and conduct a sensitivity analysis (*Results* tab).

2.4 In the black box of AHP

2.4.1 Problem structuring

In most cases, the problem is not as well defined as in Case Study 2.1. The decision maker may have a vague idea of wanting to open a shop but without knowing the precise alternatives and criteria. A structure must be formed through brainstorming sessions, analysing similar problem studies and organizing focus groups etc. Saaty and Forman (1992) have written a book describing hierarchical structures in various AHP applications, which may be of use in the structuring process.

This hierarchization of decision elements is important because a different structure may lead to a different final ranking. Several authors (Pöyhönen et al. 1997; Stillwell et al. 1987; Weber et al. 1988) have observed that criteria with a large number of sub-criteria tend to receive more weight than when they are less detailed.

Table 2.4 Food and drink quantities in two menus.

	Food [kg]	Drinks [l]
Menu A	0.80	1
Menu B	2	0.5

2.4.2 Judgement scales

The use of verbal comparisons (Table 2.2) is intuitively appealing, user-friendly and more common in our everyday lives than numbers. It may also allow for some fuzziness in difficult comparisons – a verbal comparison is not as precise as a number. However, this ambiguity in the English language has also been criticized (Donegan et al. 1992).

AHP, due to its pairwise comparisons, needs ratio scales, which, contrary to methods using interval scales (Kainulainen et al. 2009), require no units of comparison. The judgement is a relative value or a quotient a / b of two quantities a and b having the same units (intensity, utility, etc.). Barzilai (2005) claims that preferences cannot be represented with ratio scales, because in his opinion an absolute zero does not exist, for example, temperature or electrical tension. Similarly, Dodd and Donegan (1995) have criticized the absence of zero in the preference scale in Table 2.2. On the contrary, Saaty (1994a) states that ratio scales are the only possibility for aggregating measurements in a commensurate (i.e. same units) way (Example 2.2).

Example 2.2 Consider two lunch menus evaluated on two criteria of the quantity of food and quantity of drinks (Table 2.4). The food quantity is considered twice as important as the drinks quantity. The two menus can be compared on a ratio scale:

$$\frac{\text{Menu B}}{\text{Menu A}} = 2 \cdot \frac{2}{0.8} + \frac{0.5}{1} = 5.5.$$

Therefore, menu B is five and half times as good as menu A. On an interval scale,

$$\text{Menu B} - \text{Menu A} = 2 \cdot (2 - 0.8) + (0.5 - 1) = 1.9.$$

This result also indicates that menu B is better.

However, if the scale is changed from litres to decilitres, the results change for the interval scale,

$$\text{Menu B} - \text{Menu A} = 2 \cdot (2 - 0.8) + (5 - 10) = -2.6,$$

but not for the ratio scale,

$$\frac{\text{Menu B}}{\text{Menu A}} = 2 \cdot \frac{2}{0.8} + \frac{5}{10} = 5.5.$$

In order to correct this change, the weights should be adjusted as well:

$$\text{Menu B} - \text{Menu A} = 2 \cdot (2 - 0.8) + 0.1\,(5 - 10) = 1.9.$$

If mathematically the weight adjustment is feasible, can the decision maker be expected to adjust the weight preferences when a change on the scale is adopted?

To derive priorities, verbal comparisons must be converted to numerical ones. In Saaty's AHP the verbal statements are converted into integers 1–9. Theoretically, there is no reason to be restricted to these numbers and this verbal gradation. If the verbal gradation has been little investigated, various other numerical scales have been proposed (Table 2.5, Figure 2.12 and Figure 2.13). Harker and Vargas (1987) evaluated a quadratic and a square root scale in only one simple example and argued in favour of Saaty's 1–9 scale. However, one example is not enough to conclude the superiority of the 1–9 linear scale. Lootsma (1989) argued that the geometric scale is preferable to the 1–9 linear scale. Salo and Hämäläinen (1997) point out that the integers 1–9 yield local weights that are unevenly dispersed so that there is lack of sensitivity

Table 2.5 Different scales for comparing two alternatives.

Scale types	Equal importance	Weak	Moderate importance	Moderate plus	Strong importance	Strong plus	Very strong importance	Very very strong	Extreme importance
Linear (Saaty 1977)	1	2	3	4	5	6	7	8	9
Power (Harker and Vargas 1987)	1	4	9	16	25	36	49	64	81
Geometric (Lootsma 1989)	1	2	4	8	16	32	64	128	256
Logarithmic (Ishizaka et al. 2006)	1	1.58	2	2.32	2.58	2.81	3	3.17	3.32
Square root (Harker and Vargas 1987)	1	1.41	1.73	2	2.23	2.45	2.65	2.83	3
Asymptotical (Dodd and Donegan 1995)	0	0.12	0.24	0.36	0.46	0.55	0.63	0.70	0.76
Inverse linear (Ma and Zheng 1991)	1	1.13	1.29	1.5	1.8	2.25	3	4.5	9
Balanced (Salo and Hämäläinen 1997)	1	1.22	1.5	1.86	2.33	3	4	5.67	9

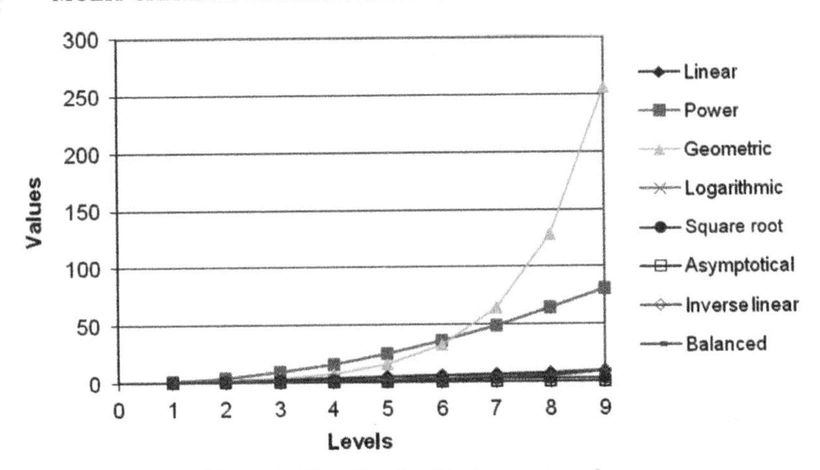

Figure 2.12 Graph of judgement scales.

when comparing elements which are preferentially close to each other. Based on this observation, they propose a balanced scale where the local weights are evenly dispersed over the weight range [0.1, 0.9]. Earlier, Ma and Zheng (1991) calculated a scale where the inverse elements x of the scale $1/x$ are linear instead of the x in the Saaty scale. Donegan et al. (1992) proposed an asymptotic scale avoiding the boundary problem (e.g. if the decision maker enters the pairwise comparison $a_{ij} = 3$ and $a_{jk} = 4$, they are forced into an intransitive relation because the upper limit of the scale is 9 and they cannot enter $a_{ik} = 12$). Ji and Jiang (2003) propose a mixture of verbal scale and geometric scale. The possibility of integrating negative values into the scale has also been explored (Millet and Schoner 2005; Saaty and Ozdemir 2003a).

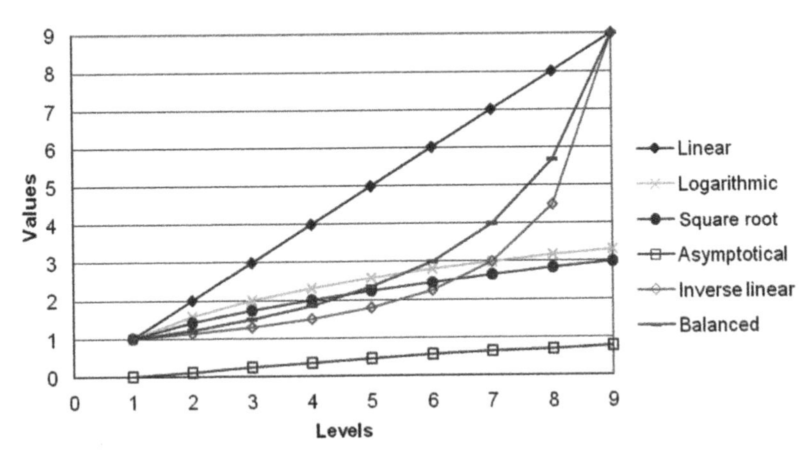

Figure 2.13 Graph of the judgement scales with the geometric and power scales omitted.

Figure 2.12 and Figure 2.13 represent graphically the different scales of Table 2.5. Large differences are noted, which imply different final results (Ishizaka et al. 2010).

Among the proposed scales, the linear scale with the integers 1–9 and their reciprocals have been used most often in applications. It is also the only one implemented in *Expert Choice* and *MakeItRational*. Saaty (1980, 1991) advocates it as the best scale to represent weight ratios. However, the cited examples deal with objective measurable alternatives like the areas of figures, whereas AHP treats mainly decision processes on subjective issues. It is technically much more difficult to verify the effectiveness of scales through subjective issues. Salo and Hämäläinen (1997) demonstrate the superiority of the balanced scale when comparing two elements. The choice of the 'best' scale is a hotly debated issue. Some scientists argue that the choice depends on the person and the decision problem (Harker and Vargas 1987; Pöyhönen et al. 1997). Therefore, other scales would be welcomed in the AHP software.

2.4.3 Consistency

In Section 2.2.3, minimal consistency was necessary to calculate meaningful priorities. A matrix filled by the pairwise comparison a_{ij} is called consistent if the transitivity (2.2) and the reciprocity (2.3) rules are respected.

Transitivity Rule:

$$a_{ij} = a_{ik} \cdot a_{kj} \tag{2.2}$$

where a_{ij} is the comparison of alternative i with j.

Suppose a person likes an apple twice as much as an orange ($a_{12} = 2$) and an orange three times as much as a banana ($a_{23} = 3$). If the person likes an apple six times as much as a banana ($a_{13} = 6$), the transitivity rule is respected.

Reciprocity Rule:

$$a_{ij} = \frac{1}{a_{ji}} \tag{2.3}$$

where i, j and k are any alternatives of the matrix.

If a person likes an apple twice as much as an orange ($a_{12} = 2$), then they like an orange half as much as an apple ($a_{21} = 1/2$).

If we suppose that preferences p_i are known, a perfectly consistent matrix

$$\mathbf{A} = \begin{bmatrix} p_1/p_1 & \cdots & p_1/p_j & \cdots & p_1/p_n \\ \cdots & 1 & \cdots & \cdots & \cdots \\ p_i/p_1 & \cdots & 1 & \cdots & p_i/p_n \\ \cdots & \cdots & \cdots & 1 & \cdots \\ p_n/p_1 & \cdots & p_n/p_j & \cdots & p_n/p_n \end{bmatrix} \tag{2.4}$$

can be constructed because all the comparisons a_{ij} obey the equality

$$a_{ij} = \frac{p_i}{p_j}, \tag{2.5}$$

where p_i is the priority of the alternative i.

Priorities are not known in advance in AHP. As priorities only make sense if derived from consistent or near-consistent matrices, a consistency check must be applied. The threshold for defining an intolerably inconsistent matrix is not clear. Several methods have been proposed to measure consistency. Peláez and Lamata (2003) describe a method based on the determinant of the matrix. Crawford and Williams (1985) prefer to add the difference between the ratio of the calculated priorities and the given comparisons. The transitivity rule (2.2) was used by Salo and Hämäläinen (1997) and later by Ji and Jiang (2003). Stein and Mizzi (2007) use the normalized column of the comparison matrix. However, the most commonly used method (including in *MakeItRational*) was developed by Saaty (1977), who proposed a consistency index (CI), which is related to the eigenvalue method (Section 2.4.4):

$$\text{CI} = \frac{\lambda_{\max} - n}{n - 1}, \tag{2.6}$$

where $\lambda_{\max}$ is the maximal eigenvalue. The consistency ratio (CR) is given by

$$\text{CR} = \text{CI}/\text{RI}, \tag{2.7}$$

where RI is the random index (the average CI of 500 randomly filled matrices). If CR is less than 10% (the inconsistency is less than 10% of 500 randomly filled matrices), then the matrix is of an acceptable consistency.

Saaty (1977) calculated the random indices shown in Table 2.6. Other researchers have run simulations with different numbers of matrices (Alonso and Lamata 2006; Lane and Verdini 1989; Tummala and Wan 1994) or incomplete matrices (Forman 1990). Their random indices are different than but close to Saaty's.

Alonso and Lamata (2006) have computed a regression of the random indices and propose the formulation

$$\lambda_{\max} < n + 0.1(1.7699n - 4.3513). \tag{2.8}$$

Table 2.6 Random indices from Saaty (1977).

n	3	4	5	6	7	8	9	10
RI	0.58	0.9	1.12	1.24	1.32	1.41	1.45	1.49

For all proposed consistency checking methods, some questions remain: where is the cut-off to declare the matrix inconsistent? Should this rule depend on the size of the matrix? How should the consistency definition be adapted when using another judgement scale?

2.4.4 Priorities derivation

Priorities derivation is the corner-stone of the mathematics behind the AHP method, otherwise rankings could not be produced. Various methods have been proposed to calculate priorities from a pairwise comparison matrix (Lin 2007; Cho and Wedley 2004). In this section, three methods will be introduced: an approximate method, the eigenvalue method and the geometric mean. The approximate method requires only additions and averages. The eigenvalue method calculates not only the priorities but also the degree of inconsistency. The geometric mean has been proposed to solve the problem of the right–left rank reversal of the eigenvalue method. Each method calculates identical priorities when matrices are consistent.

2.4.4.1 Approximate method

This method is based on two simple steps:

1. Summation of the elements of row i:

$$r_i = \sum_i a_{ij}. \tag{2.9}$$

2. Normalization of the sums:

$$pi = \frac{r_i}{\sum_i r_i}. \tag{2.10}$$

Example 2.3 Suppose the decision maker has provided the comparisons in Table 2.7. The two steps are thus as follows:

1. The rows are summed as in the final column of Table 2.7.
2. Normalization the sums gives the criteria priorities as in Table 2.8

Table 2.7 Sum of the elements of the rows.

	Industrial area	Shopping centre	City centre	Total
Industrial area	1	6	2	9.00
Shopping centre	1/6	1	1/2	1.67
City centre	1/2	2	1	3.50

Table 2.8 Criteria priorities.

	Industrial area	Shopping centre	City centre	Total
Industrial area	1	6	2	0.64
Shopping centre	1/6	1	1/2	0.12
City centre	1/2	2	1	0.25

The approximate method does not calculate the consistency of the matrices. Therefore, all the AHP software packages that have been tested prefer to use a more intensive calculation method, which allows the calculation of the inconsistency rate.

2.4.4.2 Eigenvalue method

In the eigenvalue method, the vector of the priorities **p** is calculated by solving the equation

$$\mathbf{Ap} = n\mathbf{p} \tag{2.11}$$

where n is the dimension of **A** and $\mathbf{p} = (p_1, \ldots, p_j, \ldots, p_n)$.

First, the validity of the eigenvalue method on a consistent matrix **A** is demonstrated. Let us suppose that the priorities are known. It is easy to deduce a consistent comparison matrix from the priorities as follows. Let $a_{ij} = p_i/p_j$. Multiplying **A** by the priority vector **p** gives the right-hand side of equation (2.11). To simplify the calculation, only row i of **A** is first considered:

$$\frac{p_i}{p_1}p_1 + \frac{p_i}{p_2}p_2 + \ldots + \frac{p_i}{p_j}p_j + \ldots + \frac{p_i}{p_n}p_n = p_i + p_i + \ldots + p_i + \ldots + p_i = np_i$$

or

$$\sum_j \frac{p_i}{p_j} \cdot p_j = np_i. \tag{2.12}$$

Thus the product of row i by the priority vector **p** gives n times the priority p_i. By multiplying all the elements of the comparison matrix **A** by the priority vector **p**, the following equality is obtained:

$$\begin{bmatrix} p_1/p_1 & p_1/p_2 & \cdots & p_1/p_n \\ p_2/p_1 & p_2/p_2 & \cdots & p_2/p_n \\ \cdots & \cdots & \cdots & \cdots \\ p_n/p_1 & p_n/p_2 & \cdots & p_n/p_n \end{bmatrix} \begin{bmatrix} p_1 \\ p_2 \\ \cdots \\ p_n \end{bmatrix} = n \begin{bmatrix} p_1 \\ p_2 \\ \cdots \\ p_n \end{bmatrix},$$

Table 2.9 Consistent comparison matrix.

	Industrial area	Shopping centre	City centre
Industrial area	1	6	3
Shopping centre	1/6	1	1/2
City centre	1/3	2	1

which is equation (2.11). Therefore, for a consistent matrix, the priority vector is obtained by solving equation (2.11).

Example 2.4 The eigenvalue problem is illustrated by the comparison matrix in Table 2.9. Because the comparison matrix is consistent, the priorities can be calculated by solving (2.11):

$$\mathbf{Ap} = \begin{bmatrix} 1 & 6 & 3 \\ 1/6 & 1 & 1/2 \\ 1/3 & 2 & 1 \end{bmatrix} \mathbf{p} = 3.\mathbf{p}$$

The priority vector **p** is the solution of following the linear system:

$$\begin{aligned} 1 \quad \cdot p_1 + 6 \cdot p_2 + 3 \quad \cdot p_3 &= 3 \cdot p_1 \\ 1/6 \cdot p_1 + 1 \cdot p_2 + 1/2 \cdot p_3 &= 3 \cdot p_1 \\ 1/6 \cdot p_1 + 1 \cdot p_2 + 1/2 \cdot p_3 &= 3 \cdot p_1. \end{aligned}$$

Solving this system for the unknowns p_1, p_2 and p_3 results in

$$\mathbf{p} = \begin{bmatrix} p_1 \\ p_2 \\ p_3 \end{bmatrix} = \begin{bmatrix} 0.667 \\ 0.111 \\ 0.222 \end{bmatrix}.$$

For an inconsistent matrix, this relation is no longer valid, as the comparison between element i and j is not necessarily given by formula (2.5). Therefore, the dimension n is replaced by the unknown λ. The calculation of λ and **p**, from an equation such as $\mathbf{Ap} = \lambda\mathbf{p}$ is called, in linear algebra, an eigenvalue problem. Any value λ satisfying this equation is called an eigenvalue and **p** is its associated eigenvector.

According to the Perron theorem, a positive matrix has a unique positive eigenvalue. The non-trivial eigenvalue is called the maximum eigenvalue λ_{max}. If $\lambda_{max} = n$, then the matrix is perfectly consistent. Otherwise, the difference between $\lambda_{max} - n$ is a measure of the inconsistency (Section 2.4.3).

Table 2.10 First iteration of the power method.

	Industrial area	Shopping centre	City centre	Sum of the row	Eigenvector
Industrial area	3	16	7	26	0.62
Shopping centre	0.583	3	1.333	4.916	0.12
City centre	1.333	7	3	11.333	0.27
				42.249	1

Is the eigenvalue still valid for inconsistent matrices? Saaty (1977, 1980) justifies the eigenvalue approach for slightly inconsistent matrices with perturbation theory, which says that slight variations in a consistent matrix imply only slight variations of the eigenvector and eigenvalue.

In order to calculate the eigenvector associated to the maximum eigenvalue, most software packages, including *MakeItRational*, use the power method, which is an iterative process:

1. The pairwise matrix is squared: $\mathbf{A}_{n+1} = \mathbf{A}_n\mathbf{A}_n$.
2. The row sums are then calculated and normalized. This is the first approximation of the eigenvector.
3. Using the matrix $\mathbf{A}_{n+1}$, steps 1 and 2 are repeated.
4. Step 3 is repeated until the difference between these sums in two consecutive priorities calculations is smaller than a given stop criterion.

Example 2.5 Considering the inconsistent matrix **B** of Table 2.7, the first iteration of the power method is given in Table 2.10.

The power method is not fully transparent, much less than the approximate method. Several articles have highlighted this irrationality. According to Johnson et al. (1979), the 'aggregation process embedded in the eigenvector ... is not fully understood mathematically'. Chu et al. (1979) say that 'the weighted least squares method ... is conceptually easier to understand than the eigenvalue method'.

If the matrix of Table 2.11 is considered, it is known that a comparison can be estimated indirectly by the transitivity rule (2.2). Using (2.2), the comparison a_{13} is calculated as follows:

$$a_{13} = a_{11} \cdot a_{13} = 1 \cdot 2 = 2,$$

$$a_{13} = a_{12} \cdot a_{23} = 1/7 \cdot 6 = 0.857,$$

$$a_{13} = a_{13} \cdot a_{33} = 2 \cdot 1 = 2,$$

$$a_{13} = a_{14} \cdot a_{43} = 1/2 \cdot 1 = 1/2.$$

Table 2.11 Criteria comparison matrix with illustrative values.

	Visibility	Competition	Frequency	Rental cost
Visibility	1	1/7	2	1/2
Competition	7	1	6	2
Frequency	1/5	1/6	1	1
Rental cost	2	1/2	1	1

It can be seen that the estimations of comparison a_{13} are very different. The matrix in Table 2.11 is thus inconsistent.

By squaring the matrix, the power method combines all the estimates of a_{13} due to the scalar product of the first column and the third row of the matrix. In general, each element a'_{ij} of $\mathbf{A}^2$ is given by the sum $a'_{ij} = \sum_k a_{ik} \cdot a_{kj}$. Each additional squaring refines the estimation of the comparison. Therefore, the eigenvalue method is based on a procedure of averaging the direct and indirect estimations of the comparisons.

The eigenvalue method has a major drawback: the right–left inconsistency, which leads to a rank reversal phenomenon after an inversion of the scale, was discovered two years after the publication of the original AHP method (Johnson et al. 1979). If all comparisons are replaced by their reciprocal values (e.g., 5 becomes 1/5), then the derived ranking should logically also be reversed. This is not always the case for the eigen value method; however, if a matrix is perfectly consistent or of rank $n = 3$, then a rank reversal is impossible. For inconsistent matrices of rank $n \geq 4$, rankings can be different after a scale inversion. The following example illustrates this phenomenon. Consider the matrix in Table 2.12, asking 'Which alternative is most economical?'. The calculated priorities give the following ranking of the alternatives: D > B > C > A > E.

If the question is inverted: 'Which alternative is most expensive?', then the comparisons are simply inverted (Table 2.13). The calculated priorities give the following ranking of the alternatives: B > D > C > A > E. In this case, alternative B is preferred, but before it was alternative D. This rank reversal is due to the formulation of the problem (which is different from and independent of rank reversal due to the introduction or deletion of an alternative, discussed in Section 2.4.5).

Table 2.12 Comparison matrix.

	A	B	C	D	E	Priority	Rank
A	1	1/7	1/2	1/8	2	0.061	4
B	7	1	3	1	8	0.374	2
C	2	1/3	1	1/4	5	0.134	3
D	8	1	4	1	5	0.387	**1**
E	1/2	1/8	1/5	1/5	1	0.043	5

Table 2.13 Comparisons of Table 2.11 inverted.

	A	B	C	D	E	Priority	Rank
A	1	7	2	8	1/2	0.299	4
B	1/7	1	1/3	1	1/8	0.047	**1**
C	1/2	3	1	4	1/5	0.140	3
D	1/8	1	1/4	1	1/5	0.051	2
E	2	8	5	5	1	0.462	5

2.4.4.3 Geometric mean

In order to avoid the left–right rank reversal, Crawford and Williams (1985) adopted another approach by minimizing the multiplicative error (2.13):

$$a_{ij} = \frac{p_i}{p_j} e_{ij}, \tag{2.13}$$

where a_{ij} is the comparison between object i and j, p_i is the priority of object I, and e_{ij} is the error.

The multiplicative error is commonly accepted to be log-normally distributed (similarly, the additive error would be assumed to be normally distributed). The geometric mean,

$$p_i = \sqrt[n]{\prod_{j=1}^{n} a_{ij}}, \tag{2.14}$$

is the one, which minimizes the sum of these errors,

$$\min \sum_{i=1}^{n} \sum_{j=1}^{n} \left(1\text{n}(a_{ij}) - 1\text{n}\left(\frac{p_i}{p_j}\right)\right)^2. \tag{2.15}$$

The geometric mean (also sometimes known as logarithmic least squares method) can be easily calculated by hand and has been supported by a large segment of the AHP community (Aguarón and Moreno-Jiménez 2000, 2003; Barzilai 1997; Barzilai and Lootsma 1997; Budescu 1984; Escobar and Moreno-Jiménez 2000; Fichtner 1986; Leskinen and Kangas 2005; Lootsma 1993, 1996). Its main advantage is the absence of rank reversals due to right–left inconsistency: the row and column geometric means provide the same ranking in an inverse order (this is not necessarily the case with the eigenvalue method).

Example 2.6 The priorities of Table 2.12 are

$$p_1 = \sqrt[5]{1 \cdot \tfrac{1}{7} \cdot \tfrac{1}{2} \cdot \tfrac{1}{8} \cdot 2} = 0.447, \quad p_2 = 2.787,$$
$$p_3 = .0.964,\ p_4 = 2.759,\ p_5 = 0.302.$$

Note that these priorities are in in the reverse order compared to the priorities of Table 2.13: $p_1 = 2.237$; $p_2 = 0.359$; $p_3 = 1.037$, $p_4 = 0.362$, $p_5 = 3.314$. Therefore the geometric mean does not have any rank reversal due to an inversion of the scale.

2.4.5 Aggregation

The last necessary step is the synthesis of the local priorities across all criteria in order to determine the global priority. The historical AHP approach adopts an additive aggregation with normalization of the sum of the local priorities to unity. This type of normalization is called distributive mode. This additive aggregation is expressed as

$$P_i = \sum_j w_j \cdot p_{ij} \tag{2.16}$$

where P_i is the global priority of alternative i, p_{ij} is the local priority with regard to criterion j, and w_j is the weight of the criterion j.

If priorities are known, the distributive mode is the only approach that will retrieve these priorities. However, if a copy (Belton and Gear 1983), or near-copy (the pairwise comparison is almost the same as the original) (Dyer 1990b), of an alternative is introduced or removed (Troutt 1988), a rank reversal of the alternatives may occur. This phenomenon has been criticized by some (Dyer 1990a, 1990b; Holder 1990, 1991; Stam and Duarte Silva 2003) and accepted by others (Harker and Vargas 1987, 1990; Pérez 1995; Saaty 1986, 1990, 1991, 1994b, 2006). This rank reversal phenomenon is not unique to AHP but common to all additive models having a normalization step (Triantaphyllou 2001; Wang and Luo 2009). In fact, when the number of alternatives is changed, the denominator for the normalization is also changed, which implies a change of scale and possible rank reversal.

To avoid this rank reversal problem, priorities should be normalized by dividing them by the same denominator in any configuration of the problem for which the ideal mode was proposed. This normalization is done by dividing the score of each alternative by the score of the best alternative under each criterion.

When should the distributive or ideal mode be used? Millet and Saaty (2000) give some guidance on which normalization to use. If in a *closed system* (i.e. no alternative will be added or removed), then the distributive mode should be used. If in an *open system* (i.e. alternatives can be added or removed) and the preference is allowed for alternatives to be dependent on other alternatives (in other words, the rank reversal phenomenon is accepted), then the distributive mode is indicated. If we are in an open system and do not want other alternatives to affect the outcome, then the ideal mode is recommended.

Multiplicative aggregation has been proposed to prevent the rank reversal phenomenon observed in the distributive mode (Lootsma 1993; Barzilai and Lootsma 1997). In the notation of (2.16), this form of aggregation is expressed as

$$p_i = \prod_j p_{ij}^{w_j} \tag{2.17}$$

Multiplicative aggregation has non-linear properties allowing a superior compromise to be selected; this is not the case with additive aggregation (Stam and Duarte Silva 2003; Ishizaka et al. 2006, 2010). Vargas (1997) demonstrated that additive aggregation is the only way to retrieve exact weights of known objects. *MakeItRational* offers only the additive distributive and ideal mode of aggregation.

Exercise 2.3

First you will learn to calculate priorities for one criterion step by step. Then you will be given the opportunity to complete the spreadsheet for the other criteria.

Learning Outcomes

- Understand the calculation of priorities with the approximate method in *Microsoft Excel*
- Understand the calculation of priorities with the eigenvalue method in *Excel*
- Understand the calculation of priorities with the geometric mean method in *Excel*

Tasks

Open the file Sport Shop.xls. It contains three spreadsheets with the three different priority calculation methods.

Complete the following tasks:

a) Describe the meaning of each calculation cell and its formula. Read the comments in the red squares in case of difficulties.

b) The spreadsheets are incomplete because they calculate only one local alternative. Complete them in order to calculate the other local alternatives.

2.5 Extensions of AHP

In this section, four extensions of AHP are presented. The analytic hierarchy process ordering method was introduced to separately analyse criteria that have to be minimized and maximized. The group analytic hierarchy process is used for group decisions. The clusters and pivots technique is applied to large problems to reduce the number of pairwise comparisons. AHPSort is implemented to solve sorting problems.

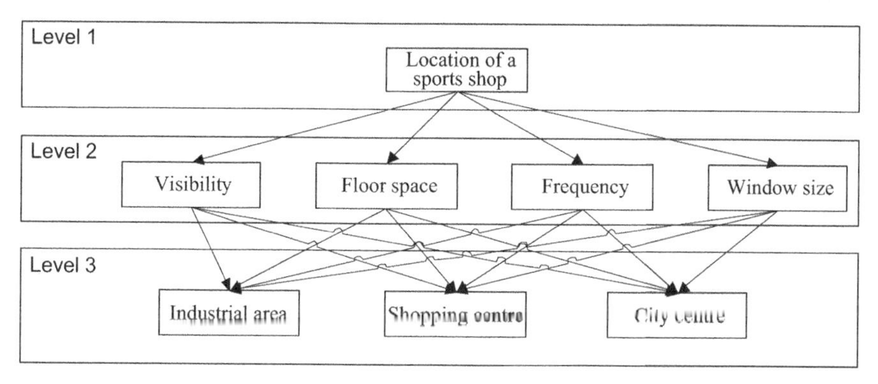

Figure 2.14 Benefit hierarchy.

2.5.1 Analytic hierarchy process ordering

The analytic hierarchy process ordering method, which considers incomparability, was first proposed by Ishizaka and Labib (2011). In line with the philosophy of AHP, some researchers have proposed deconstructing the model into sub-problems (Azis 1990; Clayton et al. 1993; Wedley et al. 2001). They propose separating the criteria in opposite directions in different hierarchies: benefits versus cost. The reason for this additional decomposition is that criteria in the same direction are much easier to compare than in opposite directions, such as a criterion to be minimized and another maximized.

For example, suppose that three more criteria are added to Case Study 2.1: vandalism, floor space and window size. We now have seven criteria in total. Some have to be maximized (visibility, floor space, frequency and window size) and others minimized (competition, rental costs, vandalism). Two hierarchies can be constructed: a benefit hierarchy (Figure 2.14) and a cost hierarchy (Figure 2.15). These two hierarchies are then solved separately. As a result the output of the method is a partial ranking (and not priorities).

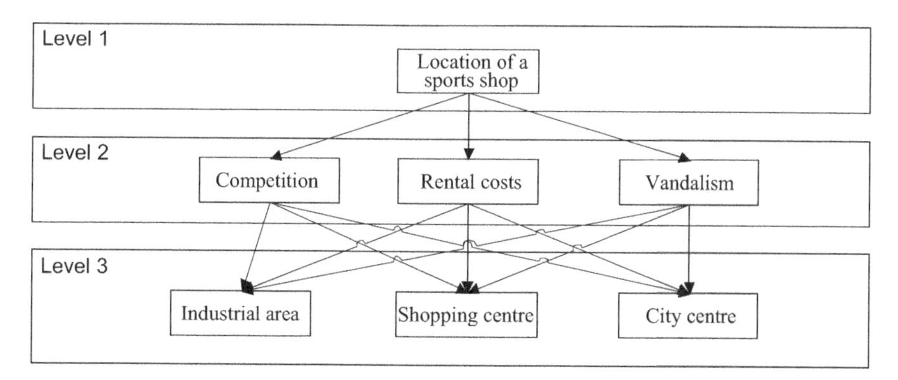

Figure 2.15 Cost hierarchy.

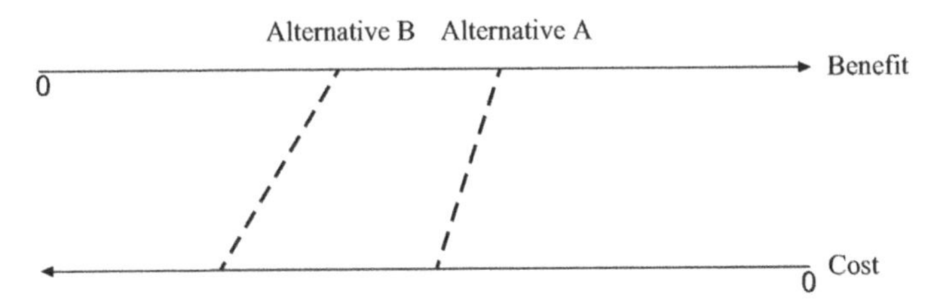

Figure 2.16 Graphical representation of the preference relation.

If cost and benefit rankings are not aggregated, we have a partial ordinal ranking (Ishizaka and Labib 2011). A partial ordinal ranking can be derived from the cost and benefit analysis, where:

1. Alternative A is *better than* alternative B if alternative A is ranked better than alternative B in the cost and benefit analysis (Figure 2.16).
2. Alternative A and alternative B are *indifferent* if alternative A has the same score as alternative B in the cost and benefit analysis (Figure 2.17).
3. Alternative A is *incomparable to* alternative B if alternative A is better in one analysis and worse in the other analysis (Figure 2.18).

Incomparability does not exist in standard AHP. This status is important as it reveals that a decision maker cannot decide which of the two alternatives is the best, while not being indifferent: an alternative is better in some aspects but worse in others. To decide which alternative is better, further discussion between the decision makers and moderation by the analyst are needed. This debate may require additional information.

However, if a debate cannot be held (e.g. if the decision maker is unavailable), the cost and benefit analysis can be merged into a complete ranking. First, the importance scores of the benefits and costs are weighted and then the weighted score of the cost analysis divides the weighted score of the benefit analysis. This produces the complete cardinal ranking.

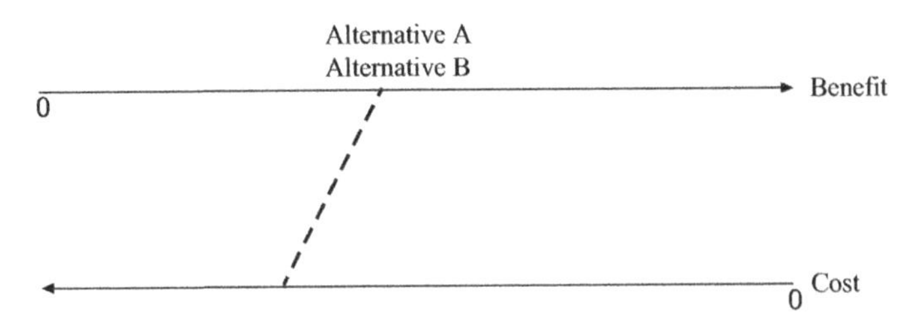

Figure 2.17 Graphical representation of the indifference relation.

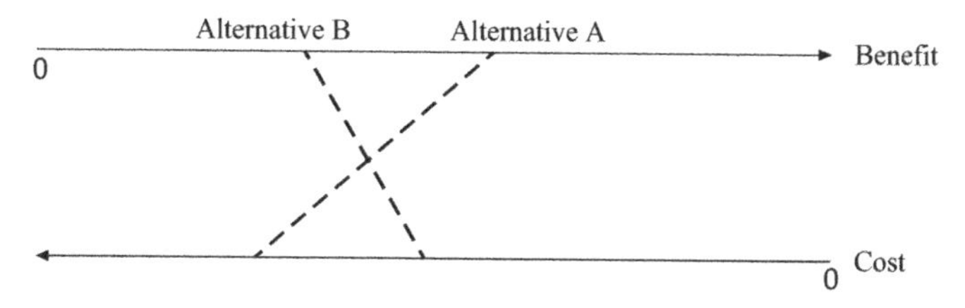

Figure 2.18 Graphical representation of incomparability.

Researchers have proposed four hierarchies: benefit, cost, opportunity and risk (Saaty and Özdemir 2003a; Saaty 1994b). Each hierarchy is solved separately and the local priorities are aggregated according to

$$p_i = \frac{a \cdot p_o + b \cdot p_b}{c \cdot p_r + d \cdot p_c} \tag{2.18}$$

where a, b, c, d are weights, p_o is the priority of the opportunity hierarchy, p_b is the priority of the benefit hierarchy, p_r is the priority of the risk hierarchy, and p_c is the priority of the cost hierarchy.

Exercise 2.4

In this exercise, you will be able to solve the sports shop problem with the Analytic Hierarchy Process Ordering.

Prerequisites

Exercise 2.2

Learning Outcomes

- Structure the benefit and cost hierarchy in *MakeItRational*
- Understand the partial ordinal ranking
- Understand the complete cardinal ranking

Tasks

a) In *MakeItRational*, construct the two hierarchies as in Figure 2.14 and Figure 2.15.

b) Evaluate the pairwise comparisons and calculate the priorities for each hierarchy.

c) Draw a graph similar to that in Figure 2.16, Figure 2.17 or Figure 2.18. Confirm whether alternatives are preferred, indifferent or incomparable.

d) Aggregate the priorities of the two alternatives to obtain a complete cardinal ranking.

Table 2.14 Four ways to combine preferences (Ishizaka and Labib 2011).

		Mathematical aggregation	
		Yes	No
Aggregation on	Judgements (Figure 2.19)	Geometric mean on comparisons	Consensus vote on comparisons
	Priorities (Figure 2.20)	Weighted arithmetic mean on priorities	Consensus vote on priorities

2.5.2 Group analytic hierarchy process

2.5.2.1 Group aggregation

As a decision often affects several people, standard AHP has been adapted so that it can be applied to group decisions. By consulting various experts, the bias often present when judgements are accepted from a single expert is eliminated. There are four ways to combine the preferences into a consensus rating (Table 2.14).

The consensus vote is used when there is a synergistic group and not a collection of individuals. In this case, the hierarchy of the problem must be the same for all decision makers. On the judgement level, this method requires the group to reach an agreement on the value of each entry in a matrix of pairwise comparisons. A consistent agreement is usually difficult to obtain, the difficulty increasing with the number of comparison matrices and related discussions. To bypass this difficulty, the consensus vote can be postponed until after the calculation of the priorities for each participant. O'Leary (1993) recommends this version because an early aggregation could result 'in a meaningless average performance measure'. An aggregation after the calculation of priorities allows decision makers with different opinions to be detected and further discussion over any disagreement.

If a consensus cannot be reached (e.g. with a large number of people or people in different locations), a mathematical aggregation can be adopted. Two synthesizing methods exist and provide the same results in the case of perfect consistency in the pairwise matrices (Saaty and Vargas 2005). In the first method, the geometric mean of the individual evaluations is used as an element in the pairwise matrices and then priorities are calculated from that (Figure 2.19). The geometric mean method must be adopted instead of the arithmetical mean to preserve the reciprocal property (Aczél and Saaty 1983). For example, if person A enters comparison 9 and person B enters 1/9, then by intuition the mathematical consensus should be $\sqrt{9 \cdot \frac{1}{9}} = 1$, which is a geometric mean and not $(9 + 1/9)/2 = 4.56$, which is an arithmetic mean. However, Ramanathan and Ganesh (1994) give an example where the Pareto optimality (i.e. if all group members prefer A to B, then the group decision should prefer A) is not satisfied with the geometric mean method. Van den Honert and Lootsma (1997) argue

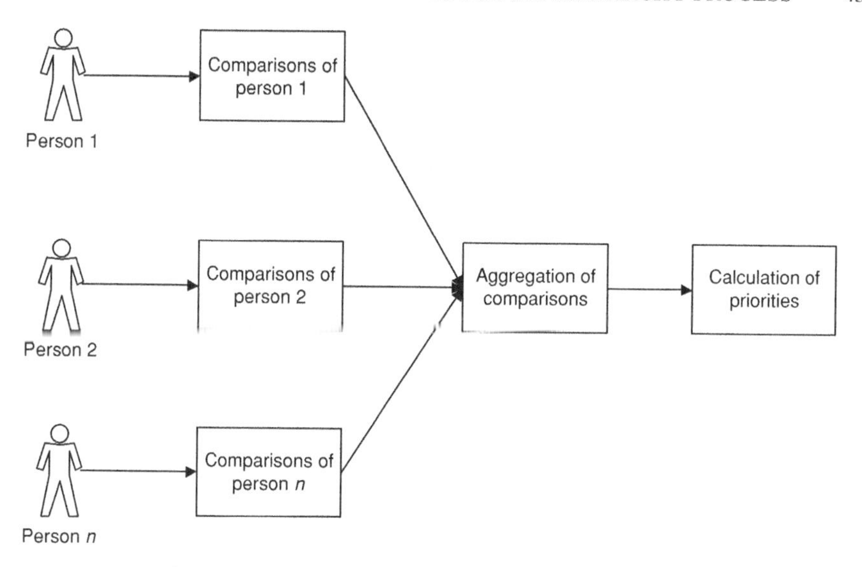

Figure 2.19 Aggregation at the comparison level.

that this violation is expected because the pairwise assessments are a compromise of all the group members' assessments and, therefore, the compromise does not represent any one opinion of the group members. Madu and Kuei (1995) and Saaty and Vargas (2007) introduce a measure of the dispersion of the judgements to avoid this problem. If the group is not homogeneous, further discussions are required to reach a consensus.

In the second method (Figure 2.20), decision makers constitute the first level below the goal of the AHP hierarchy (Figure 2.21). Priorities are computed and then aggregated using the weighted arithmetic mean method. Applications can be found in Labib and Shah (2001) and Labib et al. (1996). Arbel and Orgler (1990) introduced a level above that of the stakeholders representing the several economic scenarios. This extra level determines the priorities (weights) of the stakeholders.

In a compromise method an individual's derived priorities can be aggregated at each node. According to Forman and Peniwati (1998), this method is 'less meaningful and not commonly used'. Aggregation methods with linear programming (Mikhailov 2004) and the Bayesian approach (Altuzarra et al. 2007) have been proposed to make a decision even when comparisons are missing, for example, when a stakeholder does not have the expertise to judge a particular comparison.

A group decision may be skewed due to collusion or distortion in the judgements to secure the preferred outcome. This problem does not arise when there is a single decision maker because the first choice will always remain the first. In a group decision, a participant does not have this certitude as the results are aggregated with those of the other stakeholders. One decision maker may overweight their preferred alternative and bias the group decision. As individual identities are lost with an aggregation, early aggregation is not recommended.

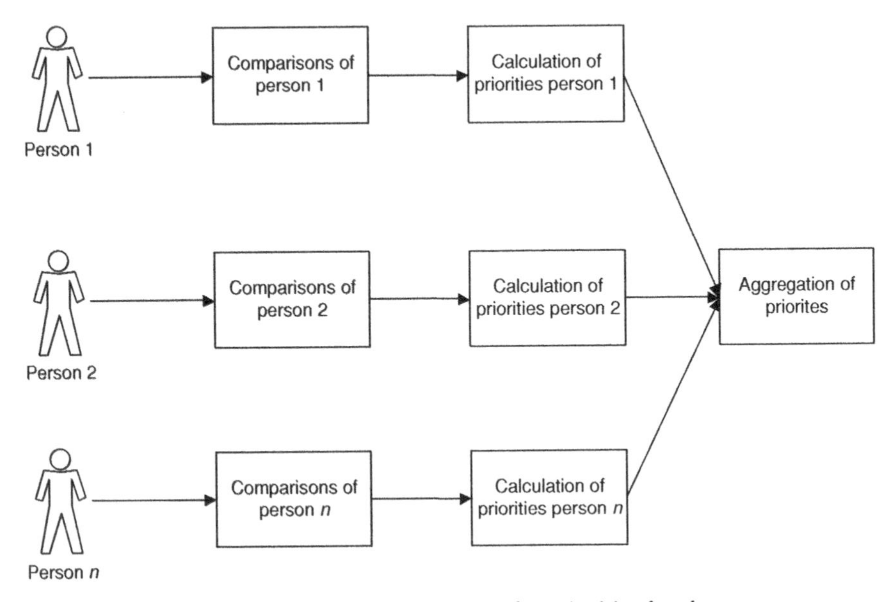

Figure 2.20 Aggregation at the priorities level.

2.5.2.2 Weight of stakeholders

If the decision makers do not have equal weight, their priorities must be determined as discussed in Ishizaka and Labib (2011). The weights should reflect the expertise of a decision maker (Weiss and Rao 1987) or the importance of the impact of the

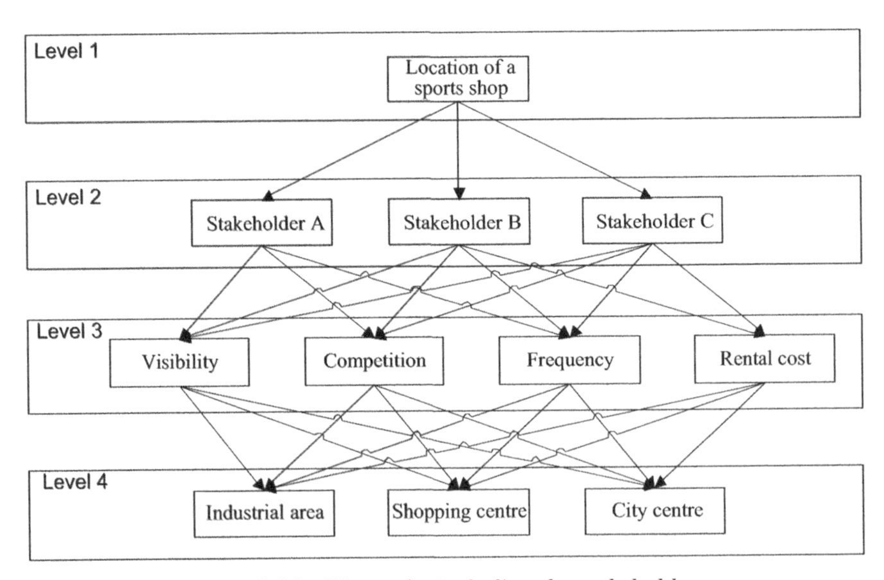

Figure 2.21 Hierarchy including the stakeholders.

decision on the decision maker. The weights can be allocated by a supra decision maker or by a participatory approach. Finding a supra decision maker or benevolent dictator, accepted by everybody, may be difficult.

Ramanathan and Ganesh (1994) proposed a method based on pairwise comparisons to calculate the weights. All n members complete a comparison matrix with the relative importance of each member. A vector of priorities is calculated for each member. The n vectors of priorities are gathered in an $n \times n$ matrix and the final weight of each member is given by the eigenvector of this matrix. To incorporate the uncertainty of the expertise of the participants, AHP has been combined with variable precision rough sets (Xie et al. 2008) and fuzzy logic (Jaganathan et al. 2007).

Ishizaka and Labib (2011) also use pairwise comparisons to judge other members of the group, with each evaluated member allowed the possibility of a veto on the received evaluation. This technique can be viewed as fairer and is applied in situations such as sporting competitions, for example ice-skating, where judges cannot evaluate competitors of the same nationality. The consistency of the weights given by the appraisers is checked with the consistency ratio formula (2.6).

Cho and Cho (2008) have a surprising way of determining weights with levels of inconsistency: decision makers with less inconsistency receive more weight. We do not support this method because inconsistency is useful feedback for the user (Section 2.2.3): it highlights the decision maker's consistency and gives a hint as to revisions of comparisons that could be manual errors in setting the comparisons, sometimes forced due to the upper limitation of the comparison scale (e.g. if the user first enters $a_{12} = 4$ and $a_{23} = 5$, they should enter $a_{13} = 20$ to be consistent, but they can only enter $a_{13} = 9$ due to the maximal value of the measurement scale). The consistency index is therefore not a measure of the quality or expertise of the decision maker.

Exercise 2.5

In this group exercise, you will be able to solve a group decision with the group analytic hierarchy process.

Prerequisites

Exercise 2.2

Learning Outcomes

- Structure a group hierarchy in *MakeItRational*
- Understand the aggregation of individual priorities
- Understand the final group ranking

Tasks

a) Form a group of three or four people.

b) In *MakeItRational* construct the hierarchy shown in Figure 2.21.

c) Each stakeholder evaluates the pairwise comparisons.

d) Evaluate pairwise the weight of each stakeholder.

e) Discuss the final ranking. Is everybody satisfied?

2.5.3 Clusters and pivots for a large number of alternatives

The main drawback of AHP is the high number of pairwise evaluations required for completing large matrices; see expression (2.1). To bypass this problem, a cluster and pivots method has been proposed (Saaty 2001; Ishizaka 2012). It is based on four steps:

a) *For each criterion, all alternatives are ordinally ranked.* If all criteria produce the same order of alternatives, they would be a replica of themselves and the problem would be a mono-criterion one.

b) *For each criterion, alternatives are divided into clusters.* The classical cluster analysis cannot be used in this case because AHP incorporates qualitative criteria in the model. It is a delicate and subjective operation, for which no algorithm exists. The decision maker must evaluate which alternatives are close enough and therefore easy to compare. A heuristic way to construct the clusters is to compare the best ordered alternative successively with the next ones, from the second best to the worst, until:

   - either the cluster contains seven elements. Psychologists have observed that it is difficult to evaluate more than seven elements (Saaty and Ozdemir 2003b). It is recommended that clusters do not contain more than seven elements.

   - or the comparison entered is 9 (if a 9-point scale is used). As no higher strength of preference is available on the comparison scale (Table 2.2), it is appropriate to close the cluster.

   The last compared alternative becomes the pivot (which becomes now the best amongst the remaining ones) at the boundary of both clusters. The same process is repeated with the pivot until seven elements are in the cluster, a comparison value of 9 is entered, or all entries are provided. In Figure 2.22, alternative D is the pivot.

c) *All alternatives of the same cluster are compared and then priorities are calculated.*

d) *Priorities of the clusters are joined with a common element: 'the pivot'.* The pivot is used for the conversion rate between two clusters.

In AHP, all alternatives are compared to each other in a unique comparison matrix, which can be perceived as a one-cluster problem. In a scoring model, direct judgements are used. Each element can be considered a separate cluster. The AHP

	A	B	C	D	E	F
A	1	----	--- ▶	9		
B		1				
C			1			
D				1	----	-- ▶
E					1	
F						1

Figure 2.22 Building clusters.

and scoring model represent the two extremes, where 1 or n clusters are used. This model is the middle way between the two methods.

Example 2.7 Suppose that in Case Study 2.1, the businessman has to choose where to situate the sports shop from a choice of 12 cities. In this example, only one criterion is considered (e.g. quality of life), as the process is identical with several criteria.

a) *Alternatives are preordered.* The 12 cities are preordered according to their quality of life: A, B, C, D, E, F, G, H, I, J, K, L.

b) *Alternatives are divided into clusters.* The best city for quality of life, A, is compared successively with the next cities until the comparison entered is 9 (Table 2.15). As city A is 9 times better than F for its quality, alternative F is declared the pivot: the last element in the first cluster and the first element in the second cluster.

c) *Comparisons are entered in clustered matrices and priorities are calculated.* Notice that the process requires 30 comparisons (highlighted in grey in

Table 2.15 Comparison matrix for the criterion quality – the unnecessary comparisons are shaded grey.

	A	B	C	D	E	F	G	H	I	J	K	L
A	1	2	3	5	8	9						
B		1	2	3	6	7						
C			1	3	5	6						
D				1	4	5						
E					1	2						
F						1	1	3	4	5	5	7
G							1	3	3	4	5	7
H								1	2	3	4	6
I									1	2	2	4
J										1	1	3
K											1	3
L												1

Table 2.16 First cluster of cities.

Cities	Priority
F	0.311
G	0.289
H	0.157
I	0.097
J	0.061
K	0.057
L	0.028

Table 2.17 Second cluster of cities.

Cities	Priorities	Priorities linked with the first cluster
A	0.404	4.333
B	0.249	2.670
C	0.178	1.909
D	0.101	1.083
E	0.040	0.429
F	0.029	0.311

Table 2.15) less than the classical AHP approach. Priorities are calculated for both clusters (Tables 2.16 and 2.17).

d) *Priorities of both clusters are joined with a common element: 'the pivot'.* Results of the second cluster (Table 2.17) are linked to the first (Table 2.16) by dividing them by the ratio of the scores of pivot F in the two clusters: 0.311/0.029. Final results are given in Table 2.18.

2.5.4 AHPSort

Whilst AHP solves ranking problems, AHPSort has been developed for sorting problems (Ishizaka et al. 2012). This method is a variant of AHP used when alternatives have to be sorted into categories predefined with central limiting profiles. Suppose that the businessman in Case Study 2.1 aims to open several sports shops in different cities. The cities will be sorted into three classes (Figure 2.23):

a) cities where sports shops will be highly profitable;

b) cities where sports shop may be profitable;

c) cities where sports shops will not be profitable.

Table 2.18 Priorities of the cities.

Cities	Priorities
A	4.333
B	2.670
C	1.909
D	1.083
E	0.429
F	**0.311**
G	0.289
H	0.157
I	0.097
J	0.061
K	0.057
L	0.028

AHPSort is based on eight steps:

A) Problem definition

1) Define the goal, criteria and alternatives of the problem.

2) Define the categories C_i, $i = 1, \ldots, n$.

3) Define the profiles of each class. This can be done with limiting profiles lp_i, which indicate the minimum performance needed on each criterion to belong to class C_i, or with central profiles cp_i, given by a typical example of an element belonging to class C_i. To define each class, $n - 1$ limiting profiles or n central profiles are needed.

B) Evaluations

4) Evaluate pairwise the importance of the criteria and derive weights (Section 2.4.4).

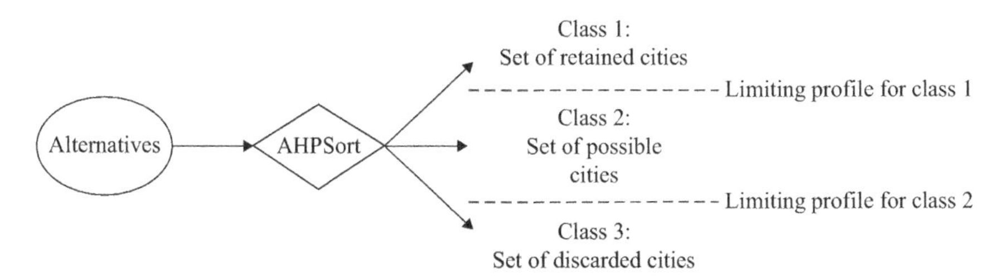

Figure 2.23 AHPSort for the sorting process.

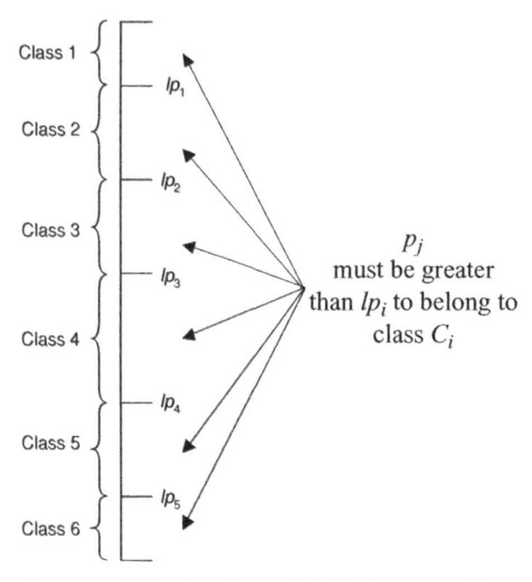

Figure 2.24 Sorting with limiting profiles.

5) Compare in a pairwise comparison matrix a single candidate to be sorted with limiting or central profiles for each criterion to derive local priorities for each criterion (Section 2.4.4).

6) Aggregate the local weighted priorities, which provide a global priority p_j of alternative a_j.

C) Assignment to classes

7) The global priority p_j is used to assign the alternative a_j to class C_i.

   a) *Limiting profiles.* If limiting profiles have been defined, then alternative a_j is assigned to class C_i which has the lp_i just below the global priority p_j (Figure 2.24).

$$\begin{array}{lcl} p_j \geq lp_1 & \Rightarrow & a_j \in C_1 \\ lp_2 \leq p_j < lp_1 & \Rightarrow & a_j \in C_2 \\ \ldots & & \\ p_j < lp_{n-1} & \Rightarrow & a_j \in C_n \end{array}$$

   b) *Central profiles.* Alternative a_j is assigned to class C_i which has the nearest central profile cp_i to p_j (Figure 2.25). In the case of equal distance between two central profiles, the optimistic assignment vision

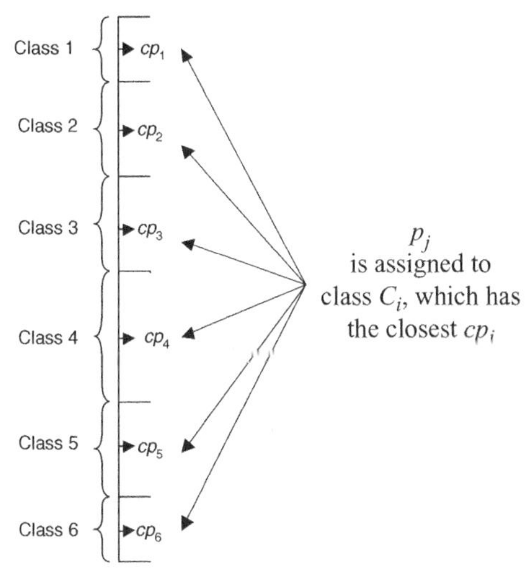

Figure 2.25 Sorting with central profiles.

allocates a_j to the upper class, whilst the pessimistic assignment vision allocates a_j to the lower class.

$$
\begin{aligned}
&p_i \geq cp_1 && \Rightarrow \mathrm{a}_j \in C_1 \\
&cp_2 \leq p_j < cp_1 \text{ AND } (cp_1 - p_j) < (cp_2 - p_j) && \Rightarrow \mathrm{a}_j \in C_1 \\
&cp_2 \leq p_j < cp_1 \text{ AND } (cp_1 - p_j) = (cp_2 - p_j) && \Rightarrow \mathrm{a}_j \in C_1 \text{ in the} \\
& && \text{optimistic vision} \\
&cp_2 \leq p_j < cp_1 \text{ AND } (cp_1 - p_j) = (cp_2 - p_j) && \Rightarrow \mathrm{a}_j \in C_2 \text{ in the} \\
& && \text{pessimistic vision} \\
&cp_2 \leq p_j < cp_1 \text{ AND } (cp_1 - p_j) > (cp_2 - p_j) && \Rightarrow \mathrm{a}_j \in C_2 \\
&\ldots \\
&p_j < cp_n \Rightarrow \mathrm{a}_j \in C_n
\end{aligned}
\tag{2.19}
$$

8) Repeat steps 4–7 for each alternative to be classified.

Exercise 2.6

In this exercise, you will sort cities into three categories:

a) cities where sports shops must be open;

b) cities where sports shops may be opened later;

c) cities where sports shops would not be profitable.

Prerequisites

Exercise 2.2

Learning Outcomes

- Structure a sorting problem in *MakeItRational*
- Define classes
- Understand the sorting results

Tasks

a) Choose 10 potential cities as alternatives for opening a sports shop.

b) Define two limiting profiles for each class.

c) Model the problem in *MakeItRational*.

d) Enter the pariwise comparison for each city.

e) Sort the city into a category according to its score.

References

Aczél, J., and Saaty, T. (1983). Procedures for synthesizing ratio judgements. *Journal of Mathematical Psychology*, *27*(1), 93–102.

Aguarón, J., and Moreno-Jiménez, J. (2000). Local stability intervals in the analytic hierarchy process. *European Journal of Operational Research*, *125*(1), 113–132.

Aguarón, J., and Moreno-Jiménez, J. (2003). The geometric consistency index: Approximated thresholds. *European Journal of Operational Research*, *147*(1), 137–145.

Alonso, J., and Lamata, T. (2006). Consistency in the analytic hierarchy process: A new approach. *International Journal of Uncertainty, Fuzziness and Knowledge-Based Systems*, *14*(4), 445–459.

Altuzarra, A., Moreno-Jiménez, J., and Salvador, M. (2007). A Bayesian priorization procedure for AHP-group decision making. *European Journal of Operational Research*, *182*(1), 367–382.

Arbel, A., and Orgler, Y. (1990). An application of the AHP to bank strategic planning: The mergers and acquisitions process. *European Journal of Operational Research*, *48*(1), 27–37.

Azis, I. (1990). Analytic hierarchy process in the benefit-cost framework: A post-evaluation of the Trans-Sumatra highway project. *European Journal of Operational Research*, *48*(1), 38–48.

Barzilai, J. (1997). Deriving weights from pairwise comparisons matrices. *Journal of the Operational Research Society*, *48*(12), 1226–1232.

Barzilai, J. (2005). Measurement and preference function modelling. *International Transactions in Operational Research*, *12*(2), 173–183.

Barzilai, J., and Lootsma, F. (1997). Power relation and group aggregation in the multiplicative AHP and SMART. *Journal of Multi-Criteria Decision Analysis*, *6*(3), 155–165.

Belton, V., and Gear, A. (1983). On a shortcoming of Saaty's method of analytical hierarchies. *Omega*, *11*(3), 228–230.

Budescu, D. (1984). Scaling binary comparison matrices: A comment on Narasimhan's proposal and other methods. *Fuzzy Sets and Systems*, *14*(2), 187–192.

Cho, E., and Wedley, W. (2004). A common framework for deriving preference values from pairwise comparison matrices. *Computers and Operations Research*, *31*(6), 893–908.

Cho, Y.-G., and Cho, K.-T. (2008). A loss function approach to group preference aggregation in the AHP. *Computers & Operations Research*, *35*(3), 884–892.

Chu, A., Kalabra, R., and Spingarn, K. (1979). A comparison of two methods for determining the weights of belonging to fuzzy sets. *Journal of Optimization Theory and Applications*, *27*(4), 531–538.

Clayton, W., Wright, M., and Sarver, W. (1993). Benefit cost analysis of riverboat gambling. *Mathematical and Computer Modelling*, *17*(4–5), 187–194.

Crawford, G., and Williams, C. (1985). A note on the analysis of subjective judgement matrices. *Journal of Mathematical Psychology*, *29*(4), 387–405.

Dodd, F., and Donegan, H. (1995). Comparison of priotization techniques using interhierarchy mappings. *Journal of the Operational Research Society*, *46*(4), 492–498.

Donegan, H., Dodd, F., and McMaster, T. (1992). A new approach to AHP decision-making. *The Statician*, *41*(3), 295–302.

Dyer, J. (1990a). A clarification of 'Remarks on the analytic hierarchy process'. *Management Science*, *36*(3), 274–275.

Dyer, J. (1990b). Remarks on the analytic hierarchy process. *Management Science*, *36*(3), 249–258.

Escobar, M., and Moreno-Jiménez, J. (2000). Reciprocal distributions in the analytic hierarchy process. *European Journal of Operational Research*, *123*(1), 154–174.

Fichtner, J. (1986). On deriving priority vectors from matrices of pairwise comparisons. *Socio-Economic Planning Sciences*, *20*(6), 341–345.

Forman, E. (1990). Random indices for incomplete pairwise comparison matrices. *European Journal of Operational Research*, *48*(1), 153–155.

Forman, E., and Peniwati, K. (1998). Aggregating individual judgements and priorities with the analytic hierarchy process. *European Journal of Operational Research*, *108*(1), 165–169.

Harker, P., and Vargas, L. (1987). The theory of ratio scale estimation: Saaty's analytic hierarchy process. *Management Science*, *33*(11), 1383–1403.

Harker, P., and Vargas, L. (1990). Reply to 'Remarks on the analytic hierarchy process'. *Management Science* *36*(3), 269–273.

Holder, R. (1990). Some comment on the analytic hierarchy process. *Journal of the Operational Research* Society, *41*(11), 1073–1076.

Holder, R. (1991). Response to Holder's comments on the analytic hierarchy process: Response to the response. *Journal of the Operational Research Society*, *42*(10), 914–918.

Ishizaka, A. (2012). A multicriteria approach with AHP and clusters for the selection among a large number of suppliers. *Pesquisa Operacional*, *32*(1), 1–15.

Ishizaka, A., Balkenborg, D., and Kaplan, T. (2006). Influence of aggregation and preference scale on ranking a compromise alternative in AHP. Paper presented at the Multidisciplinary Workshop on Advances in Preference Handling.

Ishizaka, A., Balkenborg, D., and Kaplan, T. (2010). Influence of aggregation and measurement scale on ranking a compromise alternative in AHP. *Journal of the Operational Research Society*, *62*(4), 700–710.

Ishizaka, A., and Labib, A. (2009). Analytic hierarchy process and Expert Choice: benefits and limitations. *OR Insight*, *22*(4), 201–220.

Ishizaka, A., and Labib, A. (2011). Selection of new production facilities with the group analytic hierarchy process ordering method. *Expert Systems with Applications*, *38*(6), 7317–7325.

Ishizaka, A., Nemery, P., and Pearman, C. (2012). AHPSort: An AHP based method for sorting problems. *International Journal of Production Research*, *50*(17), 4767–4784.

Jaganathan, S., Erinjeri, J., and Ker, J. (2007). Fuzzy analytic hierarchy process based group decision support system to select and evaluate new manufacturing technologies. *International Journal of Advanced Manufacturing Technology*, *32*(11), 1253–1262.

Ji, P., and Jiang, R. (2003). Scale transitivity in the AHP. *Journal of the Operational Research Society*, *54*(8), 896–905.

Johnson, C., Beine, W., and Wang, T. (1979). Right-left asymmetry in an eigenvector ranking procedure. *Journal of Mathematical Psychology*, *19*(1), 61–64.

Kainulainen, T., Leskinen, P., Korhonen, P., Haara, A., and Hujala, T. (2009). A statistical approach to assessing interval scale preferences in discrete choice problems. *Journal of the Operational Research Society*, *60*(2), 252–258.

Kaspar, R., Ossadnik, W. (2013) Evaluation of AHP software from a management accounting perspective. *Journal of Modelling in Management*, *8*(3), in press.

Labib, A., and Shah, J. (2001). Management decisions for a continuous improvement process in industry using the analytical hierarchy process. *Journal of Work Study*, *50*(5), 189–193.

Labib, A., Williams, G., and O'Connor, R. (1996). Formulation of an appropriate productive maintenance strategy using multiple criteria decision making. *Maintenance Journal*, *11*(11), 66–75.

Lane, E., and Verdini, W. (1989). A consistency test for AHP decision makers. *Decision Sciences*, *20*(3), 575–590.

Leskinen, P., and Kangas, J. (2005). Rank reversal in multi-criteria decision analysis with statistical modelling of ratio-scale pairwise comparisons. *Journal of the Operational Research Society*, *56*(7), 855–861.

Lin, C. (2007). A revised framework for deriving preference values from pairwise comparison matrices. *European Journal of Operational Research*, *176*(2), 1145–1150.

Lootsma, F. (1989). Conflict resolution via pairwise comparison of concessions. *European Journal of Operational Research*, *40*(1), 109–116.

Lootsma, F. (1993). Scale sensitivity in the multiplicative AHP and SMART. *Journal of Multi-Criteria Decision Analysis*, 2(2), 87–110.

Lootsma, F. (1996). A model for the relative importance of the criteria in the multiplicative AHP and SMART. *European Journal of Operational Research* *94*(3), 467–476.

Ma, D., and Zheng, X. (1991). 9/9-9/1 scale method of AHP. Paper presented at the 2nd International Symposium on AHP.

Madu, C., and Kuei, C.-H. (1995). Stability analyses of group decision making. *Computers & Industrial Engineering*, *28*(4), 881–892.

Mikhailov, L. (2004). Group prioritization in the AHP by fuzzy preference programming method. *Computers & Operations Research, 31*(2), 293–301.

Millet, I., and Saaty, T. (2000). On the relativity of relative measures: Accommodating both rank preservation and rank reversals in the AHP. *European Journal of Operational Research, 121*(1), 205–212.

Millet, I., and Schoner, B. (2005). Incorporating negative values into the analytic hierarchy process. *Computers and Operations Research 32*(12), 3163–3173.

O'Leary, D. (1993). Determining differences in expert judgement: Implications for knowledge acquisition and validation. *Decision Sciences, 24*(2), 395–408.

Peláez, P., and Lamata, M. (2003). A new measure of consistency for positive reciprocal matrices. *Computers & Mathematics with Applications, 46*(12), 1839–1845.

Pérez, J. (1995). Some comments on Saaty's AHP. *Management Science, 41*(6), 1091–1095.

Pöyhönen, M., Hämäläinen, R., and Salo, A. (1997). An experiment on the numerical modelling of verbal ratio statements. *Journal of Multi-Criteria Decision Analysis, 6*(1), 1–10.

Ramanathan, R., and Ganesh, L. (1994). Group preference aggregation methods employed in AHP: An evaluation and an intrinsic process for deriving members' weightages. *European Journal of Operational Research, 79*(2), 249–265.

Saaty, T. (1977). A scaling method for priorities in hierarchical structures. *Journal of Mathematical Psychology, 15*(3), 234–281.

Saaty, T. (1980). *The Analytic Hierarchy Process*. New York: McGraw-Hill.

Saaty, T. (1986). Axiomatic foundation of the analytic hierarchy process. *Management Science, 32*(7), 841–855.

Saaty, T. (1990). An exposition of the AHP in reply to the paper 'Remarks on the analytic hierarchy process'. *Management Science, 36*(3), 259–268.

Saaty, T. (1991). Response to Holder's comments on the analytic hierarchy process. *Journal of the Operational Research Society, 42*(10), 909–929.

Saaty, T. (1994a). Highlights and critical points in the theory and application of the analytic hierarchy process. *European Journal of Operational Research 74*(3), 426–447.

Saaty, T. L. (1994b). *Fundamentals of Decision Making and Priority Theory*. Pittsburgh: RWS Publications.

Saaty, T. (2001). The seven pillars of the analytic hierarchy process. Paper presented at the Multiple Criteria Decision Making in the New Millennium. Proceedings of the 15th Internatinal Conference MCDM, Istanbul.

Saaty, T. (2006). Rank from Comparisons and from ratings in the analytic hierarchy/network processes. *European Journal of Operational Research, 168*(2), 557–570.

Saaty, T., and Forman, E. (1992). *The Hierarchon: A Dictionary of Hierarchies* (Vol. V). Pittsburgh: RWS Publications.

Saaty, T., and Ozdemir, M. (2003a). Negative priorities in the analytic hierarchy process. *Mathematical and Computer Modelling, 37*(9–10), 1063–1075.

Saaty, T., and Ozdemir, M. (2003b). Why the magic number seven plus or minus two. *Mathematical and Computer Modelling, 38*(3–4), 233–244.

Saaty, T. L., and Vargas, L. G. (2005). The possibility of group welfare functions. *International Journal of Information Technology & Decision Making, 4*(2), 167–176.

Saaty, T., and Vargas, L. (2007). Dispersion of group judgements. *Mathematical and Computer Modelling*, *46*(7–8), 918–925.

Salo, A., and Hämäläinen, R. (1997). On the measurement of preference in the analytic hierarchy process. *Journal of Multi-Criteria Decision Analysis*, *6*(6), 309–319.

Stam, A., and Duarte Silva, P. (2003). On multiplicative priority rating methods for AHP. *European Journal of Operational Research*, *145*(1), 92–108.

Stein, W., and Mizzi, P. (2007). The harmonic consistency index for the analytic hierarchy process. *European Journal of Operational Research*, *177*(1), 488–497.

Stillwell, W., von Winterfeldt, D., and John, R. (1987). Comparing hierarchical and non-hierarchical weighting methods for eliciting multiattribute value models. *Management Science*, *33*(4), 442–450.

Thurstone, L. (1927). A law of comparative judgements. *Psychological Review*, *34*(4), 273–286.

Triantaphyllou, E. (2001). Two new cases of rank reversals when the AHP and some of its additive variants are used that do not occur with the multiplicative AHP. *Journal of Multi-Criteria Decision Analysis*, *10*(1), 11–25.

Troutt, M. (1988). Rank reversal and the dependence of priorities on the underlying mav function. *Omega*,*16*(4), 365–367.

Tummala, V., and Wan, Y. (1994). On the mean random inconsistency index of the analytic hierarchy process (AHP). *Computers & Industrial Engineering* *27*(1–4), 401–404.

Van Den Honert, R., and Lootsma, F. (1997). Group preference aggregation in the multiplicative AHP The model of the group decision process and Pareto optimality. *European Journal of Operational Research*, *96*(2), 363–370.

Vargas, L. (1997). Comments on Barzilai and Lootsma: Why the multiplicative AHP is invalid: A practical counterexample. *Journal of Multi-Criteria Decision Analysis*, *6*(4), 169–170.

Wang, Y., and Luo, Y. (2009). On rank reversal in decision analysis. *Mathematical and Computer Modelling*, *49*(5–6), 1221–1229.

Weber, M., Eisenführ, F., and von Winterfeldt, D. (1988). The effects of spitting attributes on weights in multiattribute utility measurement. *Management Science*, *34*(4), 431–445.

Wedley, W., Choo, E., and Schoner, B. (2001). Magnitude adjustment for AHP benefit/cost ratios. *European Journal of Operational Research*, *133*(2), 342–351.

Weiss, E., and Rao, V. (1987). AHP design issues for large-scale systems. *Decision Sciences*, *18*(1), 43–57.

Xie, G., Zhang, J., Lai, K., and Yu, L. (2008). Variable precision rough set for group decision-making: An application. *International Journal of Approximate Reasoning*, *49*(2), 331–343.

Yokoyama, M. (1921). The nature of the affective judgement in the method of paired comparison. *American Journal of Psychology*, *32*, 357–369.

3

Analytic network process

3.1 Introduction

This chapter explains the theory and practical uses of the analytic network process method (ANP). You will learn how to use *Super Decisions* (Section 3.3), a software package that helps to structure decision problems and calculate priorities (scores) based on the ANP methodology. As ANP is a generalization of the analytic hierarchy process (AHP), it is advisable to read Chapter 2 first in order to ensure better understanding. Section 3.4 is dedicated to the methodological background of ANP.

The companion website provides illustrative examples with *Microsoft Excel*, and case studies and an example with *Super Decisions*.

3.2 Essential concepts of ANP

ANP is a generalization of AHP which deals with dependencies. In AHP, as with the other methods presented in this book, we assume that criteria are independent. If they are not independent, correlated criteria would result in an overevaluated weight in the decision, as will be illustrated. For example, if we want to buy a car, the criteria of *speed* and *engine power* are correlated. In the traditional MCDA methods, this dependency implies a heavier weight of these joint criteria. The ANP method allows these dependencies, also called feedbacks, to be modelled; they are closer to reality and, as a result, yield more accurate results. As dependencies can arise between any of the elements in the decision problem (i.e. alternatives, criteria, sub-criteria, the goal), the model is no longer linear as in AHP (Figure 3.1), where the elements are arranged in levels. A hierarchy is not necessary in the ANP model, where clusters replace the levels and each cluster contains nodes or elements (see Figure 3.2). The clusters are connected by a line, which in turn means that the elements or nodes contained are connected.

Multi-Criteria Decision Analysis: Methods and Software, First Edition. Alessio Ishizaka and Philippe Nemery.

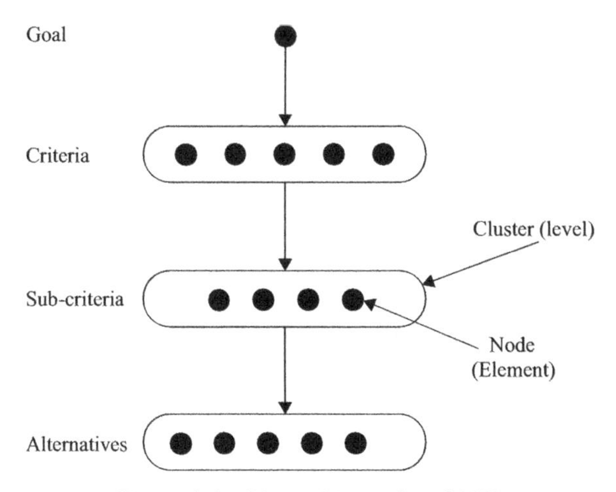

Figure 3.1 Linear hierarchy of AHP.

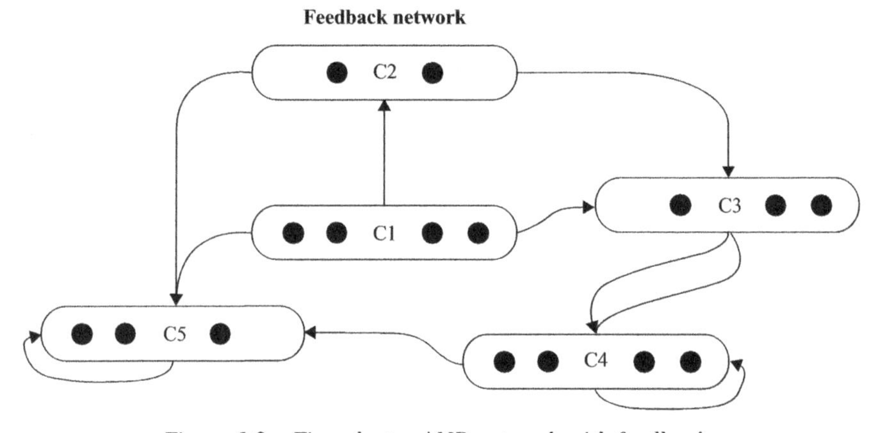

Figure 3.2 Five-cluster ANP network with feedbacks.

To illustrate the utility of ANP, we introduce in Sections 3.2.1–3.2.3 three types of dependency that cannot be solved with AHP.

3.2.1 Inner dependency in the criteria cluster

An inner dependency is a correlation of elements in the same cluster. Let us consider Case Study 3.1, which contains such a dependency.

Case Study 3.1

A school has decided to increase the educational level of the students. To accomplish this, it intends to reward the highest-achieving students in the national language, mathematics and physics. A transparent methodology for ranking students has been requested by the dean of the school.

The structure of Case Study 3.1 has three levels: goal, criteria and alternatives.

The evaluation is biased towards the scientific students, as there is a high correlation between the mathematics and physics courses. This *inner dependence* in the cluster cannot be modelled with AHP.

As an illustrative case, we consider two students: Shakespeare and Newton. Shakespeare has a talent for languages and Newton excels in scientific subjects. If the criteria are assumed to have the same weight, the result calculated by AHP would favour Newton as his strengths lie in the two scientific criteria (see Figure 3.3). ANP is necessary for modelling the inner dependency.

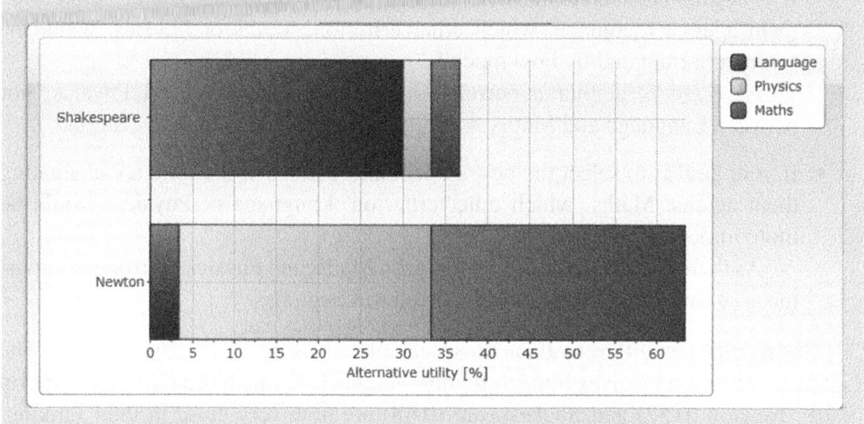

Figure 3.3 Evaluation of Shakespeare and Newton with AHP in MakeItRational. *Reproduced by permission of BS Consulting Dawid Opydo.*

The ANP problem structure is very similar to that of AHP, with three levels of clusters: goal, criteria and alternatives. The difference is the additional loop over the cluster criteria, which indicates an inner dependency (Figure 3.4).

In addition to the pairwise comparisons in traditional AHP, matrices modelling the inner dependency are required. These matrices aim to capture the relative importance

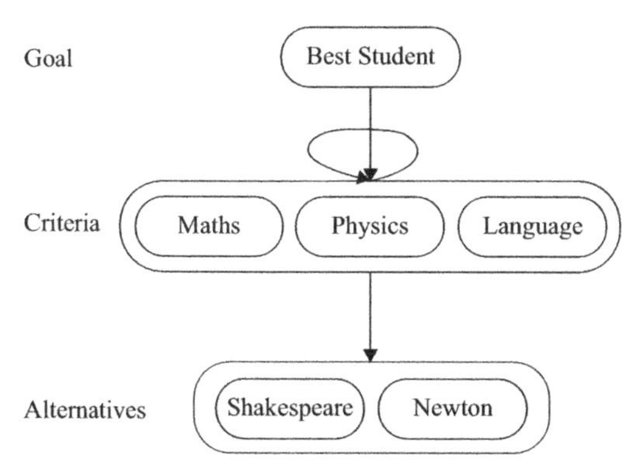

Figure 3.4 Case Study 3.1 structured in clusters.

of the criteria when another dependent criterion has already been evaluated. In Case Study 3.1, three additional matrices are required. The questions to be answered are as follows:

- If your goal is to select the best student and you know that you are evaluating them against Physics, which other criterion, Language or Maths, would be more important and by how much?

  As there is a strong correlation between Maths and Physics, a stronger importance (9 times more important) is given to Language.

- If your goal is to select the best student and you know that you are evaluating them against Language, which other criterion, Maths or Physics, would be more important and by how much?

  As there is neither a correlation between Language and Physics, nor between Language and Maths, a similar weight is given to both criteria.

- If your goal is to select the best student and you know that you are evaluating them against Maths, which other criterion, Language or Physics, would be more important and by how much?

  As there is strong correlation between Maths and Physics, a stronger importance (9 times more important) is given to Language.

The results are different than those calculated using AHP (Figure 3.5). The Language criterion receives a much higher weight (0.47) than the two other criteria (0.26). Newton (0.52) and Shakespeare (0.48) are now very close in their ranking, giving an unbiased evaluation.

If the decision maker set a weight of 0.5 for Language and 0.25 for Maths and Physics, the comparison to enter would be an infinite amount more important to Language compared to Maths and Physics. This is technically impossible to enter in the software but can be approximated with a very high number.

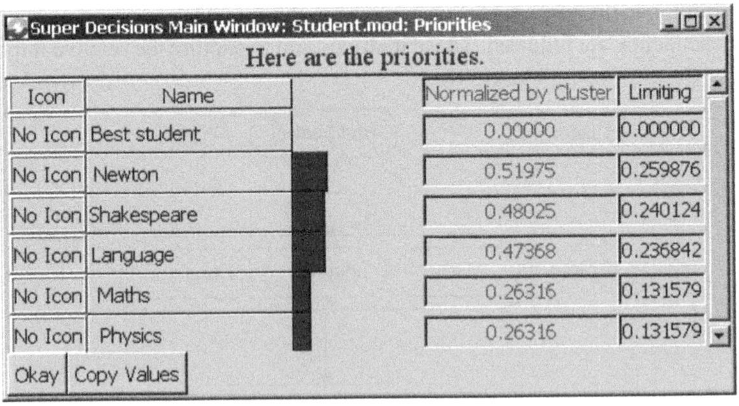

Super Decisions Main Window: Student.mod: Priorities

Here are the priorities.

Icon	Name	Normalized by Cluster	Limiting
No Icon	Best student	0.00000	0.000000
No Icon	Newton	0.51975	0.259876
No Icon	Shakespeare	0.48025	0.240124
No Icon	Language	0.47368	0.236842
No Icon	Maths	0.26316	0.131579
No Icon	Physics	0.26316	0.131579

Okay Copy Values

Figure 3.5 Ranking with ANP. © 1999/2003 Thomas L. Saaty. Reproduced by permission of Thomas L. Saaty.

3.2.2 Inner dependency in the alternative cluster

An inner dependency can also exist in an alternative cluster. This phenomenon is rare and less often considered. Two alternatives are negatively correlated in Case Study 3.2. The presence of two similar alternatives decreases their attractiveness.

Case Study 3.2

A woman wants to buy an elegant evening dress. The sales assistant suggests two dresses: A and B. The woman prefers A over B. However, the sales assistant then presents dress C as a third option, which is very similar to A. As the woman is worried about another woman wearing the same dress to the gala evening, she buys B.

If we model the problem in AHP and consider dresses A and C to be the same with regard to elegance and four times more elegant than dress B, then A and C would be given the highest priority (Figure 3.6). The inner dependency is not considered in this model and therefore ANP is needed.

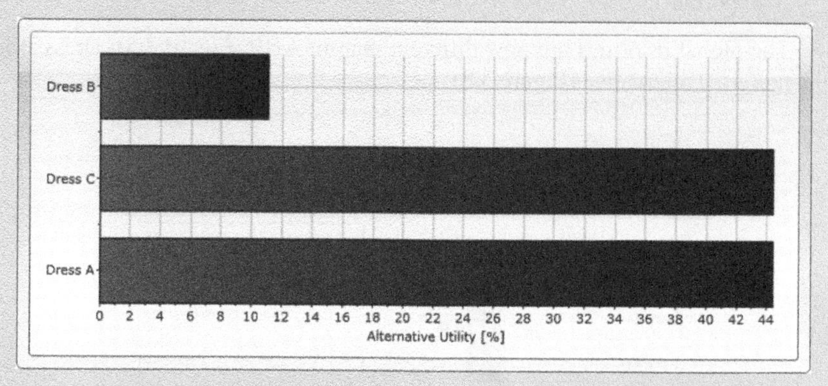

Figure 3.6 Dress evaluation with AHP in MakeItRational. *Reproduced by permission of BS Consulting Dawid Opydo.*

The ANP problem structure is similar to AHP, apart from the additional loop over the cluster alternatives, which indicates an inner dependency (Figure 3.7).

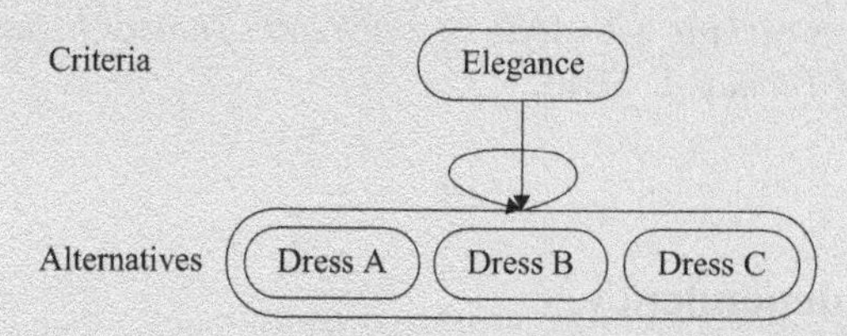

Figure 3.7 Case Study 3.2 structured in clusters.

In addition to the pairwise comparisons in traditional AHP, matrices modelling the inner dependency are required. These matrices aim to capture the relative importance of the alternatives, when it is known that another dependent alternative has already been evaluated. In Case Study 3.2, three new matrices modelling the inner dependency are required. The questions to be answered are as follows:

- If your goal is to select your preferred dress and you know that dress A is also in the evaluation cluster, which dress, B or C, is preferred?
  As dresses A and C are similar, dress B will be preferred (9 times more important).
- If your goal is to select your preferred dress and you know that dress B is also in the evaluation cluster, which dress, A or C, is preferred?
  As dresses A and C are similar and there is no correlation with dress B, they are equally preferred.
- If your goal is to select your preferred dress and you know that dress C is also in the evaluation cluster, which dress, A or B, is preferred?
  As dresses A and C are similar, dress B is preferred (9 times more important).

The global priorities are now different than in AHP (Figure 3.5). Dress B is the preferred alternative (Figure 3.8).

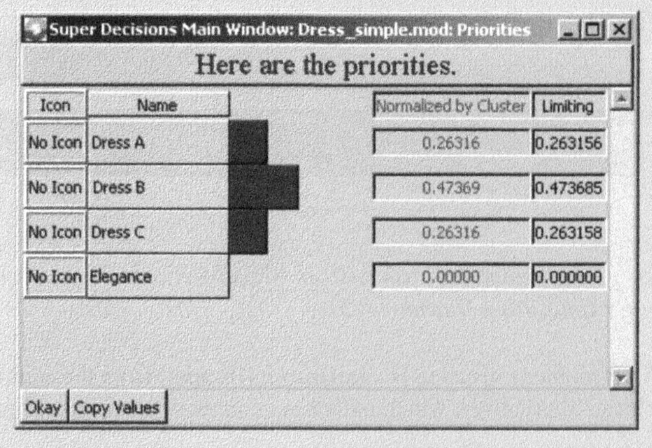

Figure 3.8 Dress ranking with ANP. © 1999/2003 Thomas L. Saaty. Reproduced by permission of Thomas L. Saaty.

3.2.3 Outer dependency

An *outer dependency* or *feedback* is a correlation between two clusters. For example, we can have an outer dependency between the cluster criteria and the cluster alternative: the weight of the criteria depends on the available alternatives. In AHP, these feedbacks can be captured through intuitive iterations. For example, in Case Study 3.3, the woman changes the weight of the criteria once she has learnt the price of the dresses. ANP uses a more formal approach. Instead of weighting the *elegance* and *price* criteria with respect to the goal, ANP assesses the relative importance of the criteria first with respect to dress A, then B.

Case Study 3.3

A woman decides to buy an evening dress. There are two criteria she considers to aid her decision: *elegance* and *price*. Elegance is considered to be much more important than price. The salesman presents the woman with two alternatives: dress B is not as elegant as dress A. The woman's preference is dress A, although it is much more expensive. In this case, the woman gives more weight to the price criterion and finally chooses dress B.

The problem structure in AHP of Case Study 3.3 has three levels: goal (buy a dress), criteria (elegance, price) and alternatives (dress A and dress B). If we consider the elegance criterion to be nine times as important as price, dress A to be twice as elegant as B, and dress B being nine times cheaper than A, dress A would be the preferred alternative (Figure 3.9).

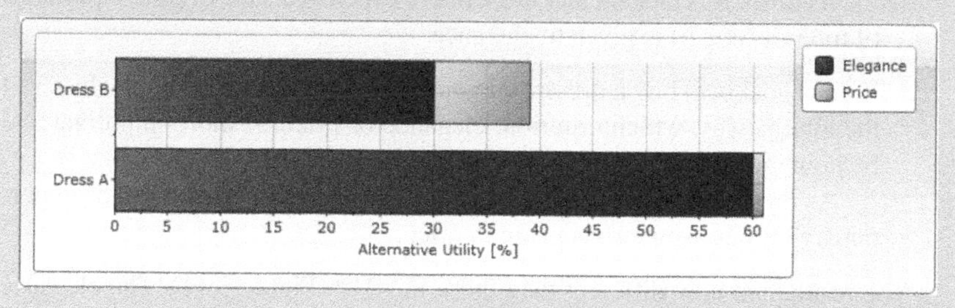

Figure 3.9 Evaluation with AHP. Reproduced by permission of BS Consulting Dawid Opydo.

Dress B would be preferred if the relative importance of elegance was twice as high as price. It is unlikely that the decision maker would take time to reconsider the judgements. Therefore, ANP is more appropriate for modelling outer dependencies.

The goal of the problem is used to establish criteria and alternatives, and it does not appear in the ANP network of Case Study 3.3 (Figure 3.10). This is

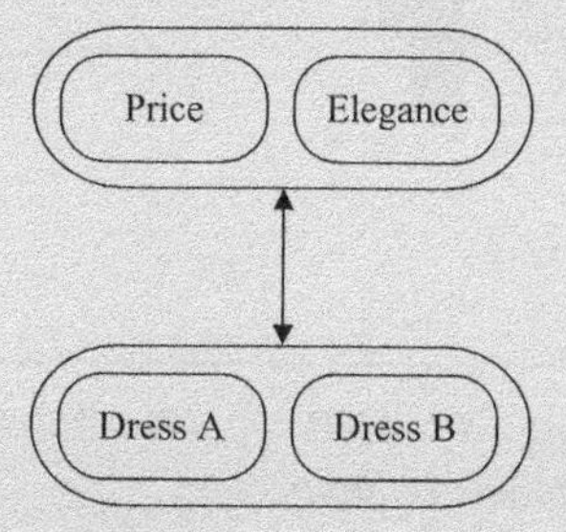

Figure 3.10 Case Study 3.3 structured in clusters.

conceptually surprising, especially for a decision maker using other MCDA methods. The justification for this absence is that no element in the problem depends on the goal: in this case, the weight of the criteria depends on the alternatives available and not on the goal.

As a result, in Figure 3.10, there is an arrow with double direction. If the weight of the criteria does not depend on the available alternatives, a goal needs to be added and the arrow between the criteria cluster and alternatives cluster would point in the direction of the alternatives.

In addition to the pairwise comparisons asked in traditional AHP, matrices modelling the outer dependency are required. These matrices aim to capture the relative importance of the criteria, with regard to the alternatives. In Case Study 3.3 two additional matrices (one for each alternative) are required.

The questions to answer are:

- If your goal is to buy a dress, and you know that dress B is in the cluster of the alternatives, which criterion, elegance or price, is more important and by how much?

  As dress B is elegant and moderately expensive, an extreme importance (9 times as much) is given to elegance.

- If your goal is to buy a dress, and you know that dress A is in the cluster of the alternatives, which criterion, elegance or price, is more important and by how much?

  As dress A is elegant but also expensive, a moderate amount of importance (3 times as much) is given to price.

The results are now different than those in AHP (Figure 3.11). The elegance criterion is given a more moderate weight and, as a result, dress B is preferred.

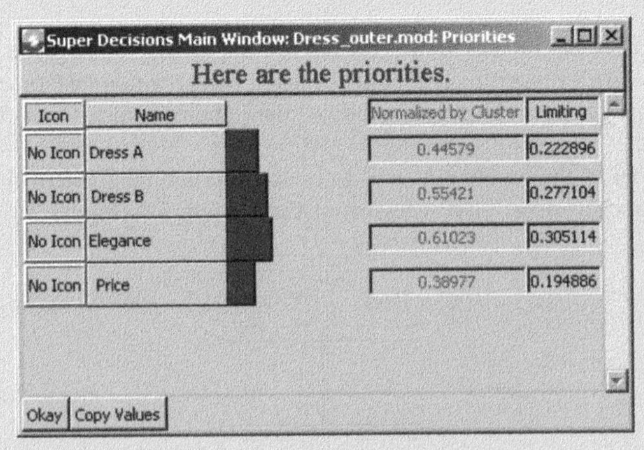

Figure 3.11 Dress ranking with ANP. © 1999/2003 Thomas L. Saaty. Reproduced by permission of Thomas L. Saaty.

Table 3.1 Influence matrix for a simple hierarchy.

		Goal	Alternatives				Criteria		
			A_1	A_2	A_3	A_4	C_2	C_3	C_4
Goal									
Alternatives	A_1						X	X	X
	A_2						X	X	X
	A_3						X	X	X
	A_4						X	X	X
Criteria	C_1	X							
	C_2	X							
	C_3	X							

3.2.4 Influence matrix

When the problem has been formulated and criteria and alternatives defined, it is good practice to complete an influence matrix. The influence matrix records with a cross any dependency between elements of the network. In AHP, the influence matrix would look like Table 3.1: the evaluation of the criteria depends on the goal, and the evaluation of the alternative depends on the criteria. In ANP, additional Xs in the influence matrix are possible to indicate the dependences between the different elements. If the criteria depend on the alternatives, the cluster goal is removed from the influence matrix (see Section 3.2.3).

Exercise 3.1

The following multiple-choice questions allow you to test your basic knowledge of ANP. Only one answer is correct. Answers can be found on the companion website.

1. What does ANP stand for?

    a) Analytic Neural Program

    b) Analytic Neural Process

    c) Analytic Network Process

    d) Analytical Network Program

2. What is an inner dependency?
    a) A correlation between two clusters
    b) A correlation between decision makers
    c) A correlation between alternatives and criteria
    d) A correlation of nodes in the same cluster
3. What is 'feedback' in an ANP structure?
    a) A correlation between two clusters
    b) A sensitivity analysis
    c) An inner dependency
    d) A master–slave dependency
4. Which of the following statements is false?
    a) ANP is more precise than AHP
    b) ANP requires more pairwise comparisons than AHP
    c) ANP is an extension of AHP
    d) ANP, unlike AHP, uses direct evaluations
5. In ANP, a goal node is not necessary when . . .
    a) Criteria depend on alternatives
    b) Alternatives depends on criteria
    c) A goal is unclear
    d) There is an inner dependency

3.3 ANP software: Super Decisions

There are two main programs that support ANP: *Super Decisions* and *Decision Lens*. *Super Decisions* has been developed for teaching purposes and can be downloaded, free of charge, from http://www.superdecisions.com. *Decision Lens* is an expensive commercial software package used in industry. Due to the cost of this software, it has not been possible to test it, so no comment can be made. In this section, the essentials for using *Super Decisions* are provided. A full tutorial can be downloaded from http://www.superdecisions.com/demos_tutorials.php3.

When using *Super Decisions* it is not necessary to understand how priorities are calculated. The decision maker identifies what should be ranked and the correlations between the elements in the problem.

To illustrate *Super Decisions*, the example of the purchase of an evening dress, as described in Case Study 3.4, will be used throughout this section.

Case Study 3.4

A woman would like to buy one of three evening dresses:

Salsaly.[1] a) Red lace ruffle dress in silk. Bodycon fit and one shoulder. Lace ruffle from shoulder to centre front and from back centre. The dress is a well-known Italian brand (same as the Tangal dress) designed by a famous German designer. It is priced at £2700.

Tangal. b) Asymmetric red draped dress in silk. Draped detail to front and back, with flower embellishment on shoulder. The dress is a well-known Italian brand (same as the Salsaly dress) designed by a famous French designer. It is priced at £2000.

Xenthila. c) Black minidress in satin. A wrap over chiffon drape design, with contrast mesh top, plunging V neck and sleeveless styling. The dress is a French brand by a famous Italian designer. It is priced at £1900.

The decision will be based on four criteria (Table 3.2).

Table 3.2 Criteria for dress purchase.

Criteria	Explanation
Brand	Brand usually determines prestige and quality
Designer	Designer is related to prestige and creativity
Fabric	Fabric of the dress
Price	Price of the dress

3.3.1 Problem structuring

Before using the software, it is important to structure the problem. All the elements (nodes) should be listed. These usually result from brainstorming and are then grouped into clusters.

In *Super Decisions*, new clusters can be created by selecting *Design/Cluster/New* from the menu. Similarly, new alternatives are added by selecting *Design/Node/New*.

The next step is to detect the influence one node could have on the others. Once detected, there are two ways to create links between the parent and child nodes.

[1] The names of the dresses are fictitious. Any resemblance to existing names is entirely coincidental.

- From the menu select *Design/Node connexions from*. From the list that appears, select a parent (or source) node. A second list will appear for all dependent nodes to be selected.
- Click on the 'three arrows' icon *Do connexions*. Left-click on the source node and right-click on all the dependent nodes, which will turn red.
  There are three types of dependencies:
- Outer dependencies:
  - ➢ Criteria influence the choice of an alternative with regard to the purchase of an evening dress, therefore all cells on the upper right-hand side of the matrix will always contain Xs.
  - ➢ The available alternatives influence the weight of the criteria, therefore all cells on the lower left-hand side of the matrix contain Xs.
- Inner dependencies cluster criteria:
  - ➢ The brand has a renowned name, which influences the price.
  - ➢ The name of the designer influences the price of the dress, as well as the type of the fabric used.
  - ➢ The fabric influences the price of the dress.
- Inner dependencies cluster alternatives:
  The availability of one alternative may have an influence on another. For example, a woman may decide not buy a dress if a similar one is available, as it might be awkward to see another woman wearing an almost identical dress on the same night (see Case Study 3.2).

There is no obligation to compile an influence matrix; the influence mapping can be done directly on the network diagram (e.g. Figure 3.12). However, an influence diagram like Table 3.3 gives a global view and a common reference for a group discussion. The nodes are grouped in clusters: criteria and alternatives. The goal node does not exist because in this case, the weight of the criteria depends on the alternatives available and not the goal. If the weight of the criteria does not depend on the available alternatives, a goal needs to be added.

3.3.2 Assessment of pairwise comparison

All the nodes to be pairwise compared are always in the same cluster. They are compared with respect to their parent node and provide local priorities. In order to start the process, you need to select *Assess/Compare* then *Node Comparisons* from the menu and click on *Do Comparison*. Four comparison modes are available: graphic, verbal, matrix and questionnaire. Figure 3.13 displays the questionnaire mode where the Salsaly and Tangal dresses are equivalent with regard to brand preference; the Salsaly brand is 5 times as preferable as that of Xenthila and Tangal is 5 times as

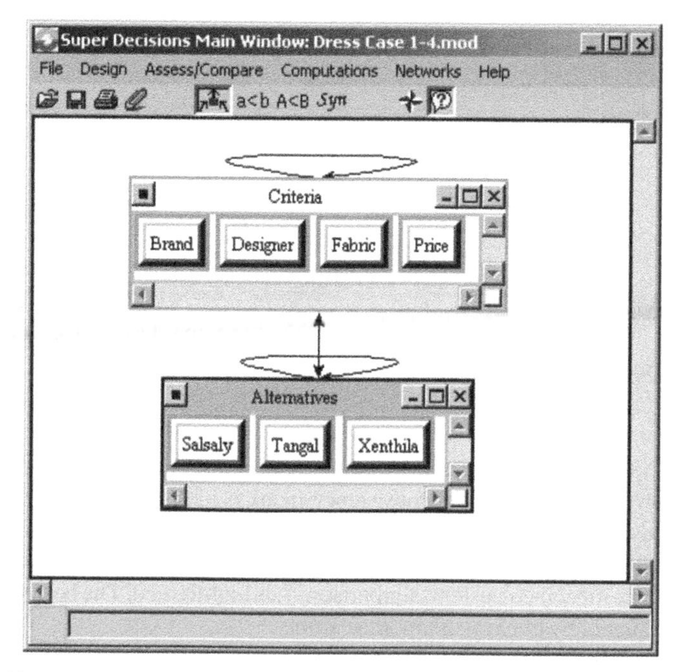

Figure 3.12 Case Study 3.4 structured with Super Decisions. *© 1999/2003 Thomas L. Saaty. Reproduced by permission of Thomas L. Saaty.*

preferable as Xenthila. To switch from one mode to another, you need to click on the top buttons.

To calculate the priorities and inconsistencies from any of the comparison modes, select *Computations/Show new Priorities*. In Figure 3.14 the local priorities of the matrix shown in Figure 3.13 are displayed. Salsaly and Tangal are the two most

Table 3.3 Influence matrix of Case Study 3.4.

		Alternatives			Criteria			
		Salsaly	Tangal	Xenthila	Brand	Designer	Fabric	Price
Alternatives	Salsaly		x	x	x	x	x	x
	Tangal	x		x	x	x	x	x
	Xenthila	x	x		x	x	x	x
Criteria	Brand	x	x	x				x
	Designer	x	x	x			x	x
	Fabric	x	x	x				x
	Price	x	x	x				

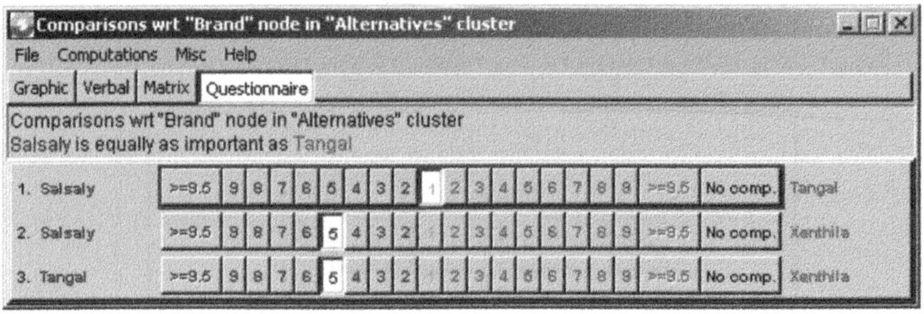

Figure 3.13 Pairwise comparisons of alternatives in Super Decisions. *© 1999/2003 Thomas L. Saaty. Reproduced by permission of Thomas L. Saaty.*

preferred alternatives with regard to brand. The matrix is consistent because the inconsistency index is 0. If the inconsistency index is higher than 0.1, it is advisable to revise the comparisons. To do this, select *Computation/Most inconsistent* from the menu where the most inconsistent comparison will be shown. For example, in Figure 3.15, the most inconsistent comparison, 2, is highlighted. The best replacement value is calculated by clicking *Show Best Value*.

Super Decisions allows a direct rating of the alternatives/criteria if they are already known. For example, in Figure 3.16, the price of each dress is known, therefore an exact amount can be entered. As a higher price is less preferable, the priorities need to be inverted, which is done by checking *Inverted*. If all evaluation preferences are rated directly, then a weighted sum is being used. *Super Decisions* supports ANP, AHP and a weighted sum because they share various common features.

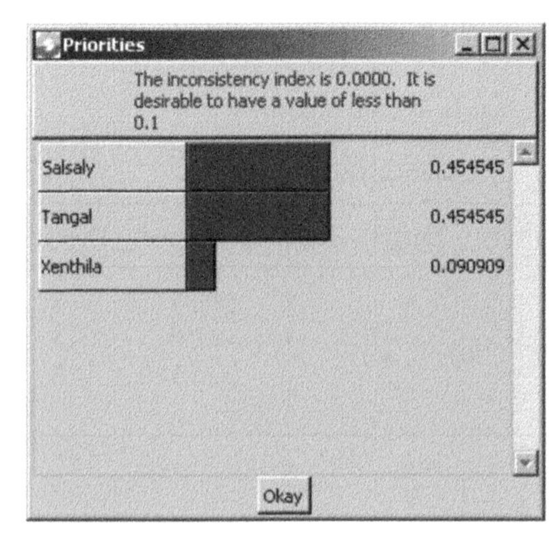

Figure 3.14 Local priorities in Super Decisions. *© 1999/2003 Thomas L. Saaty. Reproduced by permission of Thomas L. Saaty.*

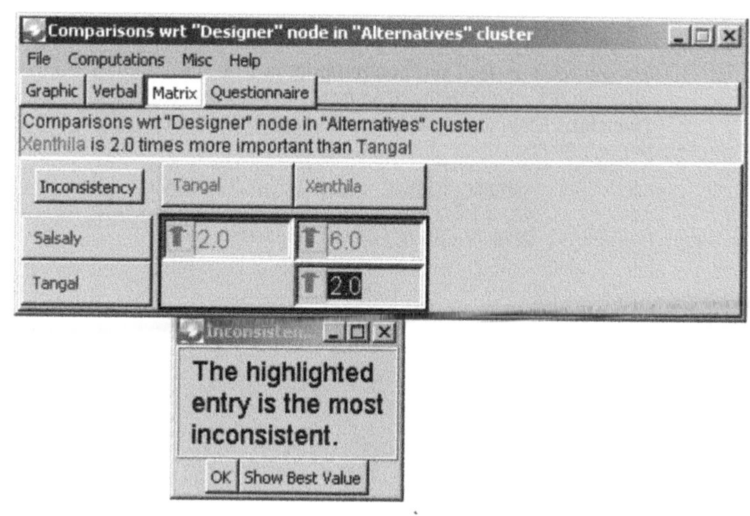

Figure 3.15 Most inconsistent comparison is highlighted. © 1999/2003 Thomas L. Saaty. Reproduced by permission of Thomas L. Saaty.

3.3.3 Results

The global priorities of the alternatives are obtained by selecting *Computations/ Synthesize* from the menu or by clicking on the Syn icon (Figure 3.12). The *Ideals* column is obtained from the *Normals* column by dividing each priority by the largest value, which means that the best alternative always has a score of 1. The *Normals* column normalizes the priorities to 1 and the *Raw* column comes from the limit supermatrix (see Section 3.4). In Figure 3.17, the Xenthila dress would be the better option.

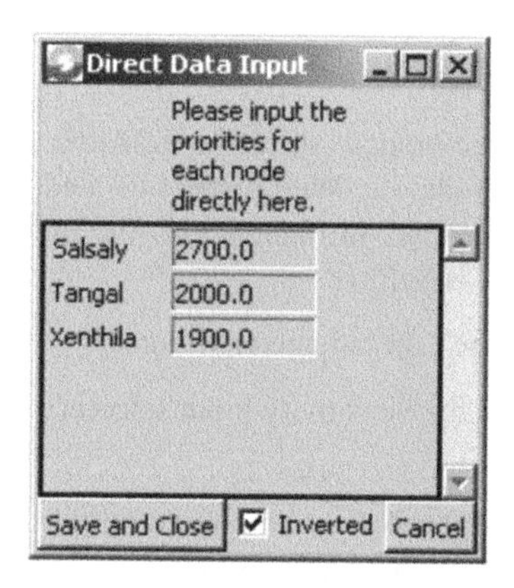

Figure 3.16 Direct rating of the alternatives. © 1999/2003 Thomas L. Saaty. Reproduced by permission of Thomas L. Saaty.

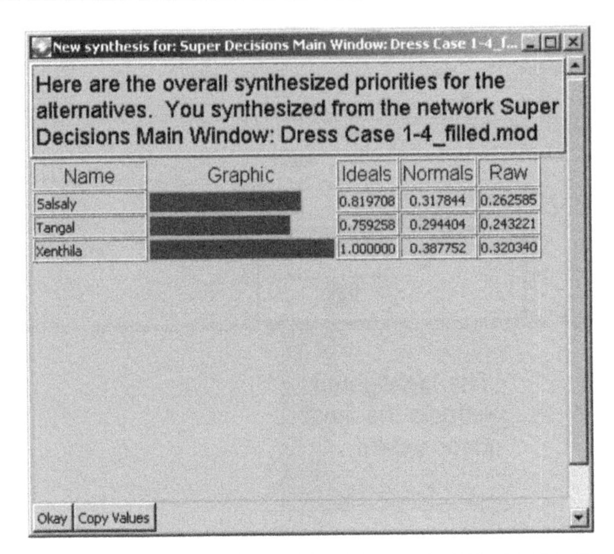

Figure 3.17 Global priorities in Super Decisions. *© 1999/2003 Thomas L. Saaty. Reproduced by permission of Thomas L. Saaty.*

3.3.4 Sensitivity analysis

The sensitivity analysis in ANP is more complicated than in AHP. In AHP, one criterion weight can be changed, whereas in ANP, each node can be linked to another node; an independent node must be defined with regard to a parent node. The following steps are used in *Super Decisions*:

- Select *Computations/Sensitivity* from the menu to access the sensitivity module.
- Select *Edit/Independent Variable* from the menu.
- In the Edit parameter window, select *SuperMatrix* for the Parameter Type, the parent node (Wrt Node, i.e. With respect to) and the independent node (1st other node). For example, in Figure 3.18, the independent variable is *Price* and the parent node is *Salsaly*.
- Click on *Done* on the Edit parameter window.
- Click on *Update* on the Sensitivity input selector.

In Figure 3.19, the weight of *Price* is plotted on the x-axis and the priorities of the alternatives are plotted on the y-axis. By clicking and dragging the black dotted vertical line it is possible to change the local priority of *Price*. We can appreciate that in any scenario, the Xenthila dress is a robust choice.

Figure 3.18 Parent and independent nodes. © 1999/2003 Thomas L. Saaty. Reproduced by permission of Thomas L. Saaty.

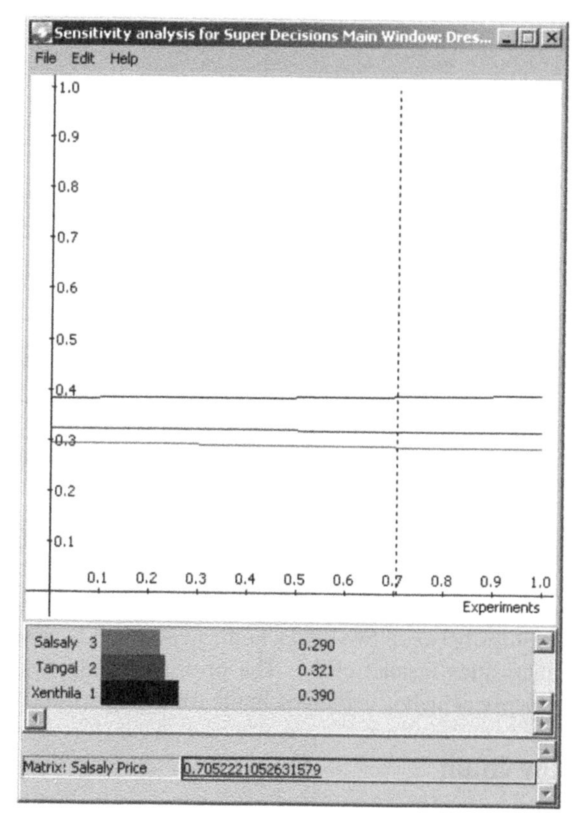

Figure 3.19 Sensitivity analysis in Super Decisions. *© 1999/2003 Thomas L. Saaty. Reproduced by permission of Thomas L. Saaty.*

Exercise 3.2

In this exercise, you will solve the evening dress problem set out in Case Study 3.4 with the *Super Decisions* software.

Learning Outcomess

- Structure a problem in *Super Decisions*
- Enter pairwise comparisons
- Understand the results
- Conduct a sensitivity analysis

Tasks

a) Open the *Super Decisions* software.

b) Read the description of Case Study 3.4 on page 69.

c) Enter the clusters (*Design/Cluster/New*).

d) Enter the nodes (*Design/Node/New*).

e) Link the dependent nodes (*Design/Node connexions from*).

f) Enter the pairwise comparisons (Menu *Assess/compare* then *Cluster Comparisons*). Are they consistent?

g) Read your global ranking (*Computations/Synthesize*).

h) Conduct a sensitivity analysis (*Computations/Sensitivity*).

3.4 In the black box of ANP

ANP rests on the same theory as AHP (Section 2.4), the only difference being the supermatrix. In this section, the priorities calculation from the supermatrix will be explained. The supermatrix is a partitioned matrix, where each submatrix is composed of a set of relationships between two nodes in the network. It is similar to Table 3.3, but consists of intensities instead of Xs. The priority calculation is based on the Markov chains process, which is explained in the following section.

3.4.1 Markov chain

A Markov chain, named after Andrey Markov, is a system that undergoes random transitions from one state to another with no memory of the past. This means that only the current state of the process can influence the next state. A more extensive presentation can be found in Norris (1997). As a simple example, we consider four children (Tom, Sam, Liz and Franz) passing a ball to each other. As they have different

	Tom	Sam	Liz	Franz
Tom	0	0.2	0.5	0.3
Sam	0.4	0	0.2	0.4
Liz	0.4	0.3	0	0.3
Franz	0.1	0.3	0.6	0

Figure 3.20 Transition probability matrix.

	Tom	Sam	Liz	Franz
Tom	0.31	0.24	0.22	0.23
Sam	0.12	0.26	0.44	0.18
Liz	0.15	0.17	0.44	0.24
Franz	0.36	0.20	0.11	0.33

Figure 3.21 Probability matrix after the second transition.

affinities, the ball is passed with unequal probabilities among them. The matrix in Figure 3.20 contains the transition probability of the ball from one child to another. For example, the probability of Tom giving the ball to Sam is 0.2, to Liz 0.5, and to Franz 0.3. The section across the diagonal will always be zero as the child will not keep the ball at each transition but pass it to another child. It is important to note that the sum of each row is 1, which indicates that the ball is always with a child. The probabilities are unchanged during the game because a previous pass has no influence on the next one.

In order to find the ball after the second transition, the matrix is squared (Figure 3.21).

If the number of transitions is infinite (i.e. high), the probabilities will stabilize and be identical in each column (Figure 3.22). This indicates that at the end of the game. After a large number of transitions, the probability of the ball being with Tom is 0.233, with Sam is 0.213, with Liz is 0.307 and with Franz is 0.247.

	Tom	Sam	Liz	Franz
Tom	0.232755	0.212865	0.307237	0.247143
Sam	0.232755	0.212865	0.307237	0.247143
Liz	0.232755	0.212865	0.307237	0.247143
Franz	0.232755	0.212865	0.307237	0.247143

Figure 3.22 Probability matrix after a large number of transitions.

Exercise 3.3

You will cover the Markov chain process step by step.

Learning Outcomes

- ➢ Understand how a probability matrix is compiled
- ➢ Understand the calculation of probabilities

Tasks

Open the file Ball Game.xls. It contains a spreadsheet with the calculation of priorities from a supermatrix.

Answer the following questions:

a) Describe the meaning of each cell in the supermatrix. (Read the comments in the red box in case of difficulty.)

b) Calculate by hand the second probability matrix.

3.4.2 Supermatrix

The influence of each node on other nodes in a network can be gathered in a supermatrix (Saaty, 2001). The supermatrix in Figure 3.23 sets out the influence on the three clusters: goal, alternatives and criteria. The order of these in the matrix is irrelevant (this is not the case with the hierarchy in AHP). If dependencies do not exist between

Cluster node model		Goal	Alternatives			Criteria		
			A_1	A_2	A_3	C_1	C_2	C_3
Goal		0	0	0	0	0	0	0
Alternatives	A_1	0	eigenvector of influence on each alternative (because of inner dependency in the alternative cluster)			local priority of alternative A_i with regard to criteria C_i		
	A_2	0						
	A_3	0						
Criteria	C_1	Weight of the criteria	0	0	0	eigenvector of influence on each criterion (because of inner dependency in the criteria cluster)		
	C_2		0	0	0			
	C_3		0	0	0			

Figure 3.23 Supermatrix with no outer dependency from the alternatives of the criteria.

Cluster node model		Alternatives			Criteria		
		A_1	A_2	A_3	C_1	C_2	C_3
Alternatives	A_1	eigenvector of influence on each alternative (because of inner dependency in the alternative cluster)			local priority of alternative A_i with regard to criteria C_i		
	A_2						
	A_3						
Criteria	C_1	weight of the criteria with regard to each alternative			eigenvector of influence on each criterion (because of inner dependency in the criteria cluster)		
	C_2						
	C_3						

Figure 3.24 Supermatrix with outer dependency from the alternatives on the criteria.

nodes, zero is entered. For example, only the grey sections in Figure 3.23 are completed. The network becomes a hierarchy and, as a result, AHP can be used; however the calculation can still be done using ANP. The results are the same in both cases. However, the disadvantage is that the calculation is more time-consuming. AHP uses a simple weighted sum for aggregation, whereas ANP requires the supermatrix to be squared many times. Because of this, ANP is not recommended if no dependency exists.

If an outer dependency from the alternatives on the criteria exists, the cluster goal can be removed and the weights with regard to each alternative can be added (section shown in grey on Figure 3.24).

The columns in the supermatrix must be normalized to 1 in order to have a stochastic matrix that can be used in a Markov chain process. To capture the transmission of influence along all possible paths of the network, the matrix is raised to powers. The matrix is squared to represent the direct influence of one element on another. The cubic power is taken to express the indirect influence of a second element, and so on. As the matrix is stochastic, it will converge to a limit supermatrix, which contains the global priorities.

Exercise 3.4

You will learn to calculate priorities from a supermatrix step by step.

Learning Outcomes

- Understand how a supermatrix is completed
- Understand the calculation of priorities from the supermatrix

Tasks

Open the file Evening Dress.xls. It contains a spreadsheet with the calculation of priorities from a supermatrix.

Answer the following questions:

a) Why does each column in the matrix have to be normalized to 1?

b) Describe the meaning of each cell in the supermatrix. (Read the comments in the red box in case of difficulty.)

References

Norris, J. (1997). *Markov Chains*. Cambridge: Cambridge University Press.

Saaty, T. (2001). *The Analytic Network Process*. Pittsburgh: RWS Publications.

4

Multi-attribute utility theory

4.1 Introduction

This chapter explains the theory and practical uses of the multi-attribute ultility theory (MAUT) method. You will learn how to use *RightChoice*, a software package that helps to structure decision problems and calculate scores based on the MAUT method. Section 4.3 is designed for readers interested in the methodological background of MAUT. As the MAUT method requires the decision maker to specify lots of input parameters, some 'elicitation' methods are presented in Section 4.4. The UTA, UTA^{GMS} and GRIP elicitation methods help to infer preference parameters from a sample ranking given by the decision maker. This section will include a brief description of the *UTA+* and *VisualUTA* software.

The companion website provides illustrative examples with *Microsoft Excel*, and case studies and an example with *RightChoice*.

4.2 Essential concepts of MAUT

MAUT is widely used in the Anglo-Saxon world and is based on the main hypothesis that every decision maker tries to optimize, consciously or implicitly, a function which aggregates all their points of view. This means that the decision maker's preferences can be represented by a function, called the utility function U (Keeney and Raiffa 1976). This function is not necessarily known at the beginning of the decision process, so the decision maker needs to construct it first.

The utility function is a way of measuring the desirability or the preference of objects, called alternatives. These can be consumer goods (cars, smartphones, etc.) or services. The utility score is the degree of well-being those alternatives provide

Multi-Criteria Decision Analysis: Methods and Software, First Edition. Alessio Ishizaka and Philippe Nemery.

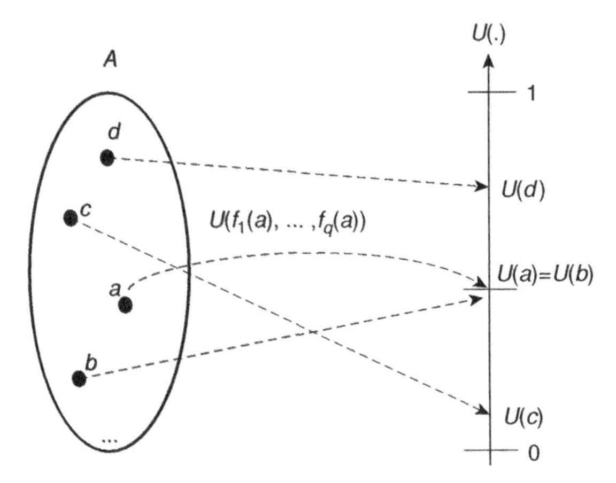

Figure 4.1 Representation of the ranking of the set A using the MAUT model.

to the decision maker. The utility function is composed of various criteria which enable the assessment of the global utility of an alternative. Consider, for instance, the purchase of a smartphone (see Case Study 4.1). In order to choose the most appropriate smartphone, one will measure the global utility of those available. The utility is usually grounded on several criteria, such as price, customer reviews, size. For each criterion, the decision maker will give a score, called the marginal utility score. The marginal utility scores of the criteria will be aggregated in a second phase to the global utility score.

Each alternative of set A is evaluated on the basis of function U and receives a 'utility score' $U(a)$ (see Figure 4.1). This utility score allows the ranking of all alternatives from best to worst. The preference and indifference relations amongst the alternatives of A are thus defined as follows:

$$\forall\, a, b \in A:\ a\mathbf{P}b \Leftrightarrow U(a) > U(b):\ a \text{ is preferred to } b, \tag{4.1}$$

$$\forall\, a, b \in A:\ a\mathbf{I}b \Leftrightarrow U(a) = U(b):\ a \text{ and } b \text{ are indifferent.} \tag{4.2}$$

The issue of incomparability between two alternatives, as in the outranking methods (see Chapters 6 and 7), does not arise since two utility scores (i.e. real numbers) are always comparable. Moreover, the preference relation on set A based on the utility scores is transitive. This means that if alternative a is better than alternative b, which in turn is better than alternative c, we can conclude that a is also better than c based on the utility score.

The utility function U can be defined in several different ways. In this chapter, the most common approach will be presented: the additive model. Moreover, only situations in which the evaluations of the alternatives are defined with certainty will be considered, although MAUT methods have been extended in the case of uncertainty and with stochastic information.

4.2.1 The additive model

Denote by F the set of q criteria f_j ($j = 1, \ldots, q$). The evaluations of the alternatives $f_j(a_i)$ are first transformed into marginal utility contributions, denoted by U_j, in order to avoid scale problems. The marginal utility scores are then aggregated with a weighted sum or addition (hence the term additive model). The additive model is the most popular and widely used model.

The general additive utility function can be written as follows:

$$\forall\, a_i \in A: U(a_i) = U(f_1(a_i), \ldots, f_q(a_i)) = \sum_{j=1}^{q} U_j(f_j(a_i)) \cdot w_j, \tag{4.3}$$

where $U_j(f_j) \geq 0$ is usually a non-decreasing function, and w_j represents the weight of criterion f_j. They generally satisfy the normalization constraint:

$$\sum_{j=1}^{q} w_j = 1 \tag{4.4}$$

The weights represent trade-offs, that is, the amount a decision maker is ready to trade on one criterion in order to gain one unit on another criterion.

When using additive functions some conditions such as the preferential independence condition between the criteria need to be respected; see the detailed explanation in Section 4.5. The 'simple' weighted sum is a special case in this model where U_j are all linear functions. The utility score corresponds to:

$$\forall\, a_i \in A: U(a_i) = \sum_{j=1}^{q} f_j(a_i) \cdot w_j. \tag{4.5}$$

Generally, the marginal utility functions are such that the best alternative (virtual or real) on a specific criterion has a marginal utility score of 1 and the worst alternative (virtual or real), on the same criterion, a score of 0. If the weights are normalized, the utility score of an alternative is always between 0 and 1.

Some examples of marginal utility functions are shown in Figures 4.2–4.4. The shapes of the marginal utility functions are determined by the decision maker and correspond to different attitudes with respect to risk or preference. If the decision maker estimates that small differences on low criteria performances are significant, he/she will opt for concave functions (risk-averse attitude). Figure 4.4 shows such a utility function. This is, for instance the case for the price criterion (which has to be minimized), where the decision maker expresses those small differences on low prices that are significant. This is given by the fact that the function decreases rapidly even if the price rises slowly.

On the other hand, if the decision maker considers small differences on high performances as important, he/she will opt for convex functions (risk-prone behaviour).

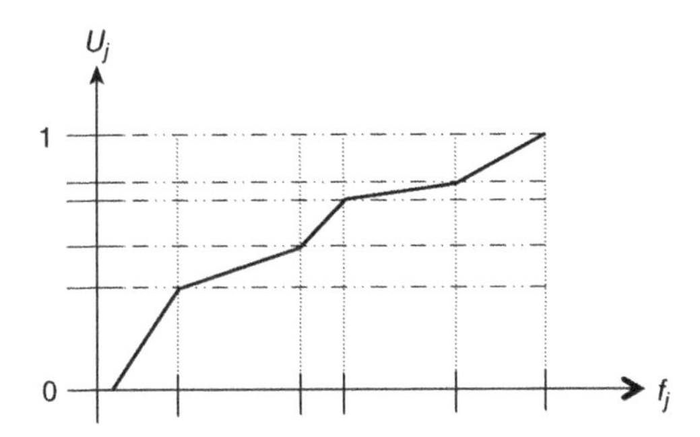

Figure 4.2 Piecewise linear marginal utility function.

In Figure 4.3, the marginal utility score of the criterion increases slowly at the beginning but rises sharply for higher values. Let us finally remark that linear functions represent a risk-neutral attitude (Zopounidis and Doumpos 2002).

Case Study 4.1

As illustrative example, consider the ranking of five fictitious smartphones (SP1, ..., SP5), which are evaluated on the following criteria:

(1) price (to be minimized);

(2) customer reviews (to be maximized);

(3) screen size (to be maximized);

(4) storage size (to be maximized).

The performance of the five smartphones on those criteria is given in Table 4.1.

Table 4.1 Performance table.

Raw data	Price (€)	Customer review	Screen size (in)	Storage size (Gb)
SP1	429	4	4.65	32
SP2	649	4	3.5	64
SP3	459	5	4.3	32
SP4	419	3.5	4.3	16
SP5	519	4.8	4.7	16

In order to compute the marginal utility functions, the decision maker first needs to rescale the raw performances between 0 and 1. This step depends on the definition of the marginal utility function. If this function takes only values between 0 and 1, then the raw performances need to be rescaled or normalized. If, on the other hand, the utility function can take any value, omit this step. The rescaling or normalization step is usually based on the minimum and maximum performance of the alternatives on each criterion:

$$f'_j(a_i) = \frac{f_j(a_i) - \min(f_j)}{\max(f_j) - \min(f_j)} \tag{4.6}$$

when maximizing the criterion, or

$$f'_j(a_i) = 1 + \left(\frac{\min(f_j) - f_j(a_i)}{\max(f_j) - \min(f_j)}\right) \tag{4.7}$$

when minimizing the criterion. Table 4.2 represents the rescaled performances of Case Study 4.1. Let us remark that this rescaling step might be performed after applying the marginal utility function in order to ensure utility scores of between 0 and 1.

Table 4.2 Rescaled performance table.

Rescaled	Price (€)	Customer Review	Screen Size (in)	Storage Size (Go)
SP1	0.957	0.333	0.958	0.333
SP2	0.000	0.333	0.000	1.000
SP3	0.826	1.000	0.667	0.333
SP4	1.000	0.000	0.667	0.000
SP5	0.565	0.867	1.000	0.000

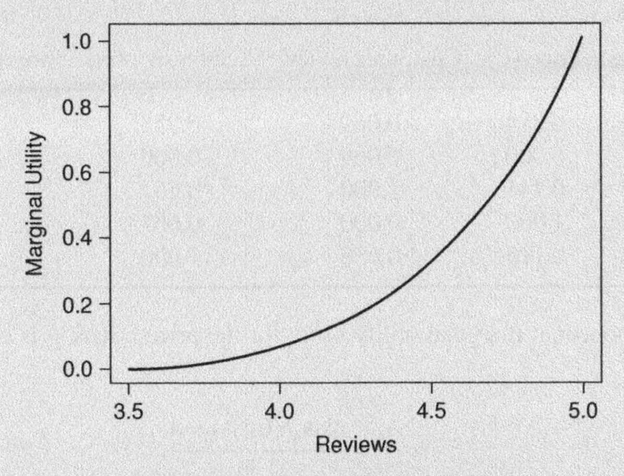

Figure 4.3 Positive exponential marginal utility function.

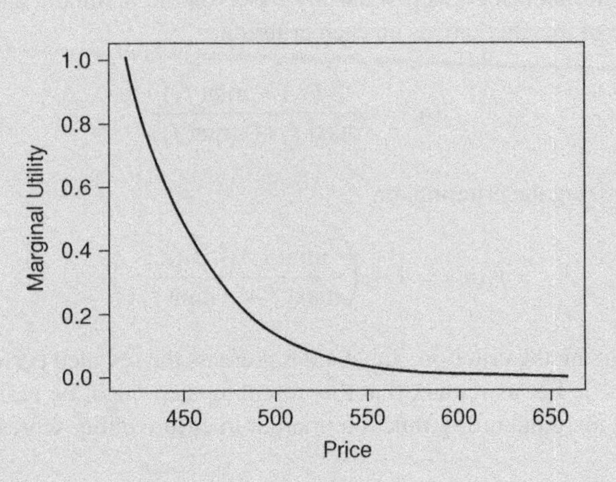

Figure 4.4 Negative exponential marginal utility function.

Suppose that the marginal utility functions of the screen size and storage size criteria are linear. On the other hand, consider an exponential marginal utility function for customer review and price as depicted in Figure 4.3 and Figure 4.4 with an exponent of 3 and 2, respectively. We remark that (4.8) uses the rescaled values (with (4.6) and (4.7)), which explains the decreasing utility function for the price utility function (Figure 4.4) despite the value of 2 for its exponent. The corresponding utility values are given in Table 4.3.

Table 4.3 Marginal utility values for the alternatives.

Marginal utility scores	Price (€)	Customer review	Screen size (in)	Storage size (Gb)
SP1	0.814	0.069	0.958	0.333
SP2	0.000	0.069	0.000	1.000
SP3	0.441	1.000	0.667	0.333
SP4	1.000	0.000	0.667	0.000
SP5	0.115	0.658	1.000	0.000

The exponential marginal utility score for the price criterion is computed as follows:[1]

$$U_1(a_j) = \frac{\exp(f_j'(a_i)^2) - 1}{1.71} \tag{4.8}$$

[1]exp(1) = 2.71, which explains the subtraction of 1 in the nominator as well as the division by 1.71 in order to obtain 1 for the best alternative.

Finally, consider the weights attached to the criteria as given in Table 4.4 and sum the marginal utility score to obtain the final utility scores. The resulting ranking is given in Table 4.5.

Table 4.4 Weights associated to the criteria.

	Price (€)	Customer review	Screen size (in)	Storage size (Gb)
Weights	0.35	0.35	0.15	0.15

Table 4.5 Global utility scores and ranking.

Final utility scores	Scores	Ranking
SP1	0.503	2
SP2	0.174	5
SP3	0.654	1
SP4	0.450	3
SP5	0.421	4

Central to the MAUT method are the marginal utility function and weights, which reflect the decision maker's preferences for each criterion. These parameters need to be designed by the decision maker with the possible help of a software package or analyst. Section 4.6 will present different analytical ways to define the utility functions and weights, whereas Section 4.4 will present the *RightChoice* Software which implements MAUT.

Exercise 4.1

You will learn how to calculate the marginal utilities of the alternatives, then aggregate the marginal utility scores to the global utility of the smartphones in Case Study 4.1.

Learning Outcomes

- ➢ Understand the normalization of data in *Microsoft Excel*
- ➢ Understand the calculation of the marginal utility functions in *Microsoft Excel*
- ➢ Understand the calculation of the global utility
- ➢ Understand the calculation of the final ranking

Tasks

Open the file Example_SmartPhones_MAUT.xls. This contains a spreadsheet with the different calculation steps of the utility scores.

Answer the following question:

Describe the meaning of each calculation cell and its formula. (Read the comments in the red squares in case of difficulties.)

Exercise 4.2

The following multiple-choice questions allow you to test your knowledge of the basics of MAUT. Only one answer is correct. Answers can be found on the companion website.

1. What does MAUT stand for?
   a) Measuring Awareness by a Utilization Technique
   b) Measuring Assurance by a Utility Technique
   c) Measuring Attractiveness by Utility Technique
   d) Multi-Attribute Utility Theory
2. Which of the following statements is correct?
   a) MAUT leads to a partial order of the alternatives
   b) A limited number of different marginal utility functions exist
   c) The utility scores lead to a complete order
   d) The utility function is always a sum of marginal utility functions
3. Which of the following statements is incorrect?
   a) The additive model is the most used aggregation model of MAUT
   b) The weighted sum is a particular MAUT model
   c) The normalization of the performances of the alternatives can be omitted in MAUT if the marginal utility functions are defined accordingly
   d) The MAUT method compare the alternatives pairwise in order to attribute them a score
4. Decreasing utility functions are generally for:
   a) Criteria to be minimized
   b) Criteria to be maximized

c) Criteria to be minimized or maximized

d) Only increasing criteria exist

5. What does not exist in MAUT?

a) Preferences

b) Indifferences

c) Incomparability

d) Sensitivity analysis

4.3 RightChoice

The additive MAUT method is easy to implement (see Exercise 4.1). However, if the decision maker wants to modify the input data easily and needs to perform a 'what-if' analysis (i.e. changing the weights, utility functions, etc. to analyse the impact on the final ranking), the use of dedicated software is recommended. *RightChoice*, which supports the additive model of MAUT, is a free software package with no limitations in usage, time or in the number of criteria or alternatives. *RightChoice* has been developed by Ventana Systems UK and can be downloaded from http://www.ventanasystems.co.uk/.

RightChoice calculates the marginal utility scores, global utility scores and the ranking of the alternatives. *RightChoice* models the decision problem as a tree of alternatives and criteria to define groups and subgroups of alternatives and criteria. *RightChoice* facilitates the sensibility analysis in a user-friendly and flexible manner. However, if a decision maker uses a similar shape of utility function (e.g. exponential) for several criteria, the parameters of those functions cannot be adapted for each criterion: they must all be the same (i.e. the exponent will always be of the same value). This is the main drawback of this software.

Furthermore, *RightChoice* allows the introduction of several scenarios of preference settings, which can be used in a group decision.

To illustrate this, Case Study 4.1 will be solved while changing some of the utility functions. The reader will find on the companion website the basic steps of the additive MAUT calculations in an *Excel* spreadsheet as well as the *RightChoice* input files.

4.3.1 Data input and utility functions

First enter the alternatives from Case Study 4.1. The user needs to define a 'root' group of alternatives by right-clicking in the left panel shown in Figure 4.5 (e.g. 'Smart Phones') and adding alternatives to the group by either right-clicking in the same place, or selecting 'Alternatives' from the menu. The user can define as many

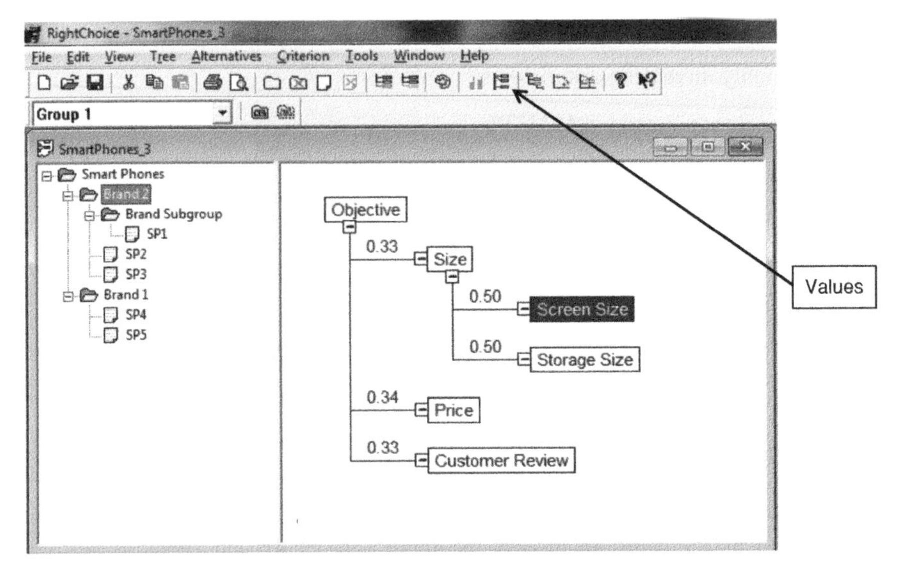

Figure 4.5 Tree for Case Study 4.1 in RightChoice. *Reproduced by permission of Ventana Systems UK Ltd.*

as groups and subgroups of alternatives as required. The menu allows the addition and removal of groups and alternatives.

Similarly, to enter the criteria of the decision problem the user needs to define a 'root' group of criteria by right-clicking in the right panel shown in Figure 4.5 (e.g. Objective). The user can define a hierarchy of criteria by adding 'sub-criteria' to a root criterion as illustrated in Figure 4.5.

The multi-level trees (compared to unique-level trees) will play a role when evaluating the alternatives (see Section 4.4.4).

Once the alternatives and criteria tree have been defined, the next step is to introduce the values of the alternatives or to evaluate them via the *Criterion* menu.

After choosing the alternative or group of alternatives, as well as a criterion (as highlighted in Figure 4.5: 'Brand 2' and 'Screen Size'), the user needs to select *Values* either using the button or menu *View/Values*. This leads to the Screen Size dialogue box shown in Figure 4.6.

If the alternatives are given in raw data (i.e. not normalized) complete the following procedure: click on the *Select* button in the dialogue window which defines the minimum and maximum values of the alternatives for this particular criterion (Units, Scale and Functions dialogue; see Figure 4.7). This step allows the normalization of data (see equations (4.6) and (4.7)). According to Table 4.1, the screen size criterion has minimum value 3.5 and maximum value 4.7. Click on *OK* and define the values of the chosen alternatives; see Figure 4.7. The raw data of the other alternatives needs to be entered in the same way.

Once all the raw data for a criterion have been entered, go back to the Units, Scale and Functions dialogue and choose from the five different marginal utility functions (linear, logarithmic, exponential, step and quadratic). The user needs to

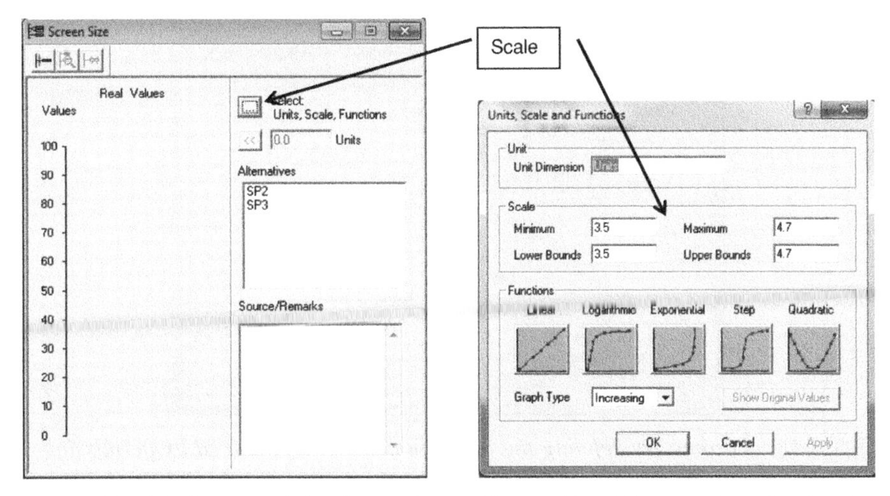

Figure 4.6 *Evaluation of the alternatives. Reproduced by permission of Ventana Systems UK Ltd.*

Figure 4.7 *Dialogue box for defining the units, scale and functions. Reproduced by permission of Ventana Systems UK Ltd.*

specify whether the criterion is to be maximized or minimized (*Graph Type* in Figure 4.8): this changes the shape of the utility function (compare Figure 4.7 and Figure 4.8).

The user can define the convexity or concavity of the marginal utility function by specifying the numerical parameter of the function (as illustrated in Figure 4.9) but cannot define piecewise linear marginal utility functions. This can be done in *Tools/Options/Functions*. However, as previously mentioned, the user cannot change the value of the parameter for different criteria once a shape is defined.

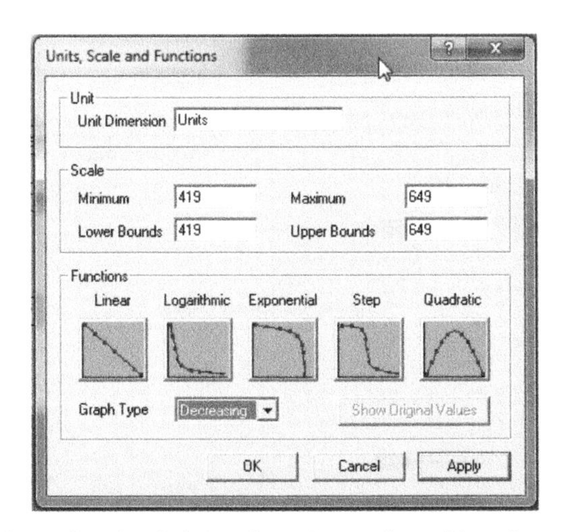

Figure 4.8 *Dialogue box for defining the units, scale and functions when minimizing the criterion. Reproduced by permission of Ventana Systems UK Ltd.*

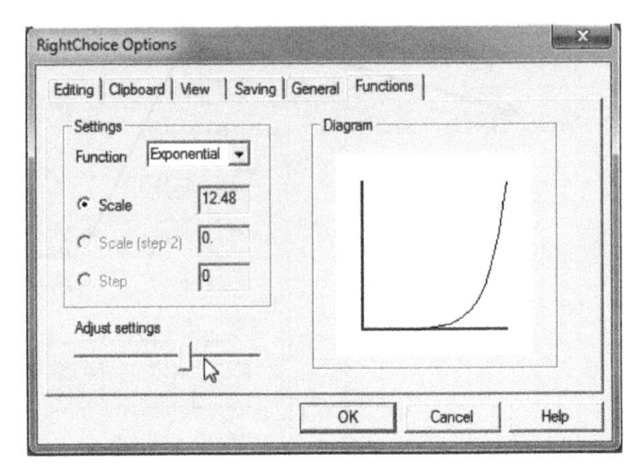

Figure 4.9 Parameters defining the utility functions. Reproduced by permission of Ventana Systems UK Ltd.

RightChoice can also be used to help the decision maker evaluate alternatives on the criteria. The dialogue box for values (Figure 4.10) for a chosen criterion permits the introduction of utility values of different alternatives. Usually, the decision maker starts by defining the worst and best alternative, to which they assign the marginal utility score of respectively 0 and 100.

The weight of the criteria can easily be modified with the Walking Weights option by clicking on the *Weights* button or via *View/Weights* (Figure 4.11). By changing the weights, the impact on the utility scores can be visualized. The user always needs to specify the group of criteria they want to act on. By selecting the root criterion, the

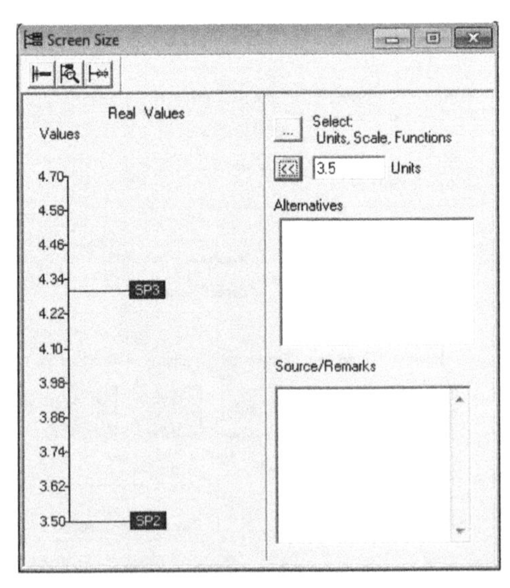

Figure 4.10 Dialogue box for the values on the screen size criterion. Reproduced by permission of Ventana Systems UK Ltd.

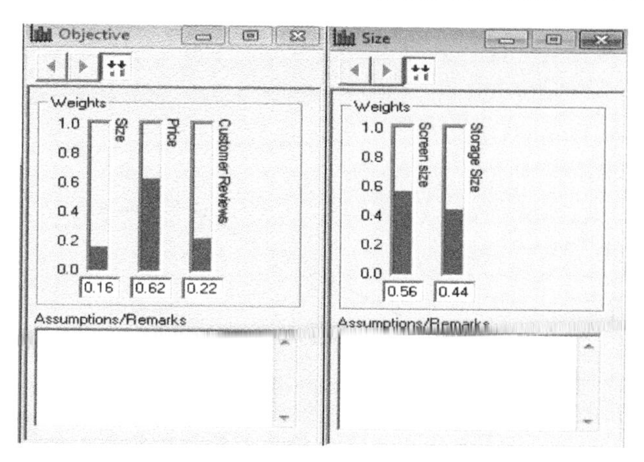

Figure 4.11 Walking weights in RightChoice. *Reproduced by permission of Ventana Systems UK Ltd.*

user will be able to modify the first-level criteria and group-criteria (see, for instance, the left panel of Figure 4.11).

4.3.2 Results

By clicking on *Frontier Analysis* or going to *Tools/Frontier Analysis*, the user can display the utility scores and the ranking of alternatives for a given criterion. The utility scores in *RightChoice* vary between 0 and 100, where 100 is the best score. To display the global utilities (Figure 4.12), the user needs to select the *root* of the criteria (i.e. Objective in Figure 4.5).

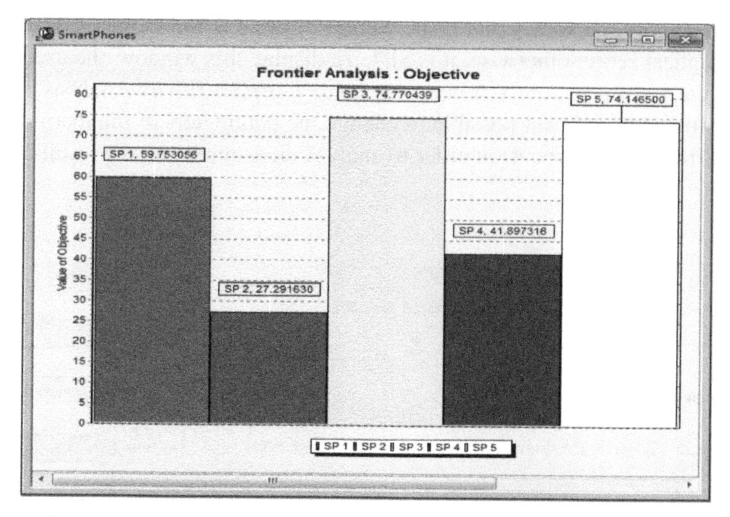

Figure 4.12 Utility scores for the alternatives in Case Study 4.1. Reproduced by permission of Ventana Systems UK Ltd.

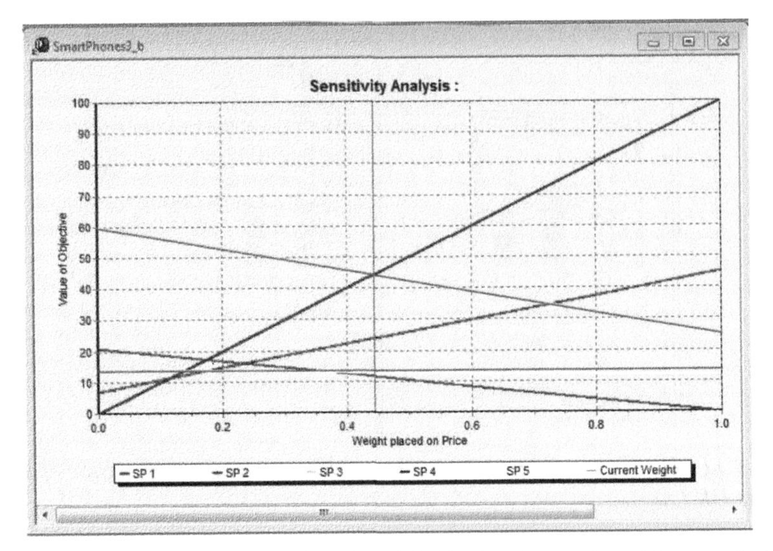

Figure 4.13 Sensitivity Analysis box in RightChoice. *Reproduced by permission of Ventana Systems UK Ltd.*

If there are groups of alternatives defined, all the alternatives need to be dragged and dropped into the root group, otherwise the user will only be able to see the scores per group and not for the complete set of alternatives.

4.3.3 Sensitivity analysis

RightChoice can perform a sensitivity analysis to illustrate when the ranking will be modified after changing a specific weight value. Figure 4.13 shows the sensitivity analysis of the global utility scores of the five smartphones when changing the price criterion. This graph shows that if the weight of price is lower than 0.44, then SP3 has the highest score; otherwise, it is SP4. To display this window, the user needs to click on the *Sensitivity Analysis* button or go to *Tools/Sensitivity* Analysis.

Unfortunately, it is not possible to change the parameters of the marginal utility functions (e.g. the exponent) in order to analyse their impact on the results.

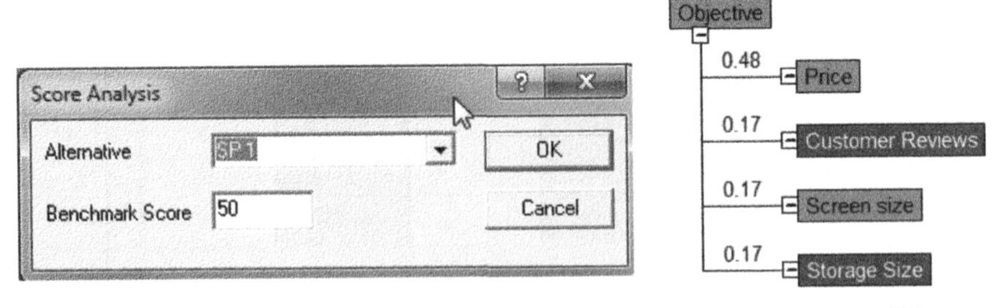

Figure 4.14 Score analysis in RightChoice. *Reproduced by permission of Ventana Systems UK Ltd.*

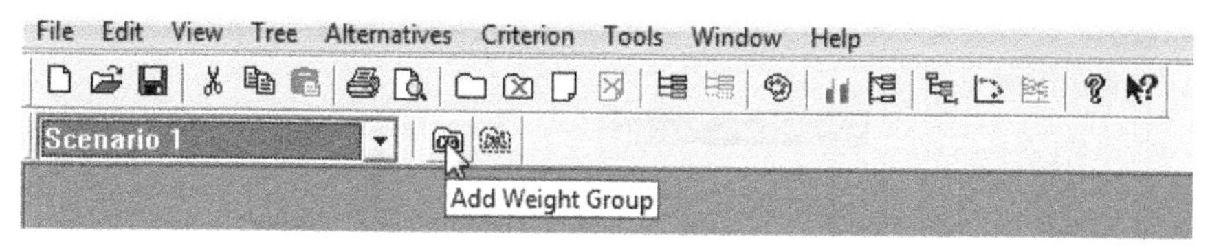

Figure 4.15 Addition/suppression of a weight group. Reproduced by permission of Ventana Systems UK Ltd.

Furthermore, the user can perform a *Score Analysis* (go to *Tools/Sensitivity Analysis*) to establish for which criteria a specific alternative has not achieved a benchmark score.

From Figure 4.14, one can see that SP1 does not reach the marginal utility score of 50 on the customer review and storage size criteria, which are marked in red.

4.3.4 Group decision and multi-scenario analysis

In some situations the decision maker might want to define several scenarios for their decision problem. For each scenario (also called a *weight group*), the decision maker can define different weight and utility functions.

The user can add/delete a new weight group via the toolbar menu as shown in Figure 4.15 and can then modify their settings by selecting the appropriate weight group.

Once the different scenarios have been set out, the user can define different weights for each scenario by opening the *Weights Series Overview*. By going to *View/Series Overview* the user can access the interactive *Weights Series* menu as well as the frontier analysis for the *Weight Series Overview* (Figure 4.16). This will allow the user to analyse the global ranking when changing the weights of the different scenarios.

If the decision problem involves several decision makers, the group facilitator can set out several scenarios corresponding to each decision maker's preference setting. Each decision maker can thus be weighted differently and the final group decision is given in the *Frontier Analysis: Weight Series Overview*. Unfortunately, no automatic insight is given about the difference in the ranking of the participants.

Exercise 4.3

In this exercise, you will solve the decision in Case Study 4.1 with the *RightChoice* software.

Learning Outcomes

- Structure a problem in *RightChoice*
- Enter the preference parameters
- Understand the results

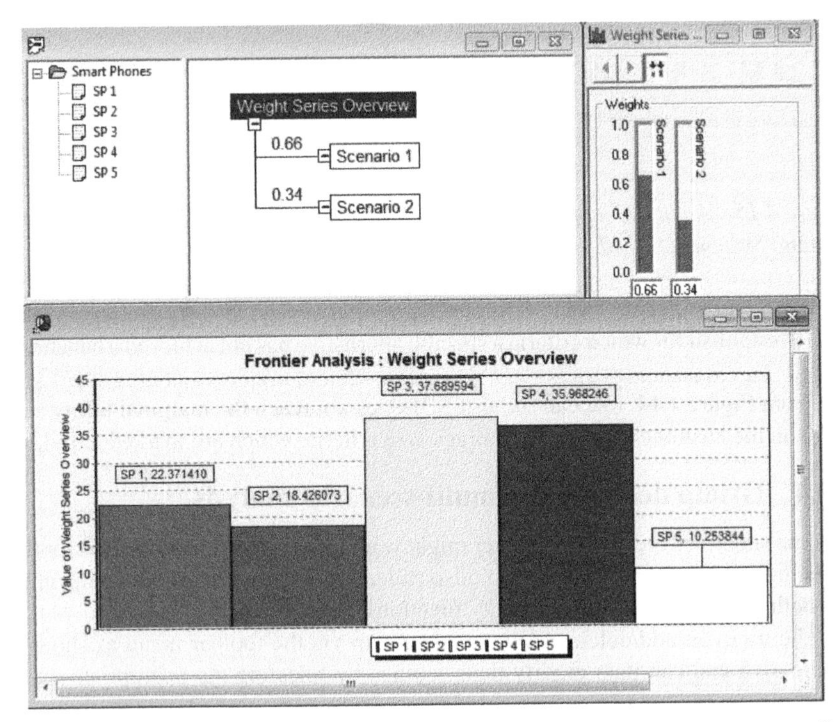

Figure 4.16 Weight Series Overview, Weights Series *menu and the corresponding* Frontier Analysis *when several scenarios have been defined. Reproduced by permission of Ventana Systems UK Ltd.*

Consider Case Study 4.1, where the performance of the alternatives is given in Table 4.1. Enter the data in the *RightChoice* software and check that the same results as in Figure 4.12 are obtained.

Tasks

a) Read the description of Case Study 4.1 and open the software. Choose *New* in the *File* menu.

b) From the *Criterion* menu, choose *Add* and insert the new criteria. For each criterion specify a name. From the *Alternatives* menu, choose *Add* and insert the new alternatives. Define some groups in the decision tree and drag and drop the alternatives and criteria into the groups (see Figure 4.5).

c) From the *View* menu, choose *Values* and edit the performance of the alternatives after choosing a specific criterion in the criterion tree. Enter the minimum and maximum values for the criterion considered, before specifying the scale transformation.

d) Choose an exponential scale for the price and customer reviews criteria with 20 as the exponent. Choose the linear scale for the two size criteria.

e) Display the marginal utility scores for each criterion and check if the marginal ranking corresponds to your intuition. Display the global utility score per group of alternatives as well as for the whole set of smartphones.

f) Is the ranking stable for any weight given to the combined size criterion?

g) At which specific weight values, given to the price criterion, will the ranking change?

h) Find the criteria for which SP2 does not reach the benchmark of 20 by performing a score analysis.

Answers can be found on the companion website.

4.4 In the black box of MAUT

MAUT is a method of total aggregation which computes the trade-off between criteria that respects the axioms of comparability, reflexivity, transitivity of choices, continuity and dominance (Beuthe and Scannella 1997). To have a family of criteria which consistently represents the decision maker's preferences, it is necessary for the criteria of F (see Section 4.3.1), defined on the set of alternatives A, to satisfy the axioms of exhaustibility, cohesion and non-redundancy. If these axioms are fulfilled, the criteria define a so-called *consistent* family of criteria (Roy 1974). The use of additive functions is only permitted if the preferential independence is respected. Refer to Keeney and Raiffa (1976) and Vincke (1992) for more information on this topic.

Having formulated the conditions that the utility functions need to fulfil, the shape of the utility functions must be considered. There are two different methods for constructing the utility functions: the direct and indirect methods. Before detailing these methods, the following considerations must be taken into account. This family of methods enables a complete pre-order of the alternatives, which is a rich result (i.e. a score is given for each alternative of A). The scores allow the preference strength to be assessed. On the other hand, the Pareto dominance relation only ascertains the dominating and dominated alternatives. Nothing is mentioned about the preference strength. Nothing can be said about efficient alternatives (i.e. alternatives that neither dominate nor are dominated by other alternatives). The dominance relation therefore gives a poor result.

The amount of information needed to build such aggregation models (through several questions asked of the decision maker) is very important and the hypotheses or axioms of this theory (existence, construction, additivity, etc. of U) are very strong. Such rich results are not always needed (e.g. a person may only want to find an ordinal ranking). These considerations have led to the development of a mid-way theory consisting of a mixture of the dominance relation and multi-attribute utility theory (Vincke, 1992): the outranking methods (see Chapters 6 and 7).

In the direct method, the marginal utility function U_j is estimated to construct U directly. The decision maker evaluates the parameters by answering direct questions about their preferences (through ratings, rankings, preferences on lotteries which generally include the choice between a certainty option and uncertain outcome but with probable higher gain). Several questioning procedures exist (Vincke, 1989), and Fishburn (1973) proposes various methods, which include probabilities, utilization of compensations between the marginal utility functions, etc.

The analyst needs to estimate this function by asking the decision maker appropriate questions. Nevertheless, two fundamental problems are inherent to this approach (Vincke, 1992):

1. How are the marginal utility functions constructed? How can their parameters be defined?
2. What properties should the decision maker's preference have? Can they be estimated analytically?

To tackle these problems, indirect methods are an alternative for constructing the marginal utility functions. They estimate the utility functions with global judgements (i.e. not for each criterion) expressed by the decision maker on a learning set L. This is achieved if the ranking of the alternatives of L is given. The elicited parameters have to respect the given ranking. These methods imply that the criteria need to be defined on a quantitative scale.

Jacquet-Lagreze and Siskos (1982) work on the basis of the utilities additives (UTA) method, which uses linear programming to obtain the parameters. The main idea of this method is that the marginal utility functions can be estimated through piecewise linear functions. Greco et al. (2008) proposed UTA^{GMS}, a generalization of the UTA method, which addresses the shortcomings of UTA. Recently Figueira et al. (2009) proposed the GRIP method, which generalizes both the UTA and UTA^{GMS} methods. GRIP has all of the features of UTA^{GMS} but takes into account the additional preference information in the form of the comparison of intensities of preference between pairs of reference actions.

UTA, UTA^{GMS} and GRIP are described in Section 4.6; other MAUT elicitation methods can be found in Krantz et al. (1971), Jacquet-Lagreze and Siskos (1982), Beuthe and Scannella (1997), Greco et al. (2008) and Figueira et al. (2009).

4.5 Extensions of the MAUT method

4.5.1 The UTA method

The aim of the UTA method (Jacquet-Lagreze and Siskos 1982) is to infer the marginal utility functions of U through the ordinal ranking given by the decision maker on the learning set L. The decision maker needs to rank the best to the worst of the alternatives of L, by giving each alternative a rank. Constraints can be imposed on the marginal utilities to respect (as much as possible) the given ranking. Properties such as transitivity will impose additional constraints. A brief introduction to the linear optimization technique is given in the Appendix.

The *UTA+* software implements the UTA method. This software has been chosen because it is not only user-friendly, but also provides a rich output such as the marginal utility functions, which can be interactively modified based on a learning set.

The software has a nice user interface but unfortunately only runs on Windows 3.1, 95, 98 and 2000. The software is available from Lamsade at Université Paris-Dauphine (http://www.lamsade.dauphine.fr/spip.php?article250). The main reason for describing this old software is to give the reader an intuitive description of the elicitation process. In Section 4.6.2, the more recent software *VisualUTA* is described, which does not currently have similar advanced functionalities.

In what follows, the weights are defined as

$$\forall\ j:\ w_j(f_j)\cdot U_j(f_j)=u_j(f_j) \tag{4.9}$$

This normalization allows us to avoid defining the weights as they are incorporated into the $u_j(f_j)$ functions.

Let f_j^+ and f_j^- be respectively the maximum and minimum values of the alternatives in set A to be ranked and the learning set L on criterion j (Figure 4.17). The interval $[f_j^-, f_j^+]$ can be subdivided into $\alpha_j - 1$ equal intervals where α_j is a parameter defined by the analyst (with $\alpha_j \geq 1$). When a criterion accepts only discrete scores, this parameter can also correspond to the number of possible values for this criterion. This allows us to define the 'endpoints' and which unknown marginal utility scores have to be determined by the linear program.

The ith endpoint f_j^i of the piecewise linear function of criterion j is defined as follows (Figure 4.17):

$$f_j^i = f_j^- + \frac{i-1\left(f_j^+ - f_j^-\right)}{(\alpha_j - 1)} \tag{4.10}$$

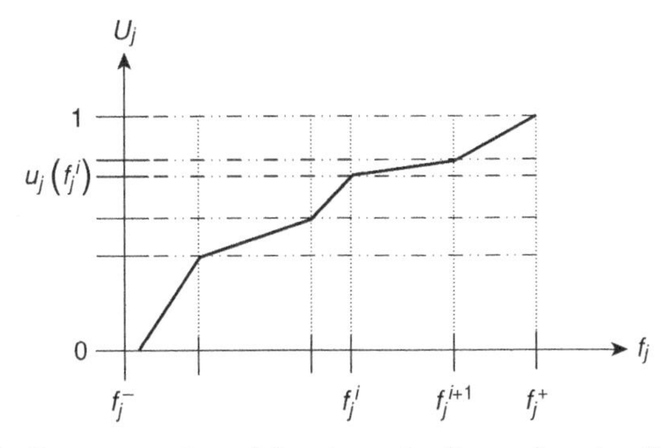

Figure 4.17 Representation of the piecewise linear function for criterion j.

If the marginal utility scores of these endpoints are known, the utility score of any point in $[f_j^-, f_j^+]$ can be inserted in the following way: if $f_j(a) \in [f_j^i, f_j^{i+1}]$, then

$$u_j(f_j(a)) = u_j\left(f_j^i\right) + \frac{f_j(a) - f_j^i}{f_j^{i+1} - f_j^i}\left[u_j\left(f_j^{i+1}\right) - u_j\left(f_j^i\right)\right] \qquad (4.11)$$

For all the alternatives of the learning set, the approximate marginal utility function U', $\forall a \in L$ can be calculated:

$$U'(a) = U(a) + \sigma(a) = \sum_{j=1}^{q} u_j(f_j(a)) + \sigma(a) \qquad (4.12)$$

where $\sigma(a)$ represents the potential error and $u_j(f_j^i)$ is unknown. The sum of all the potential errors needs to be minimized:

$$\textbf{Minimize G} = \left[\sum_{a \in L} \sigma(\text{a})\right]_{\text{minimize} \sum_{a \in L} \sigma(a)} . \qquad (4.13)$$

Since the decision maker has expressed an ordinal ranking (possibly with equal rankings) on the learning set L, some new constraints (equalities and inequalities) need to be added to the utility scores of the alternatives of L to satisfy the preference and indifference relations as defined in (4.1) and (4.2).

Given the transitivity of the utility scores, the equalities and inequalities obtained from (4.1) can be reduced to the preference relations between equivalent classes (i.e. groups of indifferent alternatives) and the indifference relations within each equivalence class. Suppose that there are Q equivalent classes with each n_p ($p = 1, \ldots, Q$) indifferent alternatives. Then the following constraints exist:

Constraint 1. Q preference relations: for all $a, b \in L$,

$$a\,\mathbf{P}\,b \Leftrightarrow U(a) > U(b) \Leftrightarrow \sum_{j=1}^{q} u_j\,(f_j(a)) - \sum_{j=1}^{q} u_j(f_j(b)) + \sigma(a) - \sigma(b) \geq \delta,$$

where δ is a small positive number.

Constraint 2. $\sum_{p=1}^{Q} n_p$ indifference relations: for all $a, b \in L$,

$$a\,\mathbf{I}\,b \Leftrightarrow U(a) = U(b) \Leftrightarrow \sum_{j=1}^{q} u_j(f_j(a)) - \sum_{j=1}^{q} u_j(f_j(b)) + \sigma(a) - \sigma(b) = 0.$$

Constraint 3. Normalized marginal utility functions are required:

$$\sum_{j=1}^{q} u_j\left(f_j^+\right) = 1.$$

Constraint 4. To fulfil the monotonicity condition, the following condition needs to be added:

$$u_j\left(f_j^{i+1}\right) - u_j\left(f_j^i\right) \geq t_j, \qquad l = 1, \ldots, \alpha_j - 1; j = 1, \ldots q$$

where $t_j \geq 0$ constitutes an indifference threshold for each criterion j.

Constraint 5. To express the fact that potential errors are positive, feasibility conditions on the error variables and marginal utilities are given: $\forall a \in L,\ \forall i, j$,

$$\sigma(a) \geq 0.$$

Constraint 6. The utility value of the minimum value on the criterion equals zero:

$$u_j\left(f_j^-\right) = 0.$$

Constraint 7. The utility values are always greater than or equal to zero:

$$u_j\left(f_j^i\right) \geq 0.$$

To find the variables $u_j(f_j^i)$, the linear program that minimizes G, the sum of the errors, needs to be solved subject to Constraints 1–7.

The optimal marginal utility functions obtained can lead to $G^* = 0$ or $G^* \neq 0$. If $G^* \neq 0$, there is no set of utility functions compatible with the given ranking by the decision maker. Jacquet-Lagreze and Siskos (1982) recommend increasing the number of endpoints (i.e. α_j, $j = 1, \ldots, q\alpha_j$) of the marginal utility functions to avoid this problem. On the other hand, several sets of utility functions may lead to the same value of G^*. In this situation, Jacquet-Lagreze and Siskos (1982) propose a further post-optimality analysis such that the utility functions lead to a ranking on L sufficiently 'close' to the reference ranking given by the user (in the sense of Kendall's rank correlation coefficient, which measures the similarity between two rankings). These steps correspond to the choice of only one set of utility functions (although several are possible), which implies that some preference information is lost (Greco et al. 2008). For more information on this topic, see Jacquet-Lagreze and Siskos (1982) and Greco et al. (2008).

In the UTA method, the marginal utility functions are limited to linear or piecewise linear marginal utility functions. This requires the analyst or decision maker to define the number of endpoints for each criterion, which is cumbersome and restrictive.

Finally, the decision maker needs to provide a complete ranking on the learning set L which avoids any incomparability amongst alternatives. A way of avoiding this problem is to ask the decision maker to define pairwise comparisons as proposed in Greco et al. (2008).

To address some of the shortcomings of UTA and in order to generalize its approach, Greco et al. (2008) proposed UTA^{GMS}, described in Section 4.6.2.

Exercise 4.4

In this exercise, you will learn how to use *UTA+*.

Learning Outcomes

- Construct the marginal utility functions using *UTA+*
- Understand the steps required in the *UTA+* software
- Obtain the ranking on the basis of a learning set
- Perform a what-if analysis

Tasks

Consider the problem of the ranking of smartphones in Case Study 4.1 based on a learning set. Define the learning set of three reference alternatives: SP3, SP2 and SP4.

Introduce the list of the alternatives of *A* and *L*. In other words, first introduce the list of all alternatives regardless of whether or not they belong to the learning set.

a) Go to *Problem/Edit*, and add the names of the alternatives (left-hand column in the Edition dialogue box in Figure 4.18).

b) Introduce the list of criteria (left-hand column in the Edition dialogue box in Figure 4.18). Define the settings *for each* criterion by clicking on *Edit* in the Edition dialogue box. Define the utility type: Gain or Cost (corresponding respectively to maximizing or minimizing the criterion) and the number of breakpoints for the utility functions.

c) Enter the performance of the alternatives (click on *Edit* in the Edition dialogue box after selecting the corresponding alternative); see Figure 4.19. Enter the

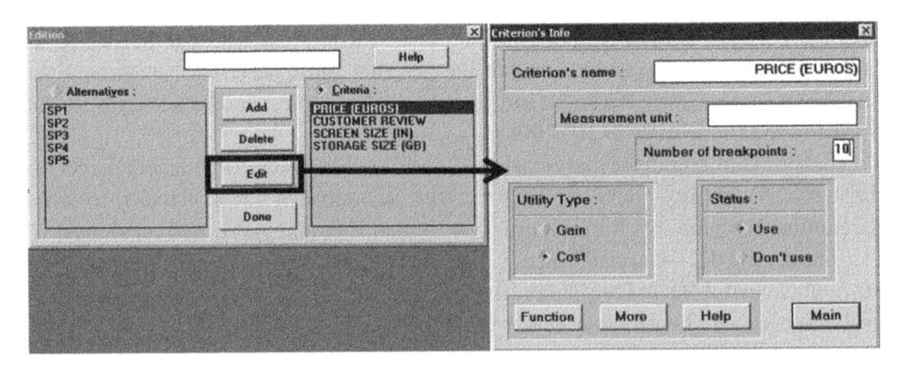

Figure 4.18 Criterion's Info dialogue box in UTA+. *Reproduced by permission of LAMSADE.*

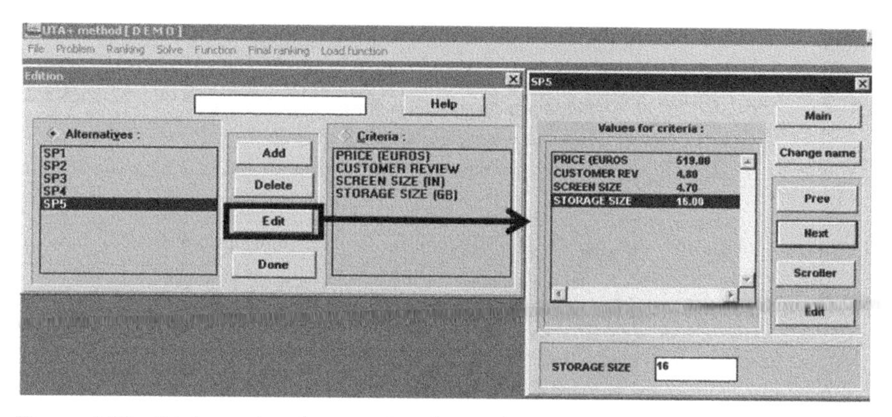

Figure 4.19 Dialogue box for entering the performance of an alternative in UTA+. *Reproduced by permission of LAMSADE.*

complete ranking on the learning set (choose *Ranking* from the menu bar); see Figure 4.20.

d) Define the settings for solving the linear program (go to *Solve*); see Figure 4.20. Generate the solution of the linear program and analyze the marginal utility functions. Do the elicited marginal utility functions lead to the given ranking on the learning set? Generate the ranking on all alternatives of A (see Figure 4.21).

e) Modify the shape of one marginal utility function and check that it defines a coherent ranking on learning set L (Figure 4.22). What is the new ranking under these new functions? (Figure 4.23).

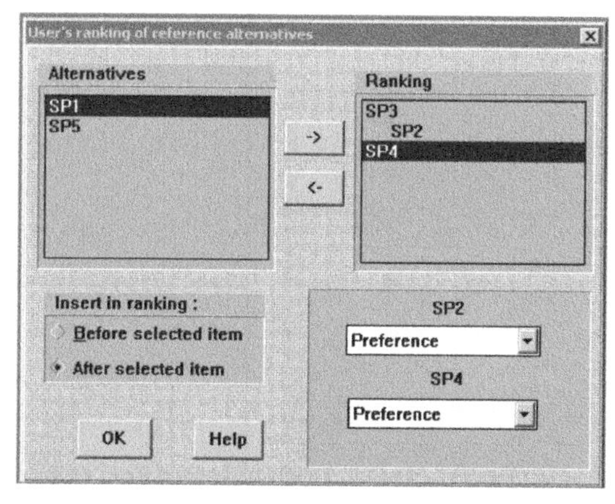

Figure 4.20 Introduction of the ranking on the learning set $L = \{\text{SP3, SP2, SP4}\}$. *Reproduced by permission of LAMSADE.*

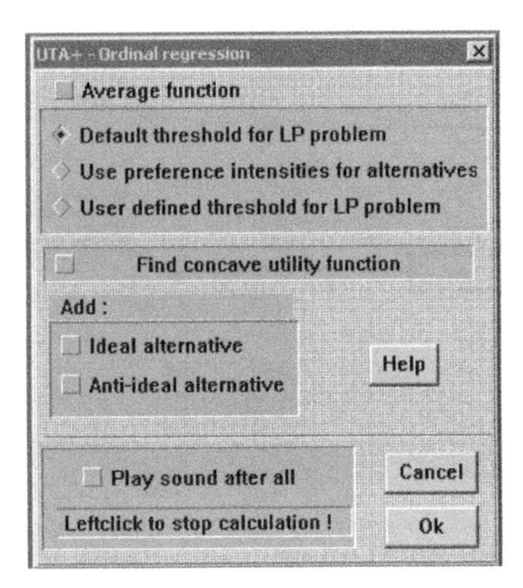

Figure 4.21 *Introduction of the parameters to solve the linear programme. Reproduced by permission of LAMSADE.*

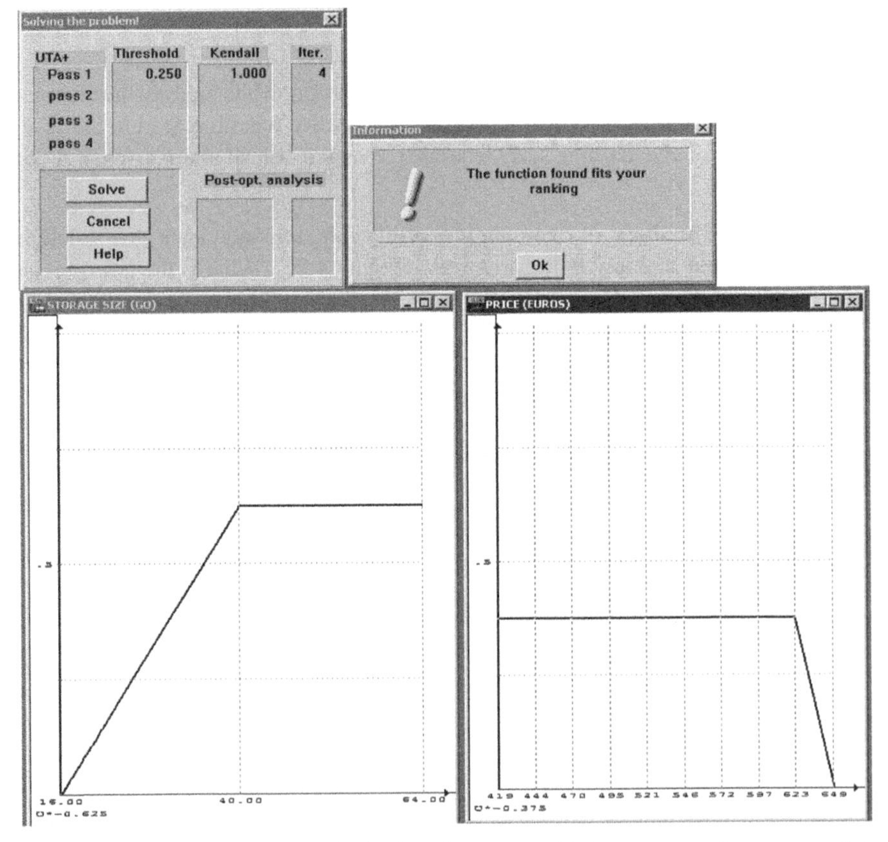

Figure 4.22 *The elicited marginal utility functions which ensure a coherent complete ranking on* L. *Reproduced by permission of LAMSADE.*

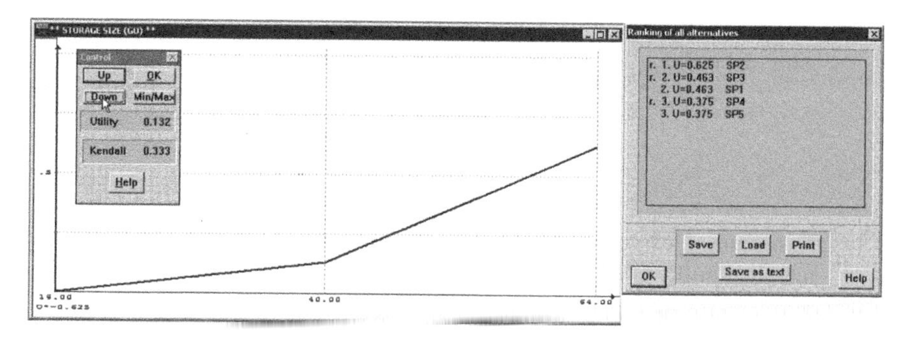

Figure 4.23 Window to modify the shape of the marginal utility function. Reproduced by permission of LAMSADE.

4.5.2 UTA^{GMS}

The UTA^{GMS} method generalizes the UTA method in three aspects (Greco et al. 2008). Unlike the UTA method which uses only one set of piecewise linear functions, the UTA^{GMS} method can take all additive value functions compatible with the preference information provided by the decision maker (see Figure 4.24). The only hypothesis on these functions is that they are non-decreasing (this generalizes the piecewise linearity). Moreover, the ranking on the learning set L does not need to be complete (i.e. there might be incomparable alternatives in L).

The software *VisualUTA* that supports UTA^{GMS} is available from http://idss.cs.put.poznan.pl/site/visualuta.html and will be described briefly in Exercise 4.5.

Similarly to UTA, the UTA^{GMS} method takes input from the decision maker in the form of an ordinal ranking on the learning set L. The ranking can be a partial pre-order: the decision maker provides pairwise comparisons between the alternatives of L. Non-decreasing additive utility functions are elicited such that they are compatible with the given preference information.

To determine the utility function values the following constraints need to be respected (analogous to Constraints 1–7):

1. Constraint 1 takes into account the user preferences on the learning set:

$$\forall a, b \in L: \begin{cases} a\text{P}b \Leftrightarrow U(a) > U(b) \\ a\text{I}b \Leftrightarrow U(a) = U(b). \end{cases}$$

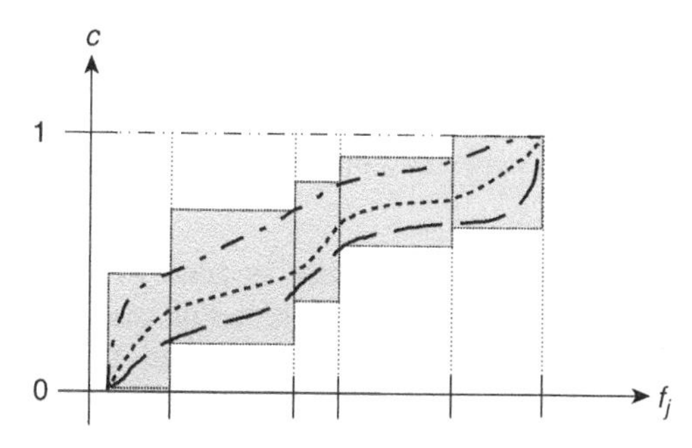

Figure 4.24 Marginal utility functions in UTA^{GMS}.

2. Constraint 2 expresses the fact that the highest value on a criterion has a marginal utility of 1, for all i:

$$\sum_{j=1}^{q} u_j(f_j^+) = 1.$$

3. Constraint 3 expresses the fact that the lowest value on a criterion has a marginal utility of 0:

$$\forall j : u_j\left(f_j^-\right) = 0.$$

4. Constraint 4 requires that the utility values are always greater than or equal to 0:

$$\forall i, j : u_j\left(f_j\left(a_{\tau_j(i-1)}\right)\right) \geq 0$$

5. Constraint 5 ensures the monotonicity condition, that is, that higher criterion values lead to higher utility values:

$$u_j\left(f_j\left(a_{\tau_j(i)}\right)\right) - u_j\left(f_j\left(a_{\tau_j(i-1)}\right)\right) \geq 0, \quad i - 1, \ldots, m; j = 1, \ldots, q,$$
$$u_j\left(f_j^+\right) - u_j\left(f_j\left(a_{\tau_j(m)}\right)\right) \geq 0, \quad j = 1, \ldots, q,$$

where $m = |L|$ and τ_j is the permutation on the set of indices of alternatives from L that reorders the alternatives with respect to the increasing evaluation on criterion f_j. In other words,

$$f_j\left(a_{\tau_j(1)}\right) \leq f_j\left(a_{\tau_j(2)}\right) \leq \ldots \leq f_j\left(a_{\tau_j(m)}\right).$$

Compatible value functions with the preference information given by the decision maker do not necessarily exist (i.e. the linear program does not necessarily have a solution). This might happen if the decision maker has made an error in their statements or if the statements are inconsistent. On the other hand, it might be possible that an additive model cannot model the preference information. Greco et al. (2008) suggest that, in order to overcome this impasse, incompatibilities are either accepted or not.

Based on the elicited functions, two preference relations can be defined on the alternatives of A: the *weak necessary* preference relation and the *possible weak* preference relation.

1. In the necessary ranking, a is at least as good as b, written as $a \succsim^N b$, if $U(a) \geq U(b)$ for all the value functions. In this case, a necessary preference relation which is reflexive and transitive is needed. The necessary ranking is a partial pre-order (Greco et al. 2008).

2. In the possible ranking, a is at least as good as b, written as $a \succsim^P b$, if $U(a) \geq U(b)$ for at least one value function compatible with the ranking on L. The possible ranking is 'strongly' complete (Greco et al. 2008): there is no incomparability in the possible ranking.

Consider the notation $a \succ b \Leftrightarrow U(a) \geq U(b)$. This leads to

$$\forall a, b \in L : a \succ b \Rightarrow a \succ^N b \quad \text{and} \quad aPb \Rightarrow not\,(b \succ^P a).$$

The necessary preference relation is included in the possible preference relation ($\succ^P \supseteq \succ^N$): If $U(a) \geq U(b)$ for all possible value functions U, then there is at least one compatible value function U' such that $U'(a) \geq U'(b)$.

The necessary ranking can be considered as robust with respect to the provided ranking on L, since the preference relations between two alternatives hold for any value function. When no preference relation exists between a pair of alternatives, the necessary ranking corresponds to the weak dominance relation and the preference relation is a complete relation.

The authors suggest using the UTA^{GMS} method interactively with the decision maker by adding alternatives to learning set L, since this will be taken into account by the model in the next iteration.

In order to compute the binary relations $\succ^{\text{N}}$ and $\succ^{\text{P}}$ for two alternatives a, b of A, proceed as follows (Greco et al. 2008):

1. Reorder the alternatives from $L \cup \{a, b\}$ from the worst to the best on criterion f_j, and note this permutation π_j:

$$f_j\left(a_{\pi_j(1)}\right) \leq f_j\left(a_{\pi_j(2)}\right) \leq \ldots \leq f_j\left(a_{\pi_j(\eta)}\right), \tag{4.14}$$

   where $\eta = |L \cup \{a, b\}|$. This value can be equal to $m+2$, $m+1$ or m according to the union operator.

2. The endpoints of the utility functions are defined as follows: $\forall_j = 1, \ldots, q$,

$$f_j^0 = f_j^-; f_j^i = f_j\left(a_{\pi_j(i)}\right); \ldots; f_j^{\eta+1} = f_j^+. \tag{4.15}$$

3. The following constraints, depending on a and b, need to be satisfied for $i = 1, \ldots, \eta+1$ and $j = 1, \ldots, q$.

Constraint 8

$$\forall a, b \in L : \begin{cases} a\text{P}b \Leftrightarrow U(a) > U(b) + \sigma \\ a\text{I}b \Leftrightarrow U(a) = U(b) \end{cases} \qquad \forall i, j : u_j\left(f_j^0\right) = 0;\ j = 1, \ldots, q,$$

where σ is random small positive number.

Constraint 9

$$u_j\left(f_j^i\right) - u_j\left(f_j^{i-1}\right) \geq 0;\ i = 1, \ldots, \eta+1.$$

Constraint 10

$$\sum_{j=1}^{q} u_j(f_j^{\eta+1}) = 1.$$

1. If the solution space is not empty:
   - $a \succsim^N b \Leftrightarrow d(a,b) \geq 0$ where $d(a,b)\min\{U(a) - U(b)\}$ respecting Constraint 10;
   - $a \succsim^P b \Leftrightarrow D(a,b) \geq 0$ where $D(a,b) = \max\{U(a) - U(b)\}$ respecting Constraint 10.

   Let us remark that not all the distances $d(a,b)$ and $D(a,b)$ need to be computed, as explained in Greco et al. (2008).

The authors of the UTA^{GMS} method have proposed an extension where the decision maker can assign confidence levels to pairwise comparisons. This yields valued necessary preference relations and valued possible preference relations.

Exercise 4.5

In this exercise, you will learn how to use *VisualUTA*.

Learning Outcomes

- Input the data into *VisualUTA*
- Understand the results of *VisualUTA*

Tasks

Consider the problem of the ranking of smartphones in Case Study 4.1 based on a learning set. As in Exercise 4.4, define the learning set of three reference alternatives: SP3, SP2 and SP4.

Input the alternatives in the same list regardless of whether they belong to ranking set A or learning set L.

1. For this task, go to the *Alternatives* panel on the left of the screen, input the name of the alternative and then click on the '+' button (see Figure 4.25).
2. Input the list of criteria (at the bottom left of the screen). Click on the '+' button in the *Criteria Set* panel, specify the name, the description and the type of criterion. Choose *Gain* if the criterion has to be maximized or *Cost* otherwise (see Figure 4.26).

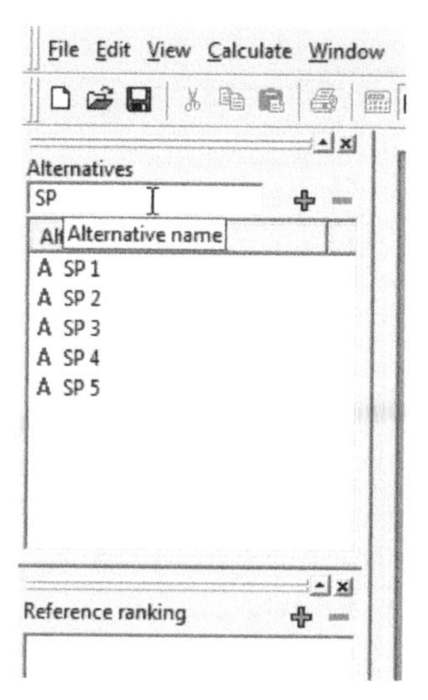

Figure 4.25 Inputting the set of alternatives A *and* L *in* VisualUTA. *Reproduced by permission of Roman Słowiński.*

3. Enter the performance of the alternatives by clicking on the toolbar menu, which opens the evaluation grid (see Figure 4.26). Define the set of reference alternatives as well as their corresponding ranking by dragging and dropping the reference alternatives from the *Alternatives Set* panel to the *Reference ranking* panel. Modify the rank of a chosen reference alternative with the '+' and '–' button (Figure 4.27) if necessary.

4. Solve the problem by pressing F5, going to *Calculate/Solve* or by clicking on the *calculator* icon in the toolbar. This leads to the 'Final Ranking worked out with UTA MD' (dominance model), the 'Alternatives comparison matrix' and the 'Resulting preference graphs worked out with the UTA GMS' (Figure 4.28). Is the ranking similar to the deduced ranking obtained in

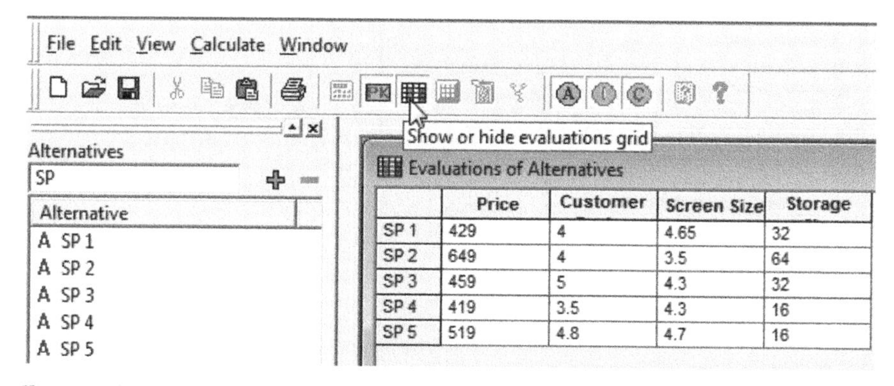

	Price	Customer	Screen Size	Storage
SP 1	429	4	4.65	32
SP 2	649	4	3.5	64
SP 3	459	5	4.3	32
SP 4	419	3.5	4.3	16
SP 5	519	4.8	4.7	16

Figure 4.26 Inputting the set of criteria in VisualUTA. *Reproduced by permission of Roman Słowiński.*

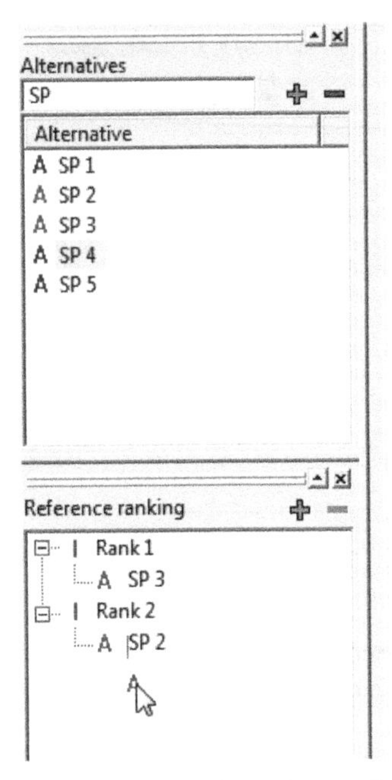

Figure 4.27 Inputting the ranking on the learning set L = {SP3,SP2,SP4}. *Reproduced by permission of Roman Słowiński.*

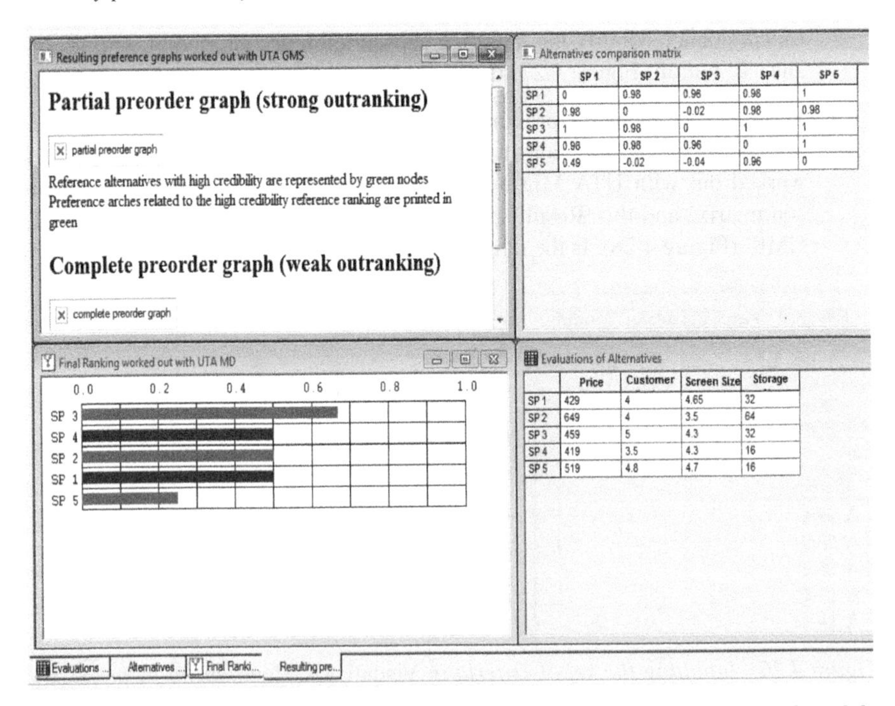

	SP 1	SP 2	SP 3	SP 4	SP 5
SP 1	0	0.98	0.96	0.98	1
SP 2	0.98	0	-0.02	0.98	0.98
SP 3	1	0.98	0	1	1
SP 4	0.98	0.98	0.96	0	1
SP 5	0.49	-0.02	-0.04	0.96	0

	Price	Customer	Screen Size	Storage
SP 1	429	4	4.65	32
SP 2	649	4	3.5	64
SP 3	459	5	4.3	32
SP 4	419	3.5	4.3	16
SP 5	519	4.8	4.7	16

Figure 4.28 Display of the results of the Visual UTA software. Reproduced by permission of Roman Słowiński.

Exercise 4.4? What are the relationships between the alternatives in the preference graphs?

5. Change the settings (the delta threshold of the linear programming corresponding to σ in Constraint 8) and epsilon used when computing the 'distances' $d(a, b)$ and $D(a, b)$ by going to *Calculate/Options*. What is the impact of those parameters on the ranking?

Members of the Laboratory of Intelligent Decision Support of the Institute of Computing Science of the Poznań University of Technology have recommended setting the value of epsilon slightly lower than delta due to the calculation errors which usually occur when solving linear programs. If the delta and epsilon thresholds are small enough, an outranking relation is defined for (almost) all compatible value functions. For greater values of these thresholds, a model is developed concerning just a set of value functions, for which the relations of preference between reference alternatives are clear (the difference between the value of two alternatives is greater than delta).

4.5.3 GRIP

The *generalized regression with intensities of preference* (GRIP) method belongs to the class of methods based on indirect preference information and the ordinal regression paradigm. In other words, GRIP enables the ranking of a finite set of alternatives A based on a partial pre-order and intensities of preferences on a subset of reference actions L given by the decision maker. Currently, there is no dedicated software that implements the GRIP method, but see Figueira et al. (2009) for more technical information and implementation steps of the method.

In contrast to other elicitation methods, the given preference information in GRIP does not need to be complete. The decision maker can provide *some* comparisons between the reference actions. A complete pre-order as in UTA is thus not required. Moreover, these comparisons can be considered on *all* criteria (i.e. comprehensively) or on *specific* criteria (i.e. partially). The decision maker can specify comparisons of strengths of preferences between some pairs of reference actions: for example, 'x is preferred to y at least as much as w is to z' or 'x is preferred to y at least as much as w is to z, on criterion f_i'. GRIP can be seen as a generalization of both the UTA and UTA^{GMS} methods.

The indirect preference information given by the decision maker is interpreted as a set of ordinal regression constraints (analogously as in UTA and UTA^{GMS}), which define a compatible set of additive value functions. The obtained additive value functions lead to the following preference relations in A and intensities of preference for these relations:

- The necessary and possible rankings for all pairs of actions $(x, y) \in A \times A$.
- The necessary and possible rankings with respect to the comprehensive intensities of preferences for all $((x, y), (w, z)) \in A \times A \times A \times A$.

- The necessary and possible rankings with respect to the partial intensities of preference for all $((x, y), (w, z)) \in A \times A \times A \times A$ and for all criteria $f_i \in F$.

In practice, the necessary and possible rankings are the most useful outputs for the decision maker.

The input given by the decision maker in GRIP can be considered as similar to the information required by AHP (see Chapter 2). In both methods, the decision maker can compare pairwise alternatives on a particular criterion with respect to strength of preferences. This permits the comparison of AHP with GRIP in respect to a single criterion. In AHP the strength of preference (given on a scale from 1 to 9) is translated into quantitative terms by for instance the principal eigenvectors, which is not always easy to justify. In GRIP, the marginal value functions are a numerical representation of the original qualitative-ordinal information without any intermediate transformation into quantitative terms (Figueira et al. 2009).

Furthermore, in GRIP weights do not need to be elicited (contrary to AHP) since the marginal value functions are expressed on the same preference scale and, as a result, the value functions can be summed. The elicited marginal value functions can be based on judgements which involve all the criteria simultaneously. For a more detailed comparison of GRIP and AHP, see Figueira et al. (2009).

GRIP can also be compared to MACBETH (see Chapter 5), which takes into account a preference order of actions and strength of preference comparisons on specific criteria (Figueira et al. 2009). Based on the preference information provided, MACBETH builds an interval scale for each criterion by means of linear programming models. GRIP and MACBETH both deal with qualitative judgements to build a numerical representation of the preferences. However, GRIP is more general than MACBETH since the comparisons in GRIP can involve all criteria simultaneously. As explained in Figueira et al. (2009), the structure of strength of preference in MACBETH is a particular case of that in GRIP. For a more detailed comparison of GRIP and MACBETH, see Figueira et al. (2009).

References

Beuthe, M., and Scannella, G. (1997). Comparative analysis of UTAs multicriteria methods. *European Journal of Operational Research*, 130, 246–262.

Figueira, J., Greco, S., and Slowinski, R. (2009). Building a set of additive value functions representing a reference preorder and intensities of preference: GRIP method. *European Journal of Operational Research*, 195, 460–486.

Fishburn, P. (1973). *The Theory of Social Choice*. Princeton, NJ: Princeton University Press.

Greco, S., Mousseau, V., and Slowinski, R. (2008). Ordinal regression revisited: Multiple criteria ranking using a set of additive value functions. *European Journal of Operational Research*, 191, 416–436.

Jacquet-Lagreze, E., and Siskos, J. (1982). Assessing a set of additive utility functions for multicriteria decision-making, the UTA method. *European Journal of Operational Research*, 10, 151–164.

Keeney, R., and Raiffa, H. (1976). *Decisions with Multiple Objectives: Preferences and Value Trade-offs.* New York: John Wiley & Sons, Inc.

Krantz, D., Luce, D., Stuppens, P., and Tvesky, A. (1971). *Foundations of Measurement.* New York: Academic Press.

Roy, B. (1974). Critères multiples et modélisation des préférences: l'apport des relations de surclassement. *Revue d'Economie Politique*, 1, 1–44.

Vincke, P. (1989). *L'aide multicritère à la décision.* Brussels: Edition Ellipses – Editions de L'Université Libre.

Vincke, P. (1992). *Multicriteria Decision Aid.* New York: John Wiley & Sons, Inc.

Zopounidis, M. and Doumpos, C. (2002). *Multicriteria Decision Aid Classification Methods.* Dordrecht: Kluwer Academic Publishers.

5

MACBETH

5.1 Introduction

This chapter explains the theory and practical uses of MACBETH. You will learn how to use the *M-MACBETH* software, which helps to structure problems and calculate the attractiveness (scores) of the options (choices). Section 5.3 is designed for readers interested in the methodological background of MACBETH. In order to understand this section, an understanding of linear programming is needed (see the Appendix).

The companion website provides illustrative examples with *Microsoft Excel*, and case studies and an example with *MACBETH*.

5.2 Essential concepts of MACBETH

MACBETH stands for 'Measuring Attractiveness by a Categorical Based Evaluation Technique'. From a user point of view, MACBETH has many similarities with AHP. A novice may even not see the difference. Both methods are based on pairwise comparisons entered by the user, but MACBETH uses an interval scale and AHP adopts a ratio scale (Example 5.1). The calculation process behind AHP (Section 2.4) is different from MACBETH as described in Section 5.3.

Example 5.1 Consider the price of a Chinese (£1000) and Japanese caterer (£5000) for a conference that you organize. The pairwise comparisons in AHP and MACBETH are performed differently. AHP uses a ratio scale, $\frac{5000}{1000} = 5$, which means that the price of the Japanese caterer is 5 times as expensive. MACBETH uses an interval scale: $5000 - 1000 = £4000$. The decision maker should therefore evaluate the difference (£4000) between the two prices.

Multi-Criteria Decision Analysis: Methods and Software, First Edition. Alessio Ishizaka and Philippe Nemery.

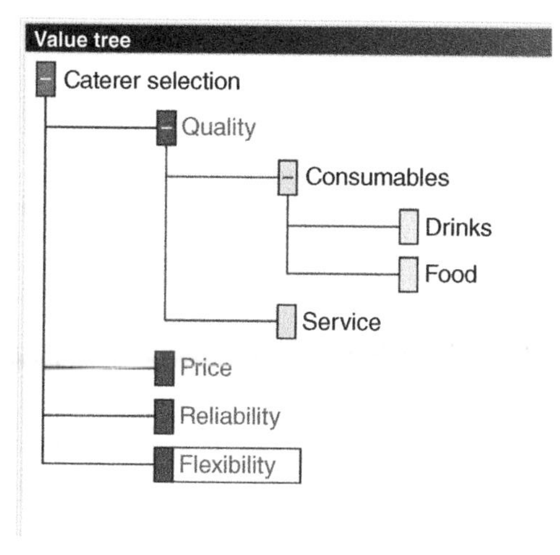

Figure 5.1 Tree of the decisions. Reproduced by permission of BANA Consulting Lda.

The user of MACBETH will need to complete three steps in order to obtain the ranking of the options. As with any MCDA method, the first step is to structure the problem, followed by entering pairwise comparisons into a judgement matrix. If the matrix is sufficiently *consistent*, the attractiveness can be calculated, otherwise the user is obliged to revise their judgements. Finally, an optional sensitivity analysis is recommended. This task is largely facilitated by the supporting software *M-MACBETH*.

5.2.1 Problem structuring: Value tree

MACBETH structures the problem in a tree or hierarchy, but makes a distinction between criteria and non-criteria nodes. Non-criteria nodes are included in the tree to help with the evaluation of criteria nodes but are not directly influential in the decision. They act only as comments to structure the problem and therefore will not be evaluated. Only one-criterion nodes can be set between the overall node (top of the tree) and the leaves (bottom of the tree). For example, in Figure 5.1, only Quality is set as a criterion between the top of the tree (Caterer selection) and the leaf (Food). It would have been impossible to set another node (Consumables, Service, Drinks or Food) as criterion under the node Quality. For those who are familiar with AHP, this is unusual. Indeed, if there is only one criterion between the overall node and leaf, then the value tree is not equivalent to the AHP criteria tree. The structure can be reduced to one level with no subcriteria.

This chapter will consider a caterer selection problem in order illustrate the different steps of the MACBETH process (Case Study 5.1).

Case Study 5.1

A university organizes a three-day conference and needs to select a caterer to provide the lunches. It is expected to have around 200 participants, but the exact number is uncertain. The schedule is very tight and a delay in the service would disrupt all the presentations in the afternoon. The organizing committee contemplates four options:

a) **Internal caterer**
The university has a catering service, which is more expensive than external caterers. It provides a large range of sandwiches. It has been used to cater for other conferences, and its service is reliable. It is very flexible in that the number of lunches can be adjusted the day before.

b) **Indian caterer**
A member of the organizing committee has used an Indian caterer for a private party. It produced high-quality food for a reasonable price. However, it does not have previous experience of catering for large numbers and may be a risky solution.

c) **Chinese caterer**
A Chinese fast food restaurant also provides catering for big events. Its service has been used successfully at past conferences as an alternative to the internal caterer. Its food quality is average and its price is low. It is very flexible with regard to the number of lunches, as the food is prepared on the spot in large woks.

d) **Japanese caterer**
A Japanese caterer has just opened. Its main dishes are sushi. It has the same price range as the internal caterer but offers higher-quality food. As it is a new business, its flexibility and reliability in providing a large number of lunches is not known.

The decision will thus be based on four criteria given in Table 5.1.

Table 5.1 List of criteria.

Criteria	Explanation
Quality	Quality of the food and service
Price	Cost of the service
Reliability	Capacity to work on a tight schedule for the conference
Flexibility	How easily they adjust the number of lunches for additional participants

Figure 5.1 represents the tree, generally referred to as a 'value tree', from Case Study 5.1. The top node of the tree contains the overall goal. Quality, Price, Reliability and Flexibility are criteria nodes (in bold in Figure 5.1). Consumables, Service, Drinks

and Food are non-criteria but essential information in order to evaluate the Quality criterion.

5.2.2 Score calculation

After the problem-structuring phase of previous section, three types of scores have to be calculated:

- **Weighting criteria.** These measure the attractiveness of each criterion in relation to the top goal.
- **Scores of options.** These represent the attractiveness of an option to one specific criterion.
- **Overall score of options.** Criteria weight and option scores are only intermediate results used to calculate the overall score of options. Whilst the score of options ranks them with regard to a single criterion, the overall score of options ranks them with regard to all criteria and consequently to the overall goal.

The relative attractiveness of each criterion is evaluated pairwise. Then, the options are compared pairwise with regard to each criterion. In Case Study 5.1, five different judgement matrices are required: four local scores of options with regard to each criterion and one weight criteria score.

MACBETH is a pairwise comparison based method on an interval scale. The user will need to provide only qualitative judgements about the difference of attractiveness between two options. The traditional MACBETH offers seven semantic categories for the evaluation (Table 5.2) but other verbal scales can be imagined. For example, a '4' would read as 'Option A is moderately more attractive than Option B'.

It is recommended to first order the elements to be evaluated. This is not a mandatory step but it helps to ensure consistency when filling the judgement matrix. Your judgements will be in an increasing order from right to left and from the bottom to the top of your matrix (see Table 5.3). Only the upper triangle of the matrix needs to be completed because the lower triangle is the reverse and can be deduced from the upper triangle. As in AHP (Chapter 2), the redundancy of information leads to a gain in precision but requires a higher effort, especially for a large number of decision elements.

Table 5.2 Seven semantic categories.

Semantic categories	Quantitative scale
no	1
very weak	2
weak	3
moderate	4
strong	5
very strong	6
extreme	7

Table 5.3 Matrix of judgements.

	Quality	Price	Reliability	Flexibility
Quality	no	very weak	weak	moderate
Price		no	very weak	weak
Reliability			no	weak
Flexibility				no

Example 5.2 In order to weight the criteria of Case Study 5.1, first rank the criteria in order of importance. In this case, quality is the most important criterion, followed by price, reliability and, finally, flexibility. Then the matrix of judgements (Table 5.3) expresses the difference in attractiveness between the criteria. For example, quality is slightly more important than price. The main diagonal has *no* values because a criterion does not need to be compared to itself.

From such a matrix of judgements, the software will calculate the weights and scores of options. First, the least attractive option/criterion is grounded to 0. Then, lists of conditions reflecting the user's judgement in Table 5.3 are set. A linear optimization is used to minimize the score of the most attractive option/criterion. However, as several solutions may exist, the scores can be readjusted in accordance with the user's feelings (see Figure 5.10), while still fulfilling the semantic category constraint conditions. This easy readjustment is done in the software with a graphical thermometer bar. Advanced readers can refer to Section 5.4 to understand how the scores are calculated.

Finally, MACBETH aggregates these weighted scores additively in order to derive the overall score of options. Unlike to AHP, MACBETH requires a high consistency in order to be able to calculate scores. Therefore, each time a judgement is provided, the software will verify the consistency and suggest changes to resolve any inconsistency. The consistency check is explained in the next section.

5.2.3 Incompatibility check

When filling a judgement matrix, the decision maker may introduce incompatible judgements. This can occur with comparative and semantic judgements. A comparative judgement is given between two actions on a semantic category (Table 5.2). It is the judgements that a decision maker enters in the matrix. A semantic judgement is the comparison between two comparative judgements. It is the difference between two judgements entered into a matrix.

Example 5.3

Comparative judgements:

- A is weakly more attractive than B.
- C is moderately more attractive than B.

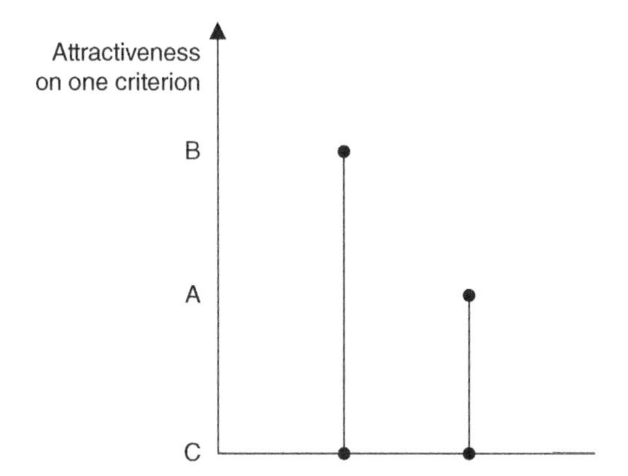

Figure 5.2 Semantic judgement difference of attractiveness B – C > A – C.

Semantic judgement:

- The difference in attractiveness between C and B (moderate) is bigger than the difference in attractiveness between A and B (weak).

There are two types of incompatibility in MACBETH: incoherence and semantic inconsistency.

Incoherence inconsistency. There are four cases where a conflict can arise between a comparative judgement and a semantic judgement.

- In Figure 5.2, the semantic judgement difference of attractiveness between B and C is higher than that of A and C (B – C > A – C). This would be 'incoherent' with the comparative judgement A better than B (A > B).
- In Figure 5.3, the semantic judgement difference of attractiveness between B and C is higher than the difference of attractiveness between B and A (B – C > B – A). This would be 'incoherent' with the comparative judgement C better than A (C > A).
- In Figure 5.2, the semantic judgement difference of attractiveness between B and C is higher than the difference of attractiveness between A and C (B – C > A – C). This would be 'incoherent' with the comparative judgement A is equal to B (A = B).
- In Figure 5.3, the semantic judgement difference of attractiveness between B and C is higher than the difference of attractiveness between B and A (B – C > B – A). This would be 'incoherent' with the comparative judgement C equal to A (C = A).

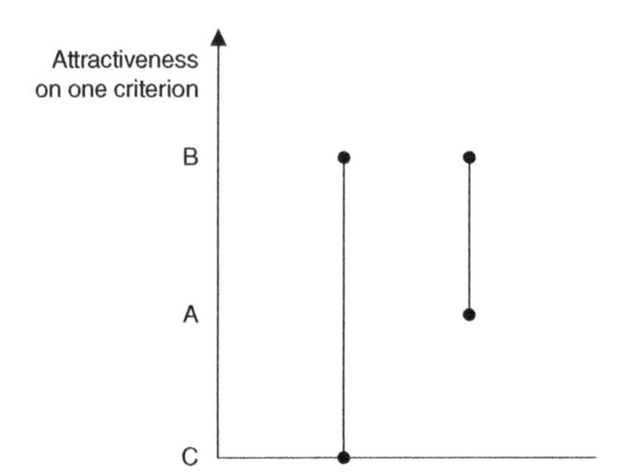

Figure 5.3 Semantic judgement difference of attractiveness B – C > B – A.

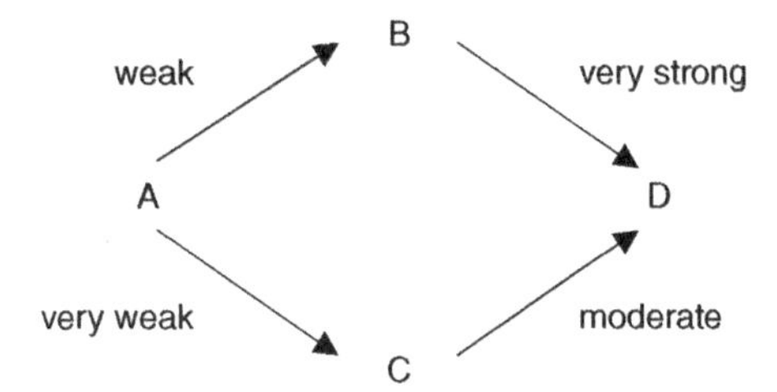

Figure 5.4 Graph of semantic inconsistency.

Semantic inconsistency. The semantic inconsistency is tested by a linear program that can be found in Bana e Costa et al. (2005). Basically, it tests that 'two paths' between two points (representing the preference strength) should have the same 'length'. For example, the graph in Figure 5.4 presents a semantic inconsistency because A – B > A – C and B – D > C – D, therefore the path A-B-D is much longer than A-C-D. However, the linear program accepts surprising cases as in Figure 5.5, where the path A-B-D is much longer than A-C-D.

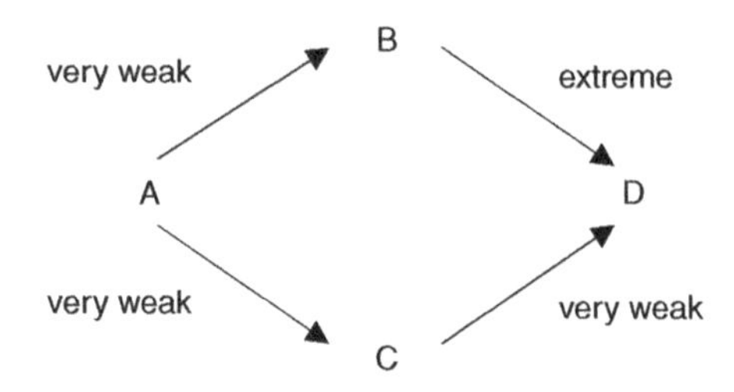

Figure 5.5 Consistent graph.

Exercise 5.1

The following multiple-choice questions allow you to test your knowledge on the basics of MACBETH. Only one answer is correct. Answers can be found on the companion website.

1. What does MACBETH stand for?
   a) Measuring Awareness by a Consistent Based Evaluation Technique
   b) Measuring Assurance by a Cooperative Based Evaluation Technique
   c) Measuring Attention by a Coherent Based Evaluation Technique
   d) Measuring Attractiveness by a Categorical Based Evaluation Technique
2. What type of scale is used in MACBETH for expressing comparisons?
   a) A ratio scale
   b) A nominal scale
   c) An interval scale
   d) A categorical scale
3. How many semantic categories are there in MACBETH?
   a) 6
   b) 7
   c) 8
   d) 9
4. If the judgements are $A < B$ and $B - C > A - C$, are they:
   a) Consistent?
   b) Incoherent?
   c) Semantic inconsistent?
   d) None of the above
5. If the judgements are: A is weakly better than B, B is strongly better than C, and A is moderately better than C, are they:
   a) Consistent?
   b) Incoherent?
   c) Semantic inconsistent?
   d) None of the above

5.3 Software description: M-MACBETH

MACBETH is supported by *M-MACBETH*. As far as we know there is no other commercial software for MACBETH, although a template in *Microsoft Excel* can be generated (Section 5.4). *M-MACBETH* is a user-friendly program with intuitive graphical user interfaces that automatically computes attractiveness and possible inconsistencies, as well as ways to process a sensitivity analysis. *M-MACBETH* has been translated into four languages (English, French, Portuguese and Spanish).

This section describes *M-MACBETH* and the essential functions of the graphical user interface. A full tutorial can be downloaded from the *M-MACBETH* website: http://www.m-macbeth.com/en/downloads.html.

A free trial version can be downloaded from http://www.m-macbeth.com/. This version is limited to five criteria and five options. It is not necessary to know *how* scores are calculated, but only *what* should be ranked.

In this section the three steps introduced in Section 5.2 will be followed.

5.3.1 Problem structuring: Value tree

Nodes are entered by right-clicking on the parent node followed by selecting *Add a node*. By default all nodes are set as *non-criteria*. In order to change a non-criterion to a criterion node, right-click on the node and select *node properties*. Select to compare the options of the problem only or the options and two benchmark references. These two benchmarks are by default the *upper* and *lower reference*. They can easily be changed, for example, to good and neutral. These two benchmarks are virtual and optional. They permit two reference points for entering judgements. The drawback is that the number of judgements required increases.

Nodes can also be directly compared on qualitative or quantitative performance-level scales. For example, in Figure 5.6, five qualitative levels for price are defined.

Options are entered via the menu *Options/Define*.

5.3.2 Evaluations and scores

Two types of evaluation are possible on a qualitative or quantitative performance-level scale: pairwise judgement of the options or direct evaluation. The choice of the evaluation technique needs to be done when setting the criterion nodes (see Section 5.3.1). In order to enter the assessments for the options or performance levels, click on the criterion node and choose *Judgements*. The evaluation of the criteria can also be done in a matrix of judgements by selecting the menu *Weighting/Judgements*.

The evaluation of the options (Figure 5.7), performance levels (Figure 5.8) and criteria follows a similar process. It is recommended to order the options by selecting and dragging them up or down in the column of the judgement matrix in decreasing order of attractiveness (Figure 5.6). Only $n - 1$ independent evaluations are required as the others can be deduced by transitivity. However, it is better to fill the upper triangle of the matrix using the semantic categories (Table 5.2). If the decision maker is unsure about the exact category, they can select two or more successive categories

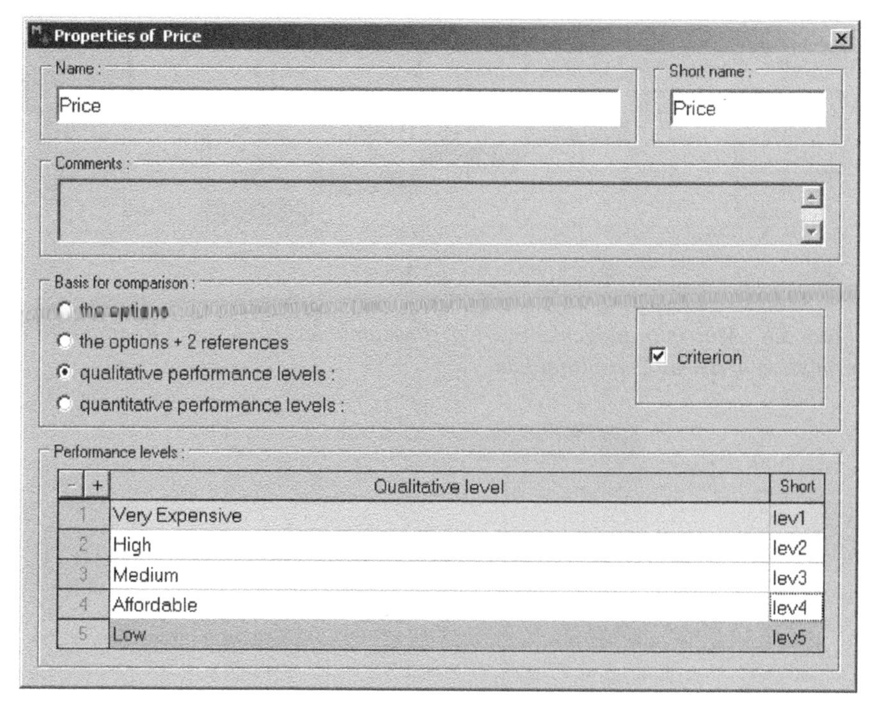

Figure 5.6 Qualitative levels. Reproduced by permission of BANA Consulting Lda.

as in Figure 5.8. Between low and high price, the difference is evaluated from *very strong* to *extreme*. The best option/level will be highlighted in green whilst the worst will be in blue.

The calculation of scores is done by clicking on the seventh button shown in Figure 5.7 and Figure 5.8. The scores are scaled so that the upper reference is 100 and the lower reference is 0 (Figure 5.9). If the two reference levels are not defined,

Quality

	Japanese	Indian	Internal	Chinese	Current scale
Japanese	no	very weak	v. strong	extreme	100.00
Indian		no	v. strong	v. strong	85.71
Internal			no	very weak	14.29
Chinese				no	0.00

extreme
v. strong
strong
moderate
weak
very weak
no

Consistent judgements

Figure 5.7 Matrix of judgements for options with regard to quality. Reproduced by permission of BANA Consulting Lda.

Price

	Low	Affordable	Medium	High	Very Expensive	Current scale
Low	no	weak	moderate	vstrg-extr	extreme	100.00
Affordable		no	moderate	strong	v. strong	88.24
Medium			no	strong	v. strong	72.84
High				no	moderate	35.29
Very Expensive					no	0.00

extreme
v. strong
strong
moderate
weak
very weak
no

Consistent judgements

Figure 5.8 Matrix of judgements for performance levels of price. Reproduced by permission of BANA Consulting Lda.

Figure 5.9 MACBETH numerical scale. Reproduced by permission of BANA Consulting Lda.

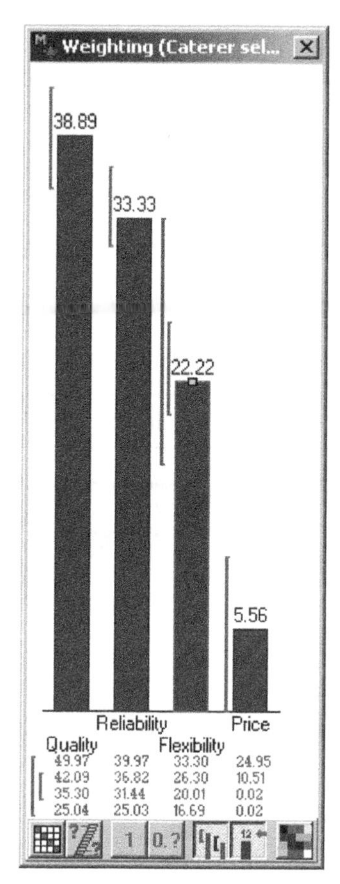

Figure 5.10 MACBETH weighting scale. Reproduced by permission of BANA Consulting Lda.

the software automatically assigns the scores 0 and 100 to the two endpoints of the scale. The normalization of the weights is different from that of the options (Figure 5.10). They are normalized to 100 (the sum of the weights adds up to 100).

M-MACBETH allows the scores to be readjusted, whilst being compatible with the judgements provided in the matrix. The permissible interval is shown in red on the left part of the thermometer bar and options can be dragged up or down the axis (Figure 5.9 and 5.10).

In the direct evaluation of options, the performance levels calculated in Figure 5.8 need to be allocated to the options by choosing *Options/Table of performances* (Figure 5.11).

5.3.3 Incompatibility check

When contradictory evaluations are entered, M-MACBETH will detect them and issue a warning. For example, in Figure 5.12, the judgement between the Indian and Chinese caterers being weak is inconsistent with the previously entered judgements.

Table of performan...

Options	Price
Indian	High
Chinese	Medium
Internal	Low
Japanese	Very Expensive

Figure 5.11 *Performance table. Reproduced by permission of BANA Consulting Lda.*

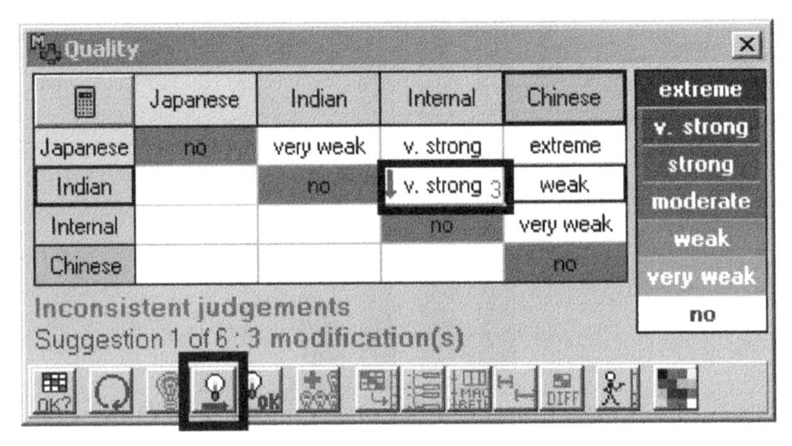

Quality

	Japanese	Indian	Internal	Chinese
Japanese	no	very weak	v. strong	extreme
Indian		no	v. strong 3	weak
Internal			no	very weak
Chinese				no

Figure 5.12 *Matrix with inconsistent judgements. Reproduced by permission of BANA Consulting Lda.*

The red arrow pointing down explains that the judgement should be decreased and the number indicates the number of categories. In Figure 5.12, the first suggestion implies that the judgement of *very strong* between the Indian and internal caterers should be reduced to *weak*. To see the other suggestions, click on the fourth button.

By clicking on the second button, the contradictions are explained. Figure 5.13 explains the incompatibility of the matrix of Figure 5.12. The first problem is an incoherence and the second one is a semantic inconsistency (see Section 5.2.3).

Quality

Problems	Diff.	Couples		Couples	Diff.
1	v. strong	Indian - Internal	>	Indian - Chinese	weak
	very weak	Internal - Chinese	>	0	
2	extreme	Japanese - Chinese	>	Japanese - Internal	v. strong
	v. strong	Indian - Internal	>	Indian - Chinese	weak

Figure 5.13 *Incompatibility explanation. Reproduced by permission of BANA Consulting Lda.*

Table of scores

Options	Overall	Quality	Price	Reliability	Flexibility
[all upper]	100.00	100.00	100.00	100.00	100.00
Internal	62.96	14.29	100.00	100.00	83.33
Indian	53.33	85.71	26.67	33.33	33.33
Chinese	50.92	0.00	66.67	75.00	100.00
Japanese	38.89	100.00	0.00	0.00	0.00
[all lower]	0.00	0.00	0.00	0.00	0.00
Weights :		0.3889	0.0556	0.3333	0.2222

Figure 5.14 Overall scores. Reproduced by permission of BANA Consulting Lda.

5.3.4 Results

Selecting *Options/Table of scores* displays the table of results (Figure 5.14). This table contains the overall scores and scores with regard to each criterion. The scores are normalized for each criterion in order that the most attractive option has a score of 100 and the least attractive option receives 0. The internal caterer is the most attractive option.

A further analysis of the options can be done graphically. The difference between the profiles for any two options can be viewed (go to *Options/Difference profiles*). For example, in Figure 5.15, the internal caterer is compared to the Indian caterer. If the quality of the Internal caterer is far less attractive than the Indian caterer, it compensates this deficit with more attractive price, reliability and flexibility.

Another way to analyze the results is to use a two-dimensional graph, where each axis represents a criterion (go to *Options/XY Map*). In Figure 5.16, the options are represented according to their attractiveness on the price and quality criteria. It can be seen that the Japanese caterer has a high quality but is not attractive on the price. On the other hand, the internal caterer has an attractive price and a low quality. The line represents the efficient frontier. In this case, the Chinese caterer is a dominated option.

5.3.5 Sensitivity analysis

A static graphical sensitivity analysis is available in *M-MACBETH* (go to *Weighting/Sensitivity analysis by weight*). It allows the impact of the change on one criterion weight on the overall score to be observed. For example, in Figure 5.17, the impact of the quality criterion weight is plotted. If the weight given to quality is over 46.2, the Indian caterer will be overall the most attractive option instead of the internal caterer. If the weight is more than 80, than the Japanese caterer becomes the most attractive option.

5.3.6 Robustness analysis

The robustness analysis allows exploration if the proposed solution remains unchanged after a variation in the weights of the criteria. This analysis is important

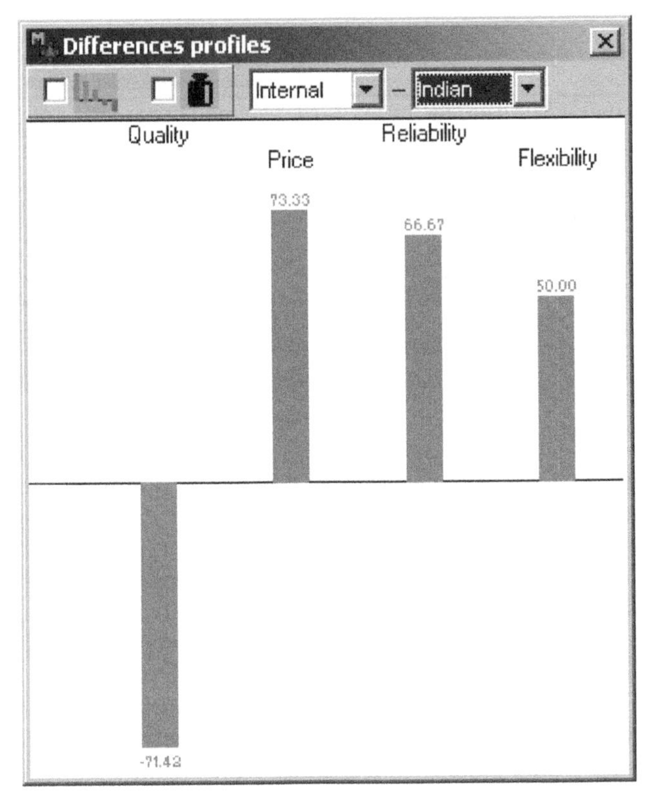

Figure 5.15 *Profiles difference. Reproduced by permission of BANA Consulting Lda.*

in an uncertain problem with imprecise information. In Figure 5.18, a 5% uncertainty is used for each weight of the criterion. Three types of symbols can be found:

- The ▲ symbol means that the global preference of the line action over the column action does not depend on any combination of weight criteria.
- The ✚ symbol means that the preference of the line action over the column action is not modified if there is less than the defined uncertainty in the weight alteration.
- The **?** symbol means that the preference of the line action over the column action can be modified if there is less (or more) than the defined uncertainty in the weight alterations.

For example, in Figure 5.18 the Indian caterer is always preferred over the Japanese caterer and is preferred over the Chinese and internal caterer if the weight alteration is below 5%.

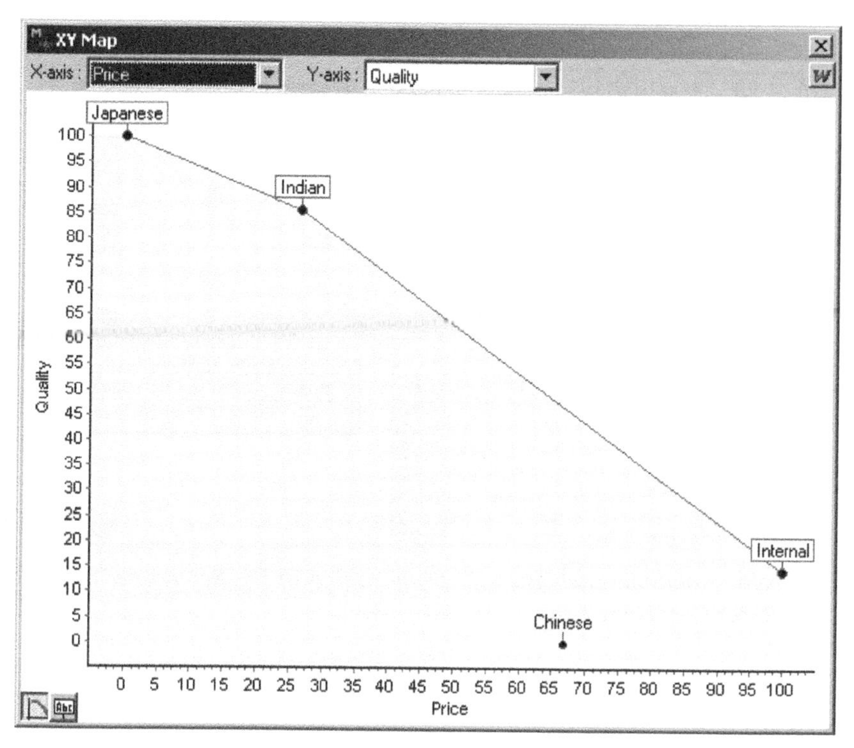

Figure 5.16 Comparison of scores on two criteria. Reproduced by permission of BANA Consulting Lda.

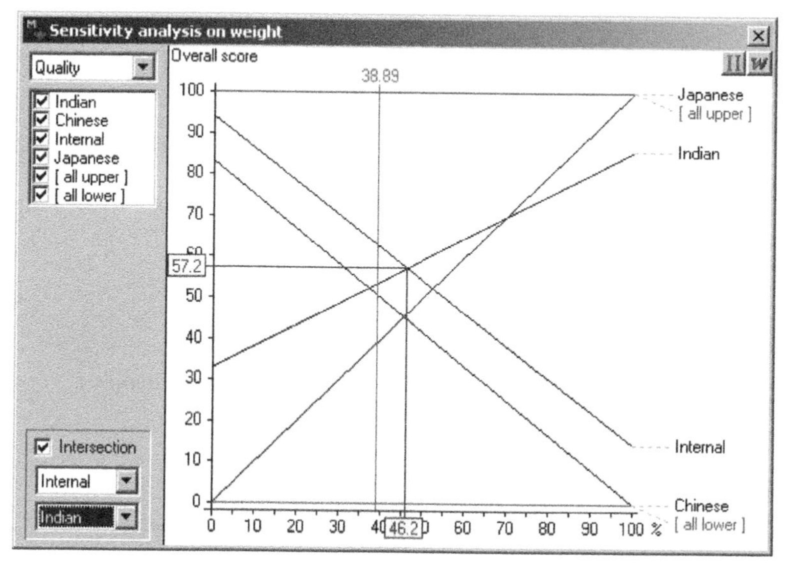

Figure 5.17 Sensitivity analysis. Reproduced by permission of BANA Consulting Lda.

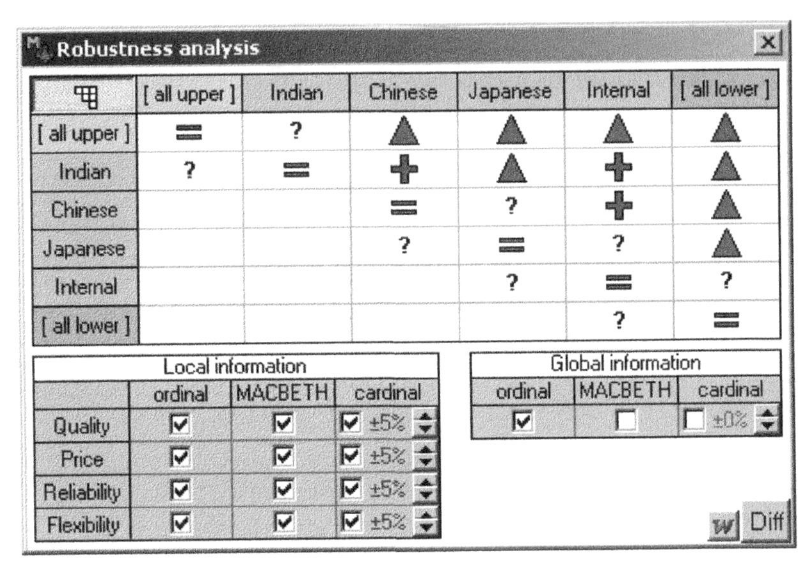

Figure 5.18 *Robustness analysis with an uncertainty of 5%. Reproduced by permission of BANA Consulting Lda.*

Exercise 5.2

In this exercise you will solve the caterer selection problem in Case Study 5.1 with the *M-MACBETH* software.

Learning Outcomes

- Structure a problem in *M-MACBETH*
- Enter pairwise judgements
- Understand the results
- Conduct a sensitivity analysis

Tasks

a) In the trial version of *M-MACBETH*, create a new file (*File/new*).

b) Read the description of Case Study 5.1, on page 116.

c) Enter the alternatives (*Options/Define*).

d) Enter the criteria (right-click on *Overall* and then *Add a node*).

e) Enter the pairwise evaluations of the options or the performance levels (click on the criterion node and then *Judgements*).

f) Enter the pairwise evaluations for the criteria (*Weighting/Judgements*).

g) Read the overall ranking (*Options/Table of scores*), analyse it (*Options/ Difference profiles* and *Options/XY Map*) and conduct a sensitivity analysis (*Weighting/ Sensitivity analysis by weight*).

5.4 In the black box of MACBETH

The basic MACBETH scale is derived from the evaluations of the decision maker by solving a linear programming problem. Therefore, if the reader is not familiar with this technique, it is advisable to first read the Appendix.

5.4.1 LP-MACBETH

LP-MACBETH is the method used to calculate scores for each judgement matrix (Bana e Costa et al. 2003; Bana e Costa and Vansnick 1999; De Corte 2002). It consists of the solution to the following linear program. We have an *objective function* of the problem:

$$\text{minimize } \Phi(o_1)$$

where $\Phi(o_1)$ is the score of the most attractive option o_1. If we maximize it, the score of the most attractive option will be infinite. Our *decision variables* are:

$$\Phi(o_i),\ i \in \{1, 2, \ldots, n\}$$

We have *constraints* as follows.

- *Ordinal conditions:*

$$\forall o_i, o_j, i, j \in \{1, 2, \ldots, n\}\text{: } o_i \text{ is preferred to } o_j \Rightarrow \Phi(o_i) \geq \Phi(o_j) + \delta(i, j)$$
$$\forall o_i, o_j, i, j \in \{1, 2, \ldots, n\}\text{: } o_i \text{ and } o_j \text{ are indifferent} \Rightarrow \Phi(o_i) = \Phi(o_j)$$

  where $\delta(i,j)$ is the difference of attractiveness between o_i and o_j

- *Semantic conditions:*

$$\forall o_i, o_j, o_k, o_l, i, j, k, l \in \{1, 2, \ldots, n\} : \Phi(o_i) - \Phi(o_j) \geq \Phi(o_k) - \Phi(o_l) + \delta(i, j, k, l)$$

  where $\delta(i, j, k, l)$ is the number of semantic categories between the difference of attractiveness between o_i and o_j, and the difference of attractiveness between o_k and o_l.

- *Grounding conditions:*

$$\Phi(o_n) = 0$$

  i.e. the score of the least attractive option is zero.

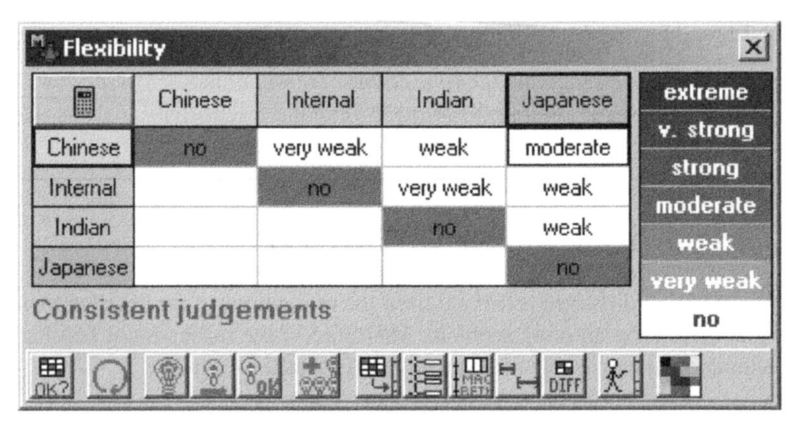

Figure 5.19 Matrix of judgements. Reproduced by permission of BANA Consulting Lda.

The optimal solution of this linear program is not always unique. The scale obtained is called a MACBETH scale. This scale is transformed in order for the highest score to be 100 and the lowest 0. This transformation allows commensurable scores between criteria, which is necessary for adding them together in an overall score. Let us reminder that this linear programme must be performed for each judgement matrix.

Exercise 5.3

You will learn to calculate the scores from a matrix of judgements step by step.

Learning Outcomes

- ➢ Understand the modelling of an LP-MACBETH problem
- ➢ Understand the configuration of *Microsoft Excel Solver*

Tasks

Open the file Caterer selection.xls. It contains a spreadsheet with the calculation of scores from Figure 5.19.

Complete the following tasks:

a) In the spreadsheet, find the objective of the problem, the decision variables and constraints.

b) Open the Solver. What is entered in the set target cell? What is entered in the box 'by changing cells'? What is entered in the box 'Subject to constraints?'

c) How do you transform the MACBETH scale in order for the most attractive score to be 100?

5.4.2 Discussion

MACBETH relies on a strong measurement theory foundation. This theoretical rigour raises practical issues.

The judgement matrix has to be consistent enough for the calculation of the attractiveness with a linear program. However, the decision maker might not always be consistent due to a lack of information or simply because the problem is inconsistent by nature (e.g. a football tournament, where the highest classified team can lose against the lowest classified). In practice, the decision maker will be forced to enter judgements proposed by *M-MACBETH* otherwise the attractiveness cannot be calculated. The final result may not reflect the real ranking the decision maker had in mind.

Another problem of MACBETH is related to linear programming. It is well known that several optimal solutions (i.e. rankings) can be obtained with the linear programming method. These different rankings can be confusing for the decision maker. The presence of a facilitator is therefore recommended in order to discuss which solution would best fit their preferences.

References

Bana e Costa, C., De Corte, J.-M., and Vansnick, J.-C. (2005). On the mathematical foundation of MACBETH. In J. Figueira, S. Greco and M. Ehrogott (eds), *Multiple Criteria Decision Analysis: State of the Art Surveys* (pp. 409–437). New York: Springer-Verlag.

Bana e Costa, C., De Corte, J.-M., and Vansnick, J.-C. (2003). MACBETH. OR Working Paper 03.56, London School of Economics and Political Science.

Bana e Costa, C., and Vansnick, J.-C. (1999). The MACBETH approach: Basic ideas, software, and an application. In N. Meskens and M. Roubens (eds), *Advances in Decision Analysis,* Mathematical Modelling: Theory and Applications (Vol. 4, pp. 131–157): Dordrecht: Kluwer Academic Publishers.

De Corte, J.-M. (2002). Un logiciel d'exploitation d'informations préférentielles pour l'aide à la décision. PhD thesis, University of Mons-Hainaut.

Part II

OUTRANKING APPROACH

6

PROMETHEE

6.1 Introduction

This chapter describes the theory along with the practical utilization of the PROMETHEE method. You will learn how to use *Smart Picker Pro*, a software package that helps to express preferences amongst actions (alternatives or choices). *Smart Picker Pro* computes, as a main output, a partial and global ranking based on the PROMETHEE methodology. Section 6.3 is designed for readers interested in the methodological background of PROMETHEE, while Section 6.4 deals with the extensions of PROMETHEE in group decision and sorting problems.

The companion website provides illustrative examples in *Microsoft Excel*, and case studies and examples with *Smart Picker Pro*.

6.2 Essential concepts of the PROMETHEE method

The acronym PROMETHEE stands for 'Preference Ranking Organization METHod for Enriched Evaluation'. Thus the PROMETHEE method will provide the decision maker with a ranking of actions (choices or alternatives) based on *preference degrees*. The method falls into three main steps:

1. the computation of preference degrees for every ordered pair of actions on each criterion;
2. the computation of unicriterion flows;
3. the computation of global flows.

Based on the global flows, a ranking of the actions will be obtained as well as a graphical representation of the decision problem. Before describing these three steps

Multi-Criteria Decision Analysis: Methods and Software, First Edition. Alessio Ishizaka and Philippe Nemery.

more in depth, let us introduce the following case study that will be used throughout this chapter.

Case Study 6.1

A decision maker wants to buy a new car and is considering five different cars of various types. He therefore defines four criteria:

- price (to be minimized);
- consumption (to be minimized);
- comfort (to be maximized);
- power (to be maximized).

After gathering data and testing the five cars, he evaluates their performances on the four criteria (Table 6.1).

Table 6.1 Performance of the five cars evaluated on four criteria.

	Price (£)	Consumption (l/km)	Comfort	Power (hp)
Economic	15 000	7.5	Very Bad	50.0
Sport	29 000	9.0	Bad	110.0
Luxury	38 000	8.5	Very Good	90.0
Touring A	24 000	8.0	Average	75.0
Touring B	25 500	7.0	Average	85.0

It can be seen from Table 6.1 that no ideal car exists: no car is the best on all four criteria. The decision maker will inevitably have to make a compromise. The PROMETHEE method will help the decision maker in his decision process.

6.2.1 Unicriterion preference degrees

The PROMETHEE method is based on the computation of preference degrees. A *preference degree* is a score (between 0 and 1) which expresses how an action is *preferred* over another action, from the decision maker's point of view.

A preference degree of 1 thus means a total or strong preference for one of the actions on the criterion considered. If there is no preference at all, then the preference degree is 0. On the other hand, if there is some preference but not a total preference, then the intensity will be somewhere between 0 and 1.

A decision maker might express a preference when looking at the prices of the cars of Table 6.1. The preference degree of Touring B compared to Luxury might be 1 since it would mean a strong preference for the cheapest car. On the other hand, the preference degrees between Touring A and Touring B will be relatively small, since they are similarly priced.

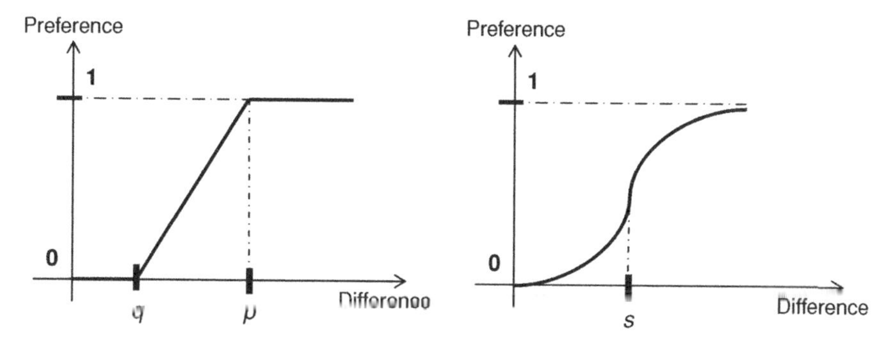

Figure 6.1 *Linear (left) and Gaussian (right) preference functions.*

We use the term *pairwise* preference degree since the preference of action A over action B cannot be deduced from the preference of action B over action A (nor vice versa). The PROMETHEE method will help the decision maker to evaluate these unicriterion pairwise preference degrees. For each criterion, this *unicriterion preference degree* is computed through *rescaling* or *enriching* the evaluations of the actions by means of *preference information.*

What matters in PROMETHEE is how the decision maker perceives the difference between the (objective) evaluations (often measured) on every specific criterion. These pairwise comparisons are based on the difference between the evaluations of the two actions (e.g. the difference in price between the actions). Therefore, he can choose between two types of preference functions: the linear and the Gaussian function. These preference functions are shown in Figure 6.1 and Figure 6.2.

If the decision maker opts for a linear function, then the preferences will gradually increase as a function of the difference between the evaluations on a particular criterion. When using the Gaussian preference function, the increase follows an exponential function.

In order to specify each preference function, one or two parameters are required. The linear preference function requires two parameters: an indifference threshold q and the preference threshold p. On the other hand, the Gaussian function requires only one parameter: the inflexion point s.

If the difference between the evaluations on a criterion is smaller than the indifference threshold, then no difference can be perceived by the decision maker between these two actions (i.e. the preference degree is 0). If the difference is higher than the preference threshold, then the preference is strong (i.e. the preference degree is 1). The preference function gives the value of the preference for differences between the indifference and preference threshold.

Figure 6.2 represents the linear preference function for particular parameters. If $q = p = 0$ (Figure 6.2(a)), then there is a strong preference for an action as soon as there is a difference (however small the difference might be). In Figure 6.2(b) there is no indifference zone (q=0) which signifies that every difference is considered proportionally. Finally, in Figure 6.2(c) the preference function is given by a 'step

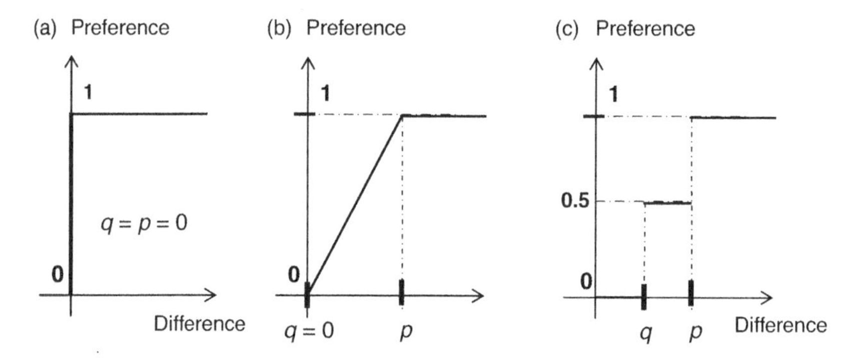

Figure 6.2 Three linear preference functions.

function'. This means that the strength of preference can only take three values: 0, 0.5 or 1.

Let us consider Table 6.1 – more specifically, the price criterion which has to be minimized. Suppose that the indifference threshold is 2000 and the preference is 5000. The unicriterion preference degrees of Figure 6.3 will be computed by the software based on this input, but it is interesting to understand its meaning. Let us therefore compute the unicriterion preference degrees given in Table 6.2.

All the elements of the diagonal are 0 because an action cannot be preferred to itself. Since the Sport car is more expensive than the Economic car, Sport cannot be preferred to Economic: the preference degree (Sport, Economic) is thus necessarily 0. Similarly, we have only zeros in the first column because Economic is the cheapest car: the first column rendering how the other actions are preferred to Economic.

Since the difference between Economic and Sport is 14 000 and thus higher than the preference threshold p (= 5000), we show that the preference of Economic over Sport is 1. Touring B is £3500 cheaper than Sport. The preference degree of Touring B on Sport can be read in Table 6.2. If you are not scared of some mathematics, this is how the preference degree is obtained: (3500 – 2000) / (5000 – 2000) = 0.5.

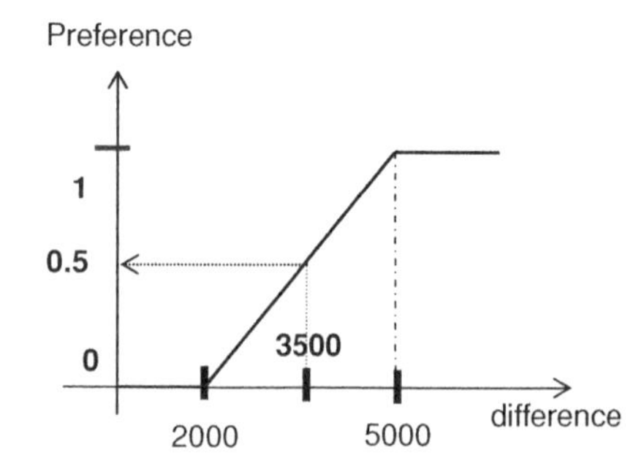

Figure 6.3 Computation of the preference degree when the difference is 3500 and q = *2000 and* p = *5000.*

Table 6.2 The unicriterion preference degrees for price with a linear function when $p = 5000$ and $q = 2000$.

Price	Economic	Sport	Luxury	Touring A	Touring B
Economic (£15 000)	**0**	1	1	1	1
Sport (£29 000)	0	**0**	1	0	0
Luxury (£38 000)	0	0	**0**	0	0
Touring A (£24 000)	0	1	1	**0**	0
Touring B (£25 500)	0	0.5	1	0	**0**

If, with the same indifference and preference threshold, the decision maker had chosen a stepwise preference function, then the difference would fall 'in the step' (Figure 6.2). The preference would also be 0.5.

Finally, let us remark that the difference between the Touring A and Touring B is 1500, which implies that there is no preference between those two cars from the price point of view. The difference in price is lower than the indifference threshold. The remaining unicriterion preference degrees are given in Table 6.2.

Figure 6.4 is a graphical representation of Table 6.2. If the preference of car A over car B is greater than 0, then there is an arrow from A to B. Thus, there is no arrow from Touring B to Economic since the preference of Touring B over Economic is 0. Moreover, the intensity or the size of the arrow indicates the strength of preference. This explains why the arrow between Touring B and Sport is dotted: the preference is only 0.5.

Moreover, from Figure 6.4 we can deduce that Economic is preferred to all the other cars since no arrow is aimed at Economic. On the other hand, Luxury is not preferred to any other car. No preference has been expressed by the decision maker between the Touring A, Touring B and Sport cars: therefore there is no arrow between them. Remember that this graph only relates to the price criterion.

It is worthwhile noting that although we considered numerical values in this example, the PROMETHEE method can deal easily with scaled values (e.g., good,

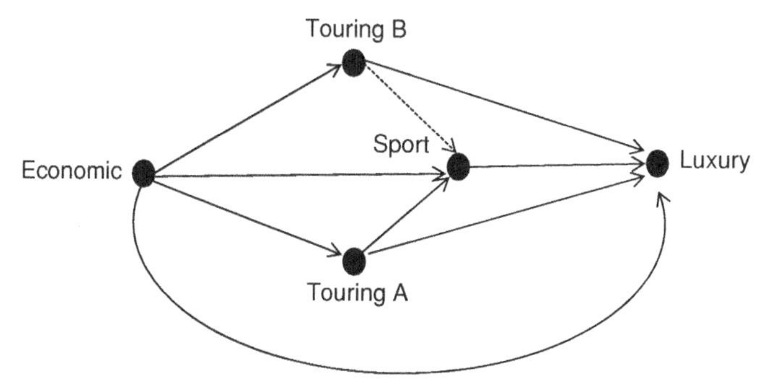

Figure 6.4 Representation of the price preference degrees.

excellent, etc.). As we will see in Section 6.3.2, the verbal values need only to be translated into numerical values.

6.2.2 Unicriterion positive, negative and net flows

It is not easy to draw conclusions from the preference degrees table (Table 6.2) or its graphical representation (Figure 6.4), especially when the number of actions is large. Therefore, the criterion pairwise preference degrees are summarized in the so-called unicriterion *leaving* or *positive flows*, the *entering* or *negative flows*, and the *net flows*. These scores measure how an action is preferred *over* all other actions or how it is preferred *by* all other actions.

6.2.2.1 Positive flows

The unicriterion positive flow (or leaving flow) of an action is a score between 0 and 1. It indicates how an action is preferred (according to the decision maker's preference) *over all other actions* on that particular criterion. The higher this positive flow is, the more preferred the action is compared to the others (i.e. the better is the action). It is in fact an average 'behaviour' obtained by the average of all the preferences of an action compared to the others (excluding the preference degree compared with itself). It is thus the normalized sum of all the row elements and always lies between 0 and 1.

Referring again to our example, the positive flow of Economic is indeed 1 since this car is preferred over all other actions. The flows thus take into account the minimizing or maximizing aspect of the criteria. The positive flow of Luxury is 0 since it is the most expensive car. The positive flow of Sport is $1/4 = 0.25$ (Table 6.3).

On the left of Figure 6.5 are the initial evaluations of the cars with respect to the price criterion. This scale is then transformed via the pairwise preferences to the positive flows depicted on the right. Although Touring A and Touring B are indifferent (preference degrees between the two cars are 0), their positive scores are slightly different. This is due to the fact that both cars 'behave' differently compared to the others cars (in particular compared to Sport). Changing the indifference and preference thresholds changes the positive scores, and their distribution as will be illustrated further in Section 6.2.5.

Table 6.3 Positive, negative and net flows for the price criterion of Case Study 6.1.

Actions\Criterion Flows	Positive	Negative	Net Price Flows
Economic (£15 000)	1	0	1
Sport (£29 000)	0.25	0.625	−0.375
Luxury (£38 000)	0	1	−1
Touring A (£24 000)	0.5	0.25	0.25
Touring B (£25 500)	0.375	0.25	0.125

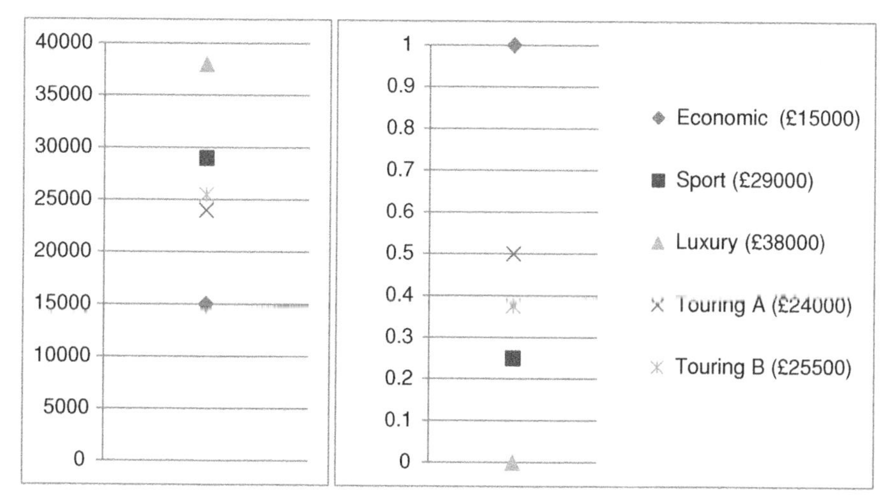

Figure 6.5 Representation of the price evaluations (on the right) and the positive flows with $q = 2000$ and $p = 5000$.

6.2.2.2 Negative flows

Analogously, the negative flows represent an average behaviour. They measure how the other actions are preferred to this action. The negative flow is thus obtained by taking an average of all the preference degrees of the actions compared to that particular action (excluding the preference degree compared with itself). It corresponds to the average of the entire column except for the diagonal element. This score thus always lies between 0 and 1.

The third column of Table 6.3 gives the negative flows of all the actions for the criterion price. Let us emphasize that this score has to be minimized since it represents the weakness of a car compared to the other cars. Since no car is cheaper than the Economic car, its negative flow is equal to zero.

6.2.2.3 Net flows

Finally, to take both the positive and the negative aspects into account, we use the net flows of an action which are obtained by subtracting the negative flows from the positive flows. These have to be maximized since they represent the balance between the global strength and the global weakness of an action. The net score of an action thus always lies between –1 and 1.

6.2.3 Global flows

In the previous section we considered only one criterion. In order to take into account all the criteria simultaneously, the decision maker needs to provide the relative importance of each criterion. For instance, the price of a car can be twice as important as the power. On the other hand, security might be more important than price. In other

Table 6.4 Preference parameters for Case Study 6.1.

Criteria	w_i	q_i	p_i
Price	0.25	2000	5000
Consumption	0.25	0.5	1
Power	0.25	1	2
Comfort	0.25	10	20

words, the decision maker specifies a weight for each criterion which permits him or her to aggregate (by means of a weighted sum) all the unicriterion positive, negative and net flows into *global positive flows*, *global negative flows* and *global net flows*. These flows thus take all the criteria into account.

The relative importance of a criterion can be depicted in several ways (verbal, pairwise comparisons, etc.) but in the end it is, in any case, transformed into a numerical value associated with each criterion (i.e. the weight of the criterion). As we will see in Section 6.3.3, the *Smart Picker Pro* software offers several ways to determine the weights of the criteria.

The global positive score indicates how an action is *globally* preferred to all the other actions when considering several criteria. Since the weights are normalized, the global positive score always lies between 0 and 1.

Analogously, the global negative score indicates how an action is preferred by the other actions. The negative score always lies between 0 and 1 and has to be minimized.

The net flows of an action, obtained by subtracting the negative flows from the positive flows, take into account both views (being preferred over and being preferred by all other actions).

If we consider equally weighted criteria and linear preference functions with parameters given in Table 6.4 we obtain the flows reflected in Table 6.5.

Note that we chose the following scale conversion for the comfort criterion: very bad, 1; bad, 4; average, 8 and very good, 10.

Table 6.5 Positive, negative and net flows for the cars of Case Study 6.1.

Actions	Positive flows	Negative flows	Net flows
A1 - Economic	0.375	0.5	–0.125
A2 - Sport	0.375	0.53125	–0.15625
A3 - Luxury	0.34375	0.4375	–0.09375
A4 - Touring A	0.375	0.28125	0.09375
A5 - Touring B	0.46875	0.1875	0.28125

6.2.3.1 The PROMETHEE I ranking

The PROMETHEE I ranking is based on the positive and the negative flows. In this ranking, there are four different scenarios when analysing the flows of two actions:

- One action has a better rank than another if its global positive and negative flows are simultaneously better (i.e. if the global positive score is higher and the global negative flow is lower). Considering the global positive and negative scores of Table 6.5, the Touring A car has a better rank than Luxury. This can be easily detected from Figure 6.6, where the positive and negative scores are shown. An action should be simultaneously the first listed on the positive and negatives axes.
- One action has a worse rank than another if both global positive and the negative scores are worse. Touring A therefore has a worse rank than Touring B: its positive scores are lower and its negative scores are higher.
- Two actions are said to be incomparable if one action has a better global positive score but worse global negative score (or vice versa). Economic and Luxury are incomparable since Luxury has a lower positive score and a lower negative score. This can be easily detected graphically as the two actions cross each other (Figure 6.6).
- Two actions are called indifferent if they have identical positive and negative flows.

6.2.3.2 The PROMETHEE II ranking

The PROMETHEE II ranking is based on the net flows only and leads to a complete ranking of the actions (i.e. the incomparable status does not exist). The actions can

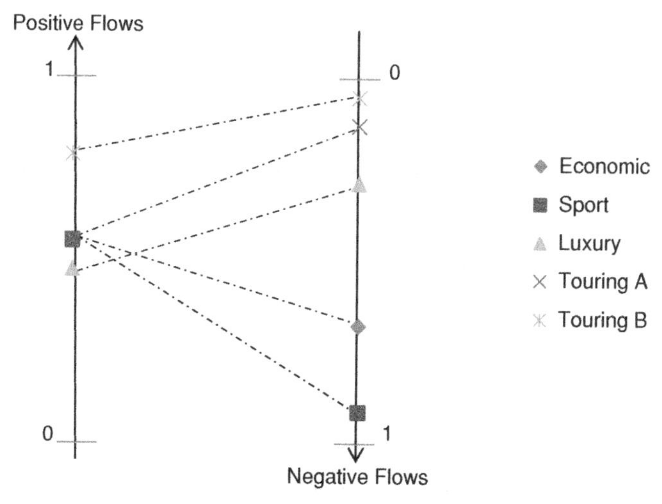

Figure 6.6 Global positive and negative scores from Case Study 6.1.

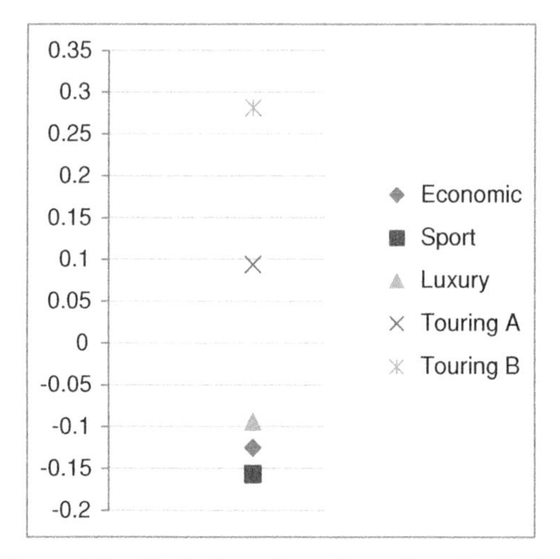

Figure 6.7 Global net flows from Case Study 6.1.

thus be ordered from the best to the worst: Table 6.5 gives the net scores of the cars and Figure 6.7 is the graphical representation.

To summarize the steps of the decision process, the decision maker needs to define the criteria taken into account in his or her decision. Then all actions to be ranked need to be evaluated according to those criteria. By specifying this preference information, the pairwise criterion preference degrees can be computed. From those preference degrees, unicriterion flows are computed. In a final step, the criterion flows are aggregated into global flows by taking into account the relative importance of each criterion. We then have a ranking.

6.2.4 The Gaia plane

The Gaia plane is a two-dimensional representation of a decision problem. It contains all the aspects of the decision problem: the actions, the criteria and the decision maker's preference information (thresholds and weights). We will not prolong the suspense any further: Figure 6.8 shows the Gaia plane for the car decision problem of Case Study 6.1.

In this Gaia plane, actions are represented by bullets and criteria by arrows. The position of the actions gives the decision maker some idea as to their similarities: the closer the actions, the more similar the actions. Touring A and Touring B are close to each other in Figure 6.8. In contrast, Sport is far away from Economic. The Sport car is thus very different from the other cars.

The similarity and non-similarity are defined by the indifference and preference threshold. This implies that the Gaia plane depends on the preference information given by the decision maker.

Analogously, the relative position of the criteria indicates to us the correlation and anti-correlation (or conflict) of criteria. The closer the arrows, the more correlated the

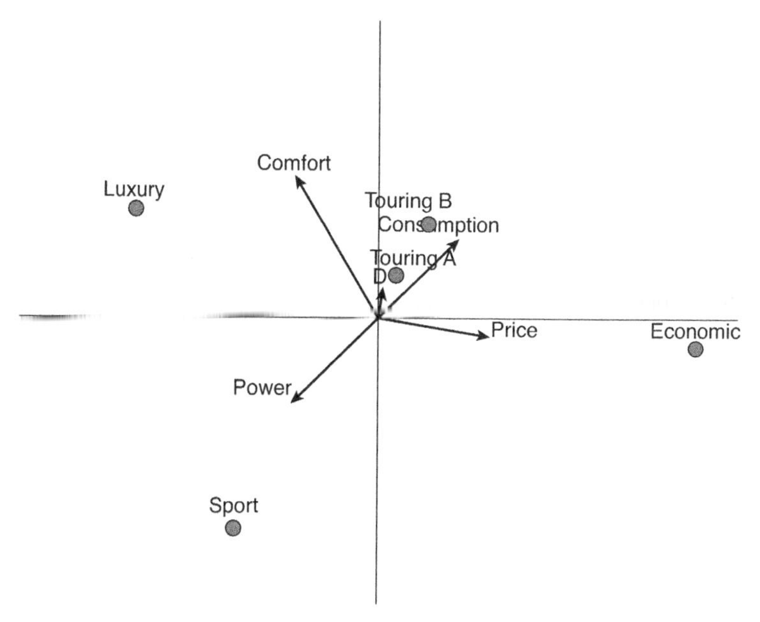

Figure 6.8 Gaia plane for the car decision problem of Case Study 6.1. Reproduced by permission of Smart Picker.

criteria in the decision problem. The greater the angle between criteria, the greater the conflict between them. In Case Study 6.1, power conflicts with consumption: the consumption has to be minimized, the power maximized, and indeed the more powerful a car, the more it consumes. The Gaia plane enables the visualization of the conflicting points of view.

As one can see, the lengths of the criteria are different. The length of a criterion measures its 'discriminating' or 'differentiating' power as a function of the data. The more different the actions on a criterion, the longer the arrow and thus the more discriminating the criterion. Power is represented by a long arrow since it has a high discriminating power, whereas consumption does not differ greatly amongst the actions. The discriminating power of a criterion depends on the chosen thresholds (the higher the indifference threshold, the less discriminating the criterion) and on the corresponding weight.

Finally, the arrow labelled D (called the *decision stick*) renders the compromise solution chosen by the decision maker, since it corresponds to his or her weight setting. The projection of the actions on this line represents their priorities. The higher the action on the stick, the better the action. However, since this is only a two-dimensional representation, projection results in a loss of information which makes the Gaia plane less accurate (or less representative of the decision problem). The amount of information preserved (the so-called *delta*) depends on the data and the number of criteria. As a consequence of the information loss, the ranking obtained by projection on the decision stick does not necessarily correspond to the PROMETHEE II ranking.

6.2.5 Sensitivity analysis

One of the main advantages of the *Smart Picker Pro* software is the ability to conduct a detailed sensitivity analysis. It is important to test the sensitivity of the ranking relative to the input parameters. Trying different parameters (e.g. dynamically changing the weights or the indifference and preference thresholds) enables us to be aware of the stability or instability of the ranking. A small variation of one single parameter inducing a complete change in the ranking indicates an unstable solution. The decision maker needs to be aware that the conclusions drawn are only valid for this precise set of chosen parameters and any small error or perturbation leads to a very different solution. If on the other hand, the ranking remains the same for a large variety of different sets of parameters, then the decision maker might be quite confident about the stability of the solution.

Exercise 6.1

The following multiple-choice questions allow you to test your knowledge on the basics of PROMETHEE. Only one answer is correct. Answers can be found on the companion website.

1. What does PROMETHEE stand for?
   a) Positive Organization METHod with Enriched Evaluation
   b) PReference Organization METHod for Enriched Evaluation
   c) PROfessional METHod for Easy Evaluation
   d) PROactive MEasurement THEory Evaluation
2. Which of these statements is incorrect?
   a) PROMETHEE can be used in a wide range of applications
   b) Every decision maker will find the same ranking
   c) The PROMETHEE method requires a lot of input parameters
   d) Results can be explained
3. What is the main purpose of PROMETHEE?
   a) PROMETHEE prioritizes actions based on criteria and constraints
   b) PROMETHEE assigns goals to actions
   c) PROMETHEE ranks actions based on criteria
   d) PROMETHEE assigns criteria to alternatives
4. Pairwise comparisons are based on which type of scale?
   a) Ratio scale
   b) Interval scale

c) Ordinal scale

d) Nominal scale

5. How many input parameters does a decision maker need to specify for a criterion (supposing he has chosen a linear preference function)?

a) 5

b) 4

c) 3

d) 2

6.3 The Smart Picker Pro software

All the PROMETHEE computations can easily be implemented in a spreadsheet, though graphs really help the decision maker to visualize the issues and key points in the decision. A number of user-friendly software packages have certainly contributed to the success of the PROMETHEE method. They incorporate intuitive graphical user interfaces, automatic calculation of preference degrees, flows, etc. One of the main advantages of the use of these software packages is the possibility to perform a sensitivity analysis. This allows the user to answer questions such as 'What if I change this parameter?' which is useful in scenario planning and dealing with risk. Among those currently available are *Decision Lab*, *D-Sight*, *Smart Picker Pro*, and *Visual Promethee*.[1]

This section will describe the *Smart Picker Pro* standalone software because of its simplicity, and because it is available as a free trial version (www.smart-picker.com) with time-unlimited use, unlike other software. The trial version is, however, restricted to five actions and four criteria, though this is ample for gaining familiarity with its application. *Smart Picker Pro* does not require much understanding of the method: as soon as the evaluations are entered, a ranking is available (by the use of default values which will be provided throughout this section).

However, this might not entirely reflect the decision maker's preferences. That is why we suggest entering all input parameters step by step.

As an illustration, we will completely resolve Case Study 6.1 according to a fictitious set of preference parameters given in Table 6.4. On the companion website the reader will find the basic steps of the PROMETHEE calculations in a spreadsheet as well as the *Smart Picker Pro* input files.

6.3.1 Data entry

The decision maker needs to first enter the *decision problem* before entering his preference parameters, although actions or criteria can be deleted or added during the

[1] At the time of writing, there was only a beta version of *Visual Promethee*.

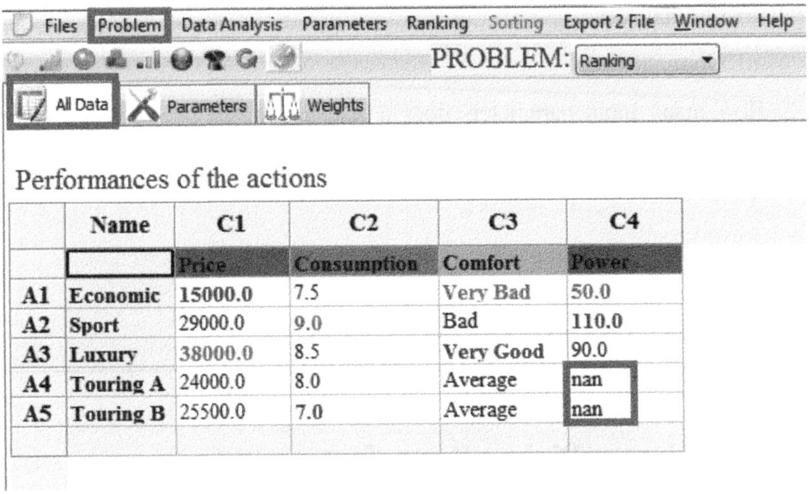

Files Problem Data Analysis Parameters Ranking Sorting Export 2 File Window Help

PROBLEM: Ranking

All Data Parameters Weights

Performances of the actions

	Name	C1	C2	C3	C4
		Price	Consumption	Comfort	Power
A1	Economic	15000.0	7.5	Very Bad	50.0
A2	Sport	29000.0	9.0	Bad	110.0
A3	Luxury	38000.0	8.5	Very Good	90.0
A4	Touring A	24000.0	8.0	Average	nan
A5	Touring B	25500.0	7.0	Average	nan

Figure 6.9 Data entry in Smart Picker Pro. *Reproduced by permission of Smart Picker.*

decision process. The spreadsheet of the *All Data* tab (Figure 6.9) shows the actions, the criteria and the initial scores of the actions on each criterion. Each row represents an action and each column represents a criterion of the decision problem. Adding or deleting actions or criteria can be easily done by:

- accessing the *Problem* menu, or
- using keyboard shortcuts (e.g., Alt+A, Alt+C, Alt+Shift+A, Alt+Shift+C), or
- placing the cell cursor on the last row or in the last column and clicking on the *Down arrow* (↓) or Alt + right arrow (Alt+ →).

The red value in a specific column indicates the worst evaluation for that criterion and blue the best value. This means that if a criterion is to be minimized, the lowest score is blue and the highest score is red. The colours are reversed when maximizing criteria.

Furthermore, the decision maker can enter numerical values as well as verbal evaluations. Sometimes, it is easier to work with a verbal scale rather than a numerical scale. For instance, it is easier to define the quality or the comfort of a car by means of words such 'good', 'average' and 'excellent'. As we will see, there is a method in the software to assess how much better 'good' is compared to 'average', etc.

The PROMETHEE method permits the use of a verbal scale provided that this scale is later translated into a numerical scale. This can be done in the Parameters tab (see Section 6.3.2).

Finally, it is worth mentioning that evaluations of all the actions are not required. In Figure 6.9 one can see that the power of the Touring A and Touring B cars are equal to 'nan' (i.e. not a number). This means that the user does not necessarily need to specify all the values. The *Smart Picker Pro* software will by default replace all the missing values on one particular criterion by the mean value for this criterion. However, the user is able to change this setting by going to *Data Analysis/Parameters Processing* and choosing either the median value or a desired default value.

6.3.2 Entering preference parameters

The Parameters tab by default displays the preference parameters of the first criterion (Figure 6.10). To enter the parameters of a specific criterion, the user can just pick the criterion from the combo-box (marked 1 in Figure 6.10).

In this tab the user can specify the following preference parameters:

- the preference function – a linear function, a stepwise function or a Gaussian function (2 in Figure 6.10);
- the preference direction – to maximize or minimize the criterion (3 in Figure 6.10);
- the values of the indifference and preference thresholds (4 in Figure 6.10).

The graph in the lower part of Figure 6.10 should help the decision maker in determining the shape of the preference function and its thresholds. Along with the

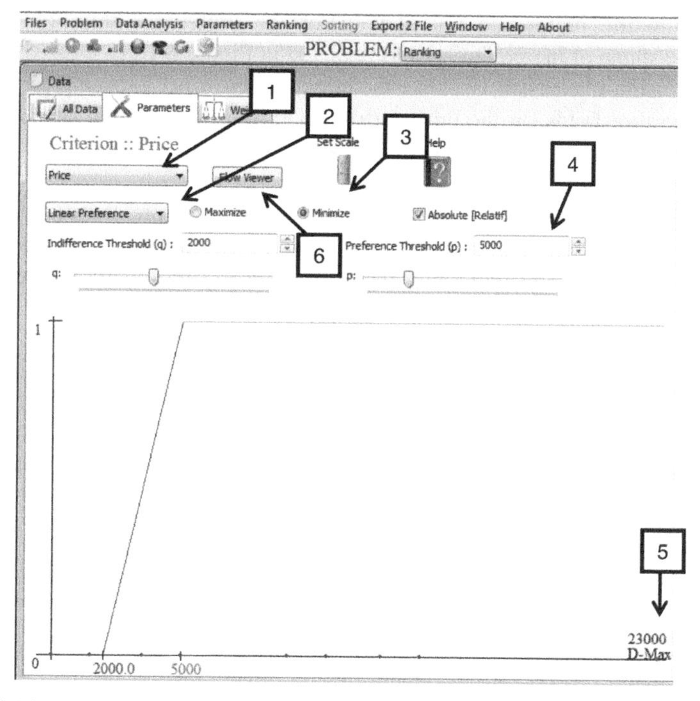

Figure 6.10 Parameters tab in Smart Picker Pro. *Reproduced by permission of Smart Picker.*

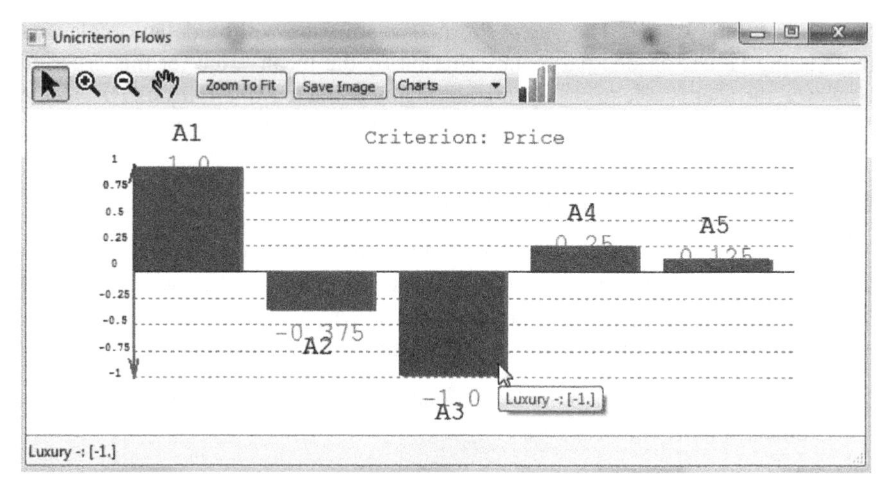

Figure 6.11 Unicriterion net flows in Smart Picker Pro. *Reproduced by permission of Smart Picker.*

preference function selected, all the differences between the actions on the selected criterion are plotted in blue on the horizontal axis. The highest difference is indicated by *D-Max* (5 in Figure 6.10).

Changing the preference function and/or the preference thresholds directly changes the shape of the preferences as depicted in Figure 6.10.

The decision maker can visualize the effect of his or her chosen preference and indifference thresholds on the unicriterion net flows/scores which are represented in (Figure 6.11), accessible via the Flow Viewer button (6 in Figure 6.10) from the Parameters tab.

Note that if the decision maker does not specify any preference information, the *Smart Picker Pro* software will assign default values so that a ranking can still be computed. The default preference parameters for a criterion are: maximizing the criterion, a linear preference function with $q = p = 0$ and a weight value equal to 1 (which will be normalized to the other criteria).

As the perception of scale values is subjective, the user will have to define his or her scale by clicking on the Scale button in the Parameters tab. This leads him or her to the view shown in Figure 6.12. *Smart Picker Pro* offers several ways to translate a verbal scale into a numerical scale:

- by specifying a direct numerical value for each verbal scale label.
- by ranking the verbal scale labels by their importance (i.e. specifying for instance that *very good* is the best, *good* is the second best, etc.). The software generates a score based on this ranking.
- by pairwise comparing the labels on a 3-, 5- or 7-level scale. This enables the user to specify for instance that *very good* is 3 times better than *good* and that *very bad* is neutral to *bad*.

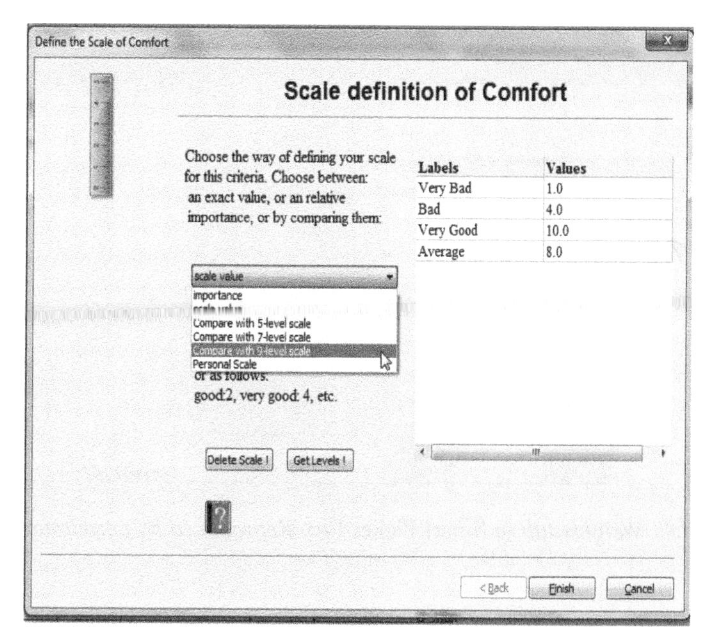

Figure 6.12 Entering scale definition in Smart Picker Pro. *Reproduced by permission of Smart Picker.*

6.3.3 Weights

The decision maker can specify the weights in the Weights tab (Figure 6.13). He or she can modify the value of the weights by directly dragging the weight bars up or down or by selecting a criterion on the drop-down menu and using the slider to its right. This can be done for a criterion relative to the others (i.e. when the weight of a criterion is increased, then the other weights will be proportionally decreased) or in absolute terms (the values of the other weights do not change). Furthermore, *Smart Picker Pro* provides the user with a context-specific assistance tool for evaluating the weights (by clicking on the '?'button – (1 in Figure 6.13).

The decision maker can choose how he or she specifies the weights (1 in Figure 6.14):

- by directly entering a value.
- by ranking the criteria according to their importance (e.g. specifying that the space criterion is the most important). The software generates a score based on this ranking.
- by pairwise comparing the weights on a 3-, 5- or 7-level scale. This facilitates specifying, for instance, that price is 3 times more important than consumption and that power is as important as space (2 in Figure 6.14).

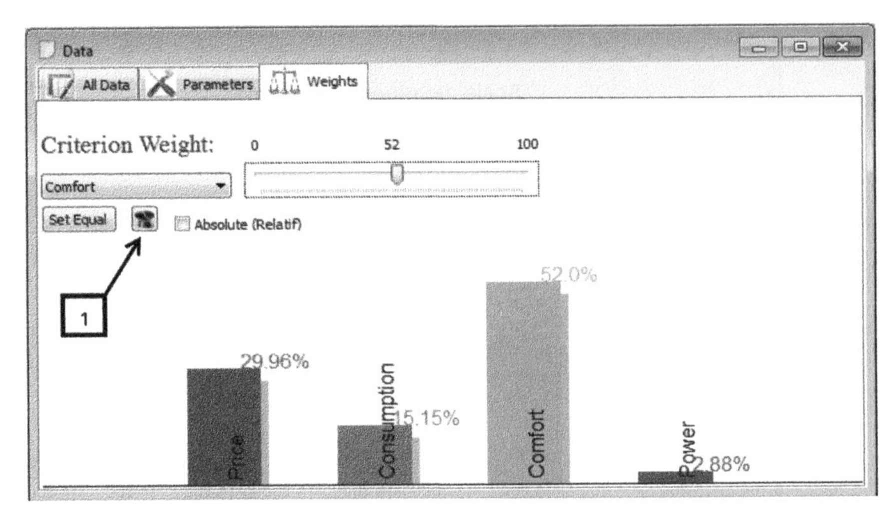

Figure 6.13 Weights tab in Smart Picker Pro. *Reproduced by permission of Smart Picker.*

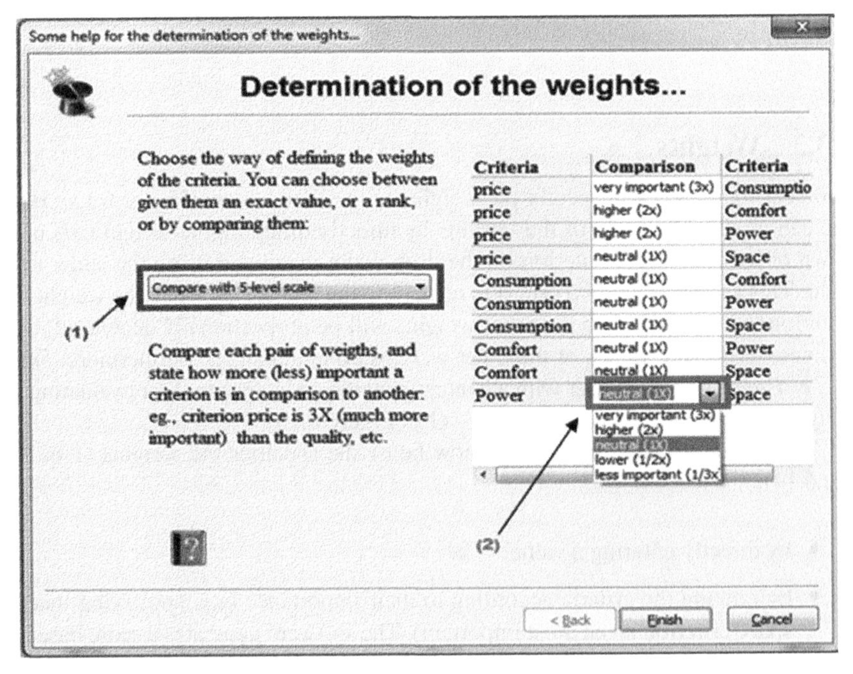

Figure 6.14 Weights assistant in Smart Picker Pro. *Reproduced by permission of Smart Picker.*

	Price	Consumption	Comfort	Power
new name	Price	Consumption	Comfort	Power
min max	min	min	max	max
weight	0.3	0.152	0.52	0.029
Type (L/S)	L	L	L	L
p	5000.0	1	2.0	20.0
q	2000.0	0.5	1.0	10.0
Abs/Rel	Abs	Abs	Abs	Abs

Figure 6.15 Summary of the preference parameters in Smart Picker Pro. *Reproduced by permission of Smart Picker.*

Note here that the user can retrieve all of the preference parameters from a summary table (shown in Figure 6.15). This table can also be used to directly enter these parameters.

6.3.4 PROMETHEE II ranking

Once the data, the preference parameters and the weights are entered, the complete PROMETHEE II ranking is computed by pressing F6 or by selecting *Ranking/ Ranking Scores*. To help the decision maker in his or her analysis, the results can be displayed as a table (Figure 6.16) or in one of several graphical representations (accessible via the button marked 1 in Figure 6.17).

The contribution of each positive flow (stacked bar above the zero line) and negative flow (stacked bar below the zero line) in Figure 6.18 to the net flows (the white dotted line and the black bullet) permit the decision maker to clearly understand the strengths and the weaknesses of each action; and thus their rank.

For example, the Economic car has a high score due to its good performance on the price criterion (indicated by the high price bar). However, the Economic car does

Results

Results | Processed Data

	Actions	Net Flows	Position
	A1 - Economic	-0.125	4.0
	A2 - Sport	-0.15625	5.0
	A3 - Luxury	-0.09375	3.0
	A4 - Touring A	0.09375	2.0
	A5 - Touring B	0.28125	1.0

Figure 6.16 Net scores in table view in Smart Picker Pro. *Reproduced by permission of Smart Picker.*

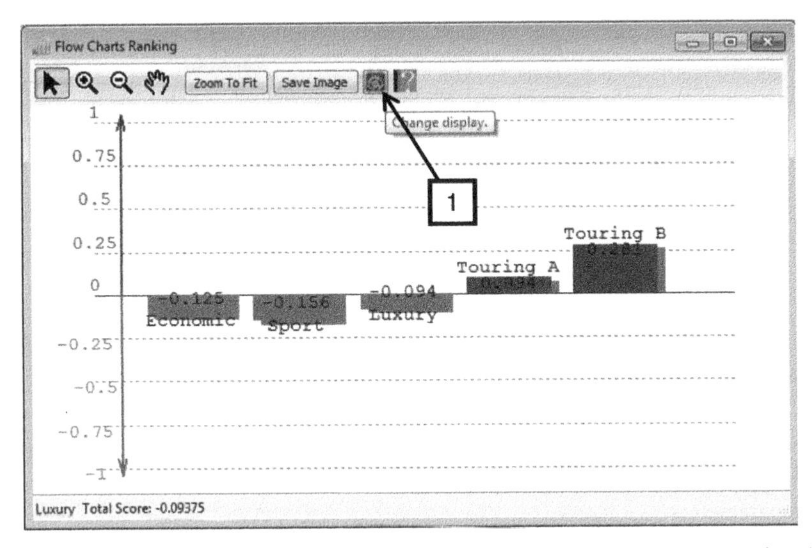

Figure 6.17 Graphical representation of net scores (PROMETHEE II ranking) in Smart Picker Pro. *Reproduced by permission of Smart Picker.*

not perform well on the comfort and power criteria. This is indicated by their negative contribution. The Luxury car performs well only on the power criterion (only positive contribution) and performs very badly on the price criterion.

Smart Picker Pro includes an easily understandable representation of the PROMETHEE I ranking via a triple *thermometer view*, where incomparability is easily detectable. The first thermometer (on the left in Figure 6.19) indicates the scores of the negative flows, the second thermometer represents the scores of the

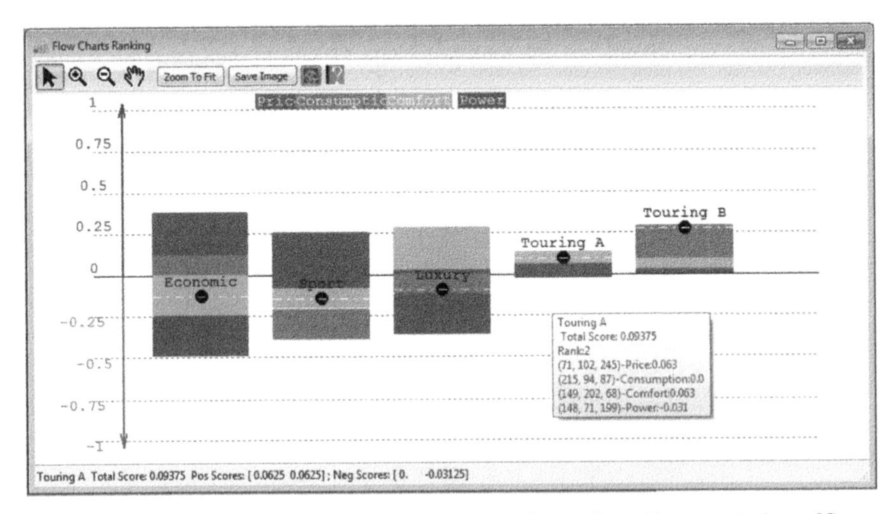

Figure 6.18 Criterion flows in Smart Picker Pro. *Reproduced by permission of Smart Picker.*

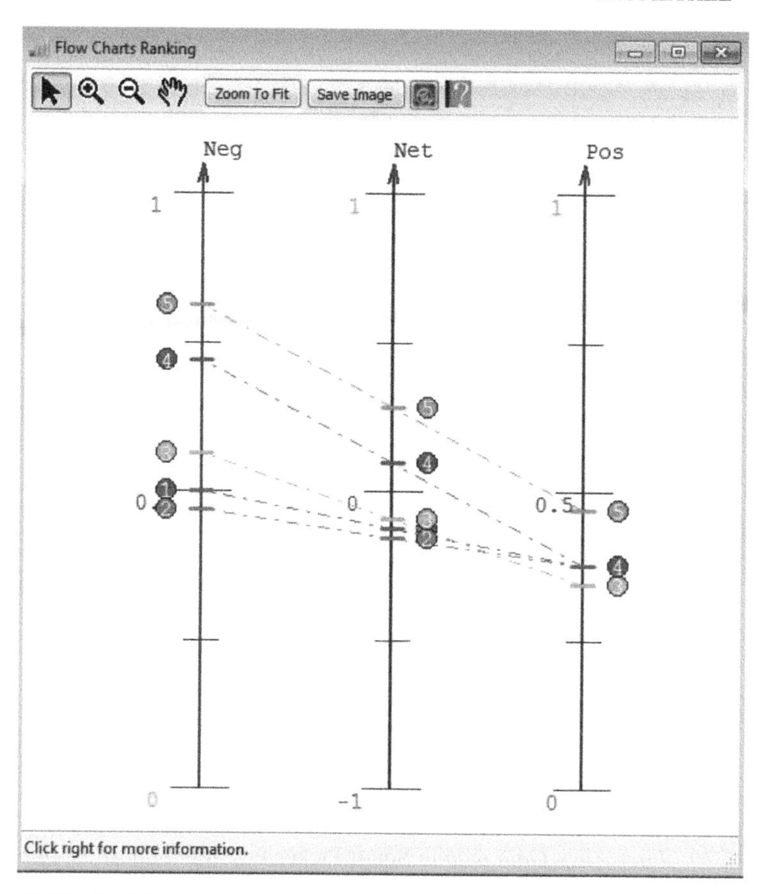

Figure 6.19 Positive, negative and net flows in Smart Picker Pro. *Reproduced by permission of Smart Picker.*

net flows, and the third thermometer shows the scores of the positive flows. All thermometers are bounded from the worst score (bottom) to the best score (top). Therefore, the negative flow thermometer starts from 1 on the left axis and finishes at 0. A dotted line represents the score of one action on the three flows. If two lines cross, there is an incomparability between the two actions.

6.3.5 Gaia plane

Smart Picker Pro provides the user with an interactive Gaia map. The projections on each criterion and on the decision stick can be visualized easily by left-clicking on the arrows. Some information can be gained by left-clicking on the bullets (representing the actions). Moreover, in the menu (see the toolbar in Figure 6.20), the user can choose to display the initial performances of the actions, to see a weighted view of the Gaia map, etc. The user can also see the consequences of changing the thresholds or the weights very easily: increasing the indifference threshold may lead to some

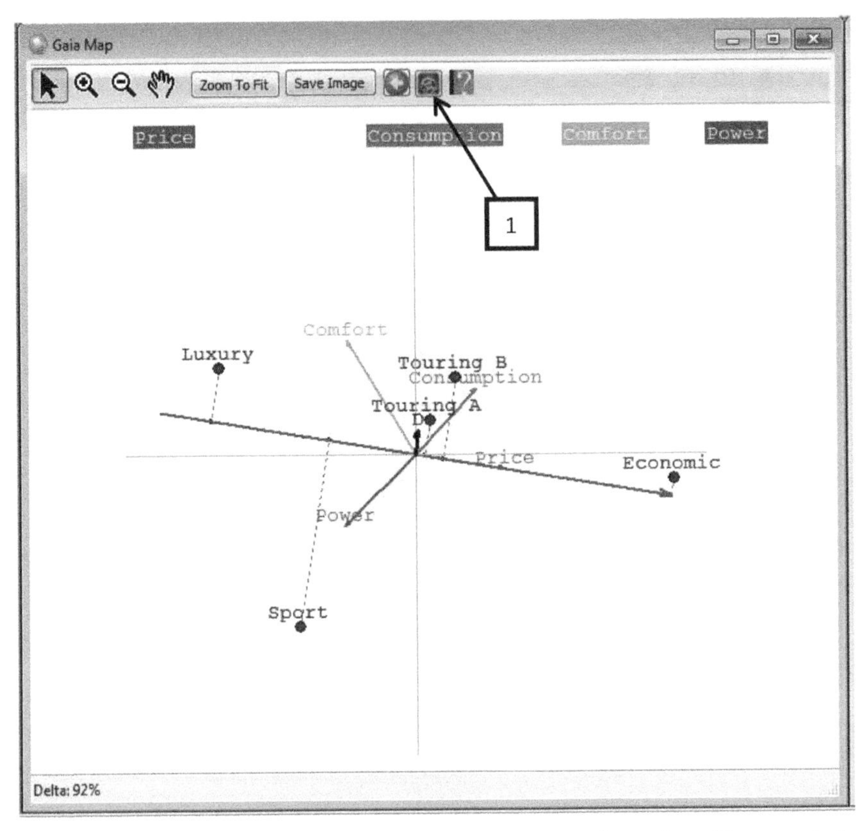

Figure 6.20 Interactive Gaia map in Smart Picker Pro. *Reproduced by permission of Smart Picker.*

actions being closer (more similar). In contrast, increasing the preference threshold or changing the preference function may lead to more spread (difference) on the Gaia map. This step helps the decision maker to better understand the decision problem as well as the sensitivity of the preference parameters.

6.3.6 Sensitivity analysis

As mentioned in Section 6.2.5, a variation in the values of the parameters may lead to a change in the scores and ranking. It is therefore crucial to perform some sensitivity tests in order to have an idea of the stability of the final decision. One of the main features of *Smart Picker Pro* is that it enables the decision maker to change all of the parameters dynamically while simultaneously analysing the different rankings. The user can thus directly modify all the preference parameters. On the other hand, *Smart Picker Pro* permits the user to change the evaluations of the actions, while keeping the same preference parameters. As shown in Figure 6.21, the user can

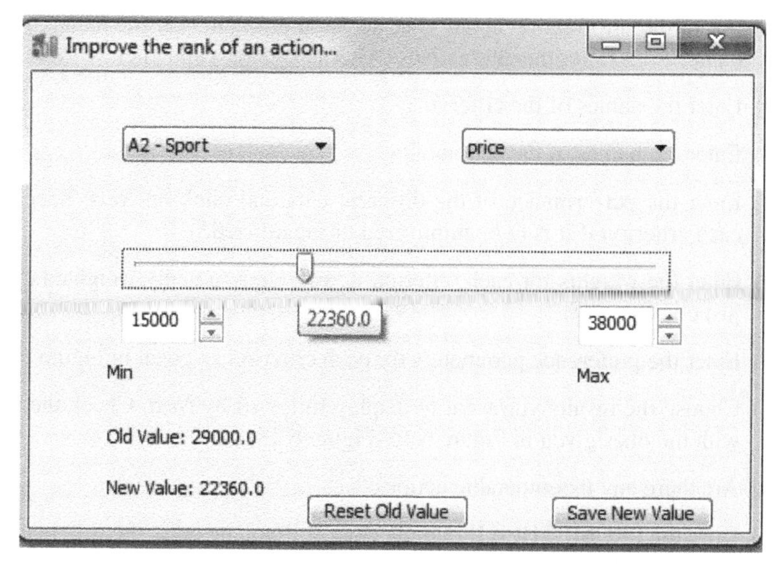

Figure 6.21 Sensitivity analysis by changing the evaluations of the actions in Smart Picker Pro. *Reproduced by permission of Smart Picker.*

gradually increase/decrease the evaluation of a specific action on a chosen criterion. This permits the measurement of the required improvement (or deterioration) of the criterion evaluation of an action in order to be higher (or lower) ranked.

Exercise 6.2

In this exercise, you will solve the car selection problem of Case Study 6.1 using *Smart Picker Pro.*

Learning Outcomes

- Structure a problem in *Smart Picker Pro*
- Enter the preference parameters
- Understand the results
- Conduct a sensitivity analysis

Tasks

a) Read the description of the Case Study 6.1, on page 138 and open *Smart Picker Pro*. Choose 'Create a new problem with the Companion' in the opening wizard and click on the *Go* button.

b) Choose the left figure or *Ranking actions*.

c) Fill in the information about your decision problem (i.e. file name, the user name, etc.). Enter the alternatives (Alternatives tab) and click on *Next*.

d) Enter the names of the criteria.

e) Enter the names of the actions.

f) Enter the performance of the different cars and click on *Next*. Specify for each criterion if it is to be minimized or maximized.

g) Enter the weights for each criterion. If you chose equally weighted criteria, just click on *Next*.

h) Enter the preference parameters for each criterion as given in Figure 6.15.

i) Choose the results you want to display followed by *Next*. Check the results with the ones given in Figure 6.16–Figure 6.18.

j) Are there any incomparable actions?

k) Does the PROMETHEE II ranking change if you increase the weight of price to 35% while maintaining the relative weight of the other criteria?

l) On which criterion does the best action perform the worst?

m) How does the decision stick move? Do the projections of the actions on the decision stick correspond to the net flow ranking?

n) Does an increase of 10% on the power criterion change the top-two ranking?

6.4 In the black box of PROMETHEE

The PROMETHEE method has already been used successfully in a lot of cases. Behzadian et al. (2010) listed 200 papers since its conception where PROMETHEE has been applied in environment management (47 papers), business and financial management (25), hydrology and water management (28), chemistry (24), logistics and transportation (19), manufacturing and assembly (19), energy management (17), social science (7), design (2), agriculture (2), education (2), sports (1), information technology (1) and medicine (1) up to 2008. Recently, PROMETHEE has been used in

- water management (Kodika 2010; Silva et al. 2010);
- banking (Doumpos and Zopounidis, 2010);
- energy management (Ghafghazi 2010; Oberschmidt 2010);
- manufacturing and assembly (Kwak and Kim 2009; Tuzkaya 2010; Venkata Rao and Patel 2010; Zhu 2010);
- logistics and transportation (Lanza and Ude 2010; Safaei Mohamadabadi 2009; Semaan 2010);

- chemistry (Cornelissen 2010; Ni 2009);
- maritime commerce (Castillo-Manzano 2009);
- strategy (Ghazinoory 2009);
- project management (Halouani 2009);
- construction (Castillo-Manzano 2009; Frenette 2010);
- urban development (Juan 2010);
- location analysis (Luk 2010; Ishizaka et al. 2013);
- environment (Nikolic 2010; Soltanmohammadi et al. 2009; Zhang 2009);
- safety (Ramzan 2009);
- engineering (Ishizaka and Nemery 2011);
- e-commerce (Andreopoulou 2009).

The PROMETHEE method belongs to the family of outranking methods, which means that the method is based on pairwise comparisons of the actions. As we will see, the use of pairwise comparisons to infer global rankings has a direct influence when an action is added to or deleted from the problem (Keyser and Peeters 1996; Mareschal et al. 2008).

Denote by $A = \{a_1, a_2, \ldots, a_n\}$ the set of actions to be ranked; and let $F=\{f_1, f_2, \ldots f_m\}$ be the set of criteria. As in most multi-criteria decision aid methods, the set of criteria is assumed to be a coherent set of criteria as defined in Vincke (1992). Without loss of generality, we will suppose in this section that all criteria have to be maximized. Furthermore, we will denote by $f_i(a_j)$ the evaluation of action a_j on criterion f_i. Let us first assume that $f_i(a_j)$ is a numeric value.

In this section we will present the PROMETHEE method in a slightly different way than in Section 6.2, where we introduced the unicriterion flows as an aggregation of the criterion preference degrees at a global action level, that is, from a pairwise behaviour to a global behaviour. These unicriterion flows are then aggregated (by means of a weighted sum) to the global positive and negative flows by taking into account the weights of the criteria. This is illustrated on the left-hand side of Figure 6.22.

In this section we will deduce the global flows from the so-called *preference matrix*. This preference matrix contains the global pairwise preference degrees, computed between all the ordered pairs of actions. The global preference degrees are deduced from the criterion preference degrees by means of the weighted sum. Based on the global preference degrees, the global flows can be easily deduced, as we will see.

The reason for introducing this preference matrix is the information in the pairwise comparisons. The preference matrix enables us, for instance, to detect 'local incomparable actions' (incomparable if you only consider these two actions) but which are 'indifferent' in the final ranking. The preference matrix contains information

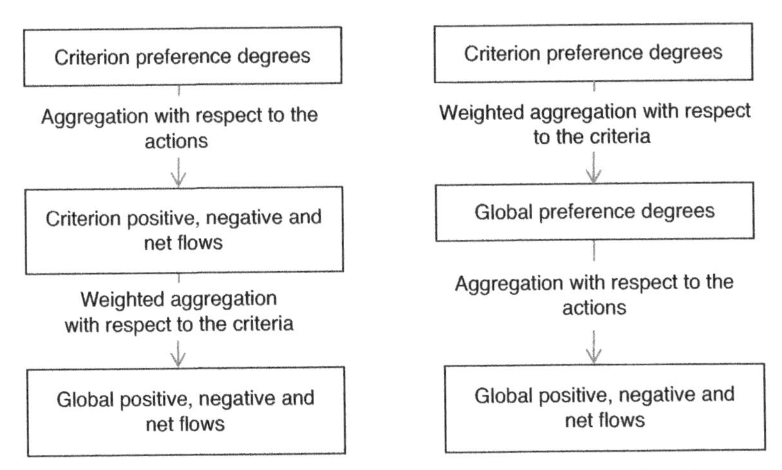

Figure 6.22 Two different ways of computing the global flows based on criterion preference degrees.

which is lost when aggregating the pairwise preference degrees to the global flows. This computation process is represented on the right-hand side of Figure 6.22.

The two approaches are totally complementary since the first approach allows one to render the global behaviour of an action compared to all the other actions on one particular criterion. On the other hand, the global preference degree represents how an action compares globally to one specific other action.

The outline of this section is as follows. First, we will introduce the unicriterion preference degree. Based on this, we will define the preference degree and the preference matrix. The global flows will be directly deduced from these pairwise preference degrees. A sensitivity analysis will then be performed on the pairwise comparisons.

6.4.1 Unicriterion preference degrees

For each ordered pair of actions (a_i,a_j) of A, the unicriterion preference degree P_{ij}^k (also noted $P_k(a_i, a_j)$ is computed and reflects how strongly action a_i is preferred to a_j based solely on criterion f_k. P_{ij}^k is a number between 0 and 1, and is a function of the difference between the evaluations (i.e. $f_k(a_i) - f_k(a_j)$): the higher this difference, the stronger the unicriterion preference degree.

The preference degree is computed based on the preference functions (illustrated in Figure 6.1 and Figure 6.2). The decision maker thus has the choice between three different types of preference functions. If we consider the linear preference function with q and p as, respectively, the indifference and preference threshold, we have formally that

$$P_{ij}^k = \begin{cases} 0 & \textbf{if} \quad f_k(a_i) - f_k(a_j) \leq q \\ \dfrac{[f_k(a_i) - f_k(a_j) - q]}{[p - q]} & \textbf{if} \quad q < f_k(a_i) - f_k(a_j) < p \\ 1 & \textbf{if} \quad f_k(a_i) - f_k(a_j) \geq p \end{cases} \tag{6.1}$$

(Brans and Mareschal 2005; Brans and Vincke 1985). On the other hand, if we consider the Gaussian preference function, where s represents the inflexion point, we have

$$P_{ij}^{k} = \begin{cases} 1 - \exp\left(\dfrac{-(f_k(a_i) - f_k(a_j))^2}{2s^2}\right) & \textbf{if } f_k(a_i) - f_k(a_j) \geq 0 \\ 0 & \textbf{otherwise}. \end{cases} \tag{6.2}$$

Conversely, the unicriterion preference degree P_{ij}^{k} expresses how a_j is preferred to a_i according to the decision maker. P_{ij}^{k} and P_{ji}^{k} are not symmetric numbers but respect the condition $0 \leq P_{ij}^{k} + P_{ji}^{k} \leq 1$.

6.4.2 Global preference degree

Having all the ordered unicriterion preference degrees, the global preference degree π_{ij} can be computed while taking into account the weights associated to each criterion. Let w_k be the weight associated to criterion f_k. If the weights respect the condition $\sum_{k=1}^{q} w_k = 1$, we have (Brans et al. 1986) that

$$\pi(a_i, a_j) = \pi_{ij} = \sum_{k=1}^{q} w_j \cdot P_{ij}^{k}. \tag{6.3}$$

The preference degree π_{ij} expresses the global preference of action a_i on a_j according to all criteria. This preference degree, which lies between 0 and 1, respects the constraint $0 \leq \pi_{ij} + \pi_{ji} \leq 1$. We observe that we have necessarily that $\forall i : \pi_{ii} = 0$.

Suppose, therefore, that the preference degree reflects a 'local' behaviour: comparison of one action to just one other action. The preference degree translates either an indifference (similar and low preference degrees between the actions, $\pi_{ij} \approx 0 \approx \pi_{ji}$), an incomparability (similar and high preference degrees, $\pi_{ij} \approx 0.5 \approx \pi_{ji}$) or a preference (high difference between the preference degrees, $\left|\pi_{ij} - \pi_{ji}\right| \gg 0$) between two actions. We note here that the definitions of indifference, incomparability and preference are not precise but left to the interpretation of the decision maker.

All the ordered pairwise comparisons are usually presented in the preference matrix Π, where the element $\Pi(i, j)$ represents π_{ij}.

As an illustration of this, let us consider the following numerical example:

Example 6.1 Table 6.6 represents the evaluation matrix of four actions on two criteria. If we consider the criteria to be equally weighted, modelled with the linear preference function where $q = 0.2$ and $p = 0.5$, we have the preference matrix given in Table 6.7, from which we can conclude that:

- Action A1 is preferred to action A2 since we have: $\pi_{12} = 1$ and $\pi_{21} = 0$.
- Action A2 is incomparable to action A1: $\pi_{32} = 0.5$ and $\pi_{32} = 0.5$. In other words, actions A1 present some strength on one criterion and some weaknesses on the other criterion, compared to action A2.

Table 6.6 Numeric example.

Actions	Criterion 1	Criterion 2
Action 1 (A1)	1	1
Action 2 (A2)	0	0.5
Action 3 (A3)	0.75	0
Action 4 (A4)	1	0.1

- Actions A3 and A4 can be considered as indifferent, since the preference degrees between both actions are similar and small ($\pi_{43} = 0.08$ and $\pi_{34} = 0$).

6.4.3 Global flows

The positive and negative flows summarize the ordered preference degrees into a unique score for each action. Let us denote by $\Phi^+(a_i)$ and $\Phi^-(a_i)$ respectively the positive and negative flows of action a_i. These can be computed as follows:

$$\Phi^+(a_i) = \frac{\sum_{j=1}^{n} \pi_{ij}}{n-1}, \tag{6.4}$$

$$\Phi^-(a_i) = \frac{\sum_{j=1}^{n} \pi_{ji}}{n-1}. \tag{6.5}$$

The positive flow of action a_i thus represents the mean preference degree of action a_i compared to all other actions. In other words, it measures the global 'preferred' behaviour of action a_i compared to the other actions. Mathematically, it corresponds to the sum of all the elements in the corresponding row divided by the number of actions reduced by 1, as it cannot be compared to itself. The positive flow always lies between 0 and 1. The higher this number is, the better the action will be.

Analogously, the negative flow of action a_i represents the mean preference of the other actions on action a_i. In other words, it measures the global 'being preferred' behaviour of action a_i compared to the other actions. Mathematically, it corresponds

Table 6.7 Preference matrix of the problem in Table 6.6.

	A1	A2	A3	A4
A1	0.0	1.0	0.58333	0.5
A2	0.0	0.0	0.5	0.33333
A3	0.0	0.5	0.0	0.0
A4	0.0	0.5	0.08333	0.0

Table 6.8 The global flows of the numerical example of Table 6.7.

Actions	Φ^+	Φ^-	Φ
Action 1 (A1)	0.69444	0	0.69444
Action 2 (A2)	0.27778	0.66667	–0.38889
Action 3 (A3)	0.16667	0.38889	–0.22222
Action 4 (A4)	0.19444	0.27778	–0.08333

to the sum of all the elements in the corresponding column divided by the number of actions minus 1.The negative flow also always lies between 0 and 1. The higher this number is, the worse the action will be.

Finally, the net flow summarizes those two perspectives in the following way:

$$\Phi(a_i) = \Phi^+(a_i) - \Phi^-(a_i). \tag{6.6}$$

It is a number between –1 and 1. The higher this number is, the better the action will be.

All the flows of actions of Example 6.1 are given in Table 6.8 and illustrated in Figure 6.23. Observe that the scales lie between 0 and 1 for the positive and negative flows (with an inverted orientation of the axis) and between –1 and 1 for the net flows.

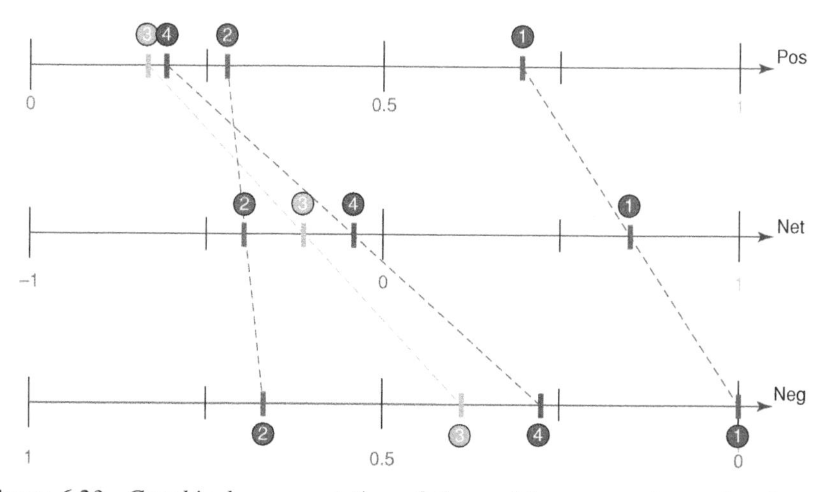

Figure 6.23 Graphical representation of the positive, net and negative flows of Table 6.8. Reproduced by permission of Smart Picker.

6.4.4 PROMETHEE I and PROMETHEE II ranking

The positive and negative flows can be used to compute the PROMETHEE I ranking, which is a partial ranking, whereas the net flows can be used for the PROMETHEE II ranking, which is a complete ranking.

Let us express the relations (S^+,I^+) and (S^-,I^-) as the two complete pre-orders induced by the positive and negative flows (Brans and Mareschal 2005):

- The positive flow of action a_i outranks the positive flow of action a_j:

$$a_i S^+ a_j \Leftrightarrow \Phi^+(a_i) > \Phi^-(a_j).$$

- The positve flow of action a_i and the positive flow of action a_j are indifferent:

$$a_i I^+ a_j \Leftrightarrow \Phi^+(a_i) = \Phi^-(a_j).$$

- The negative flow of action a_i outranks the negative flow of action a_j:

$$a_i S^- a_j \Leftrightarrow \Phi^-(a_i) < \Phi^-(a_j).$$

- The negative flow of action a_i and the negative flow of action a_j are indifferent:

$$a_i I^- a_j \Leftrightarrow \Phi^-(a_i) = \Phi^-(a_j).$$

These relations permit us to define the partial PROMETHEE I ranking which is the intersection of these two pre-orders (Brans and Mareschal 2005):

- The action a_i is preferred to the action a_j if

$$a_i \, P a_j \Leftrightarrow [a_i \, S^+ \, a_j \text{ AND } a_i \, S^- \, a_j] \text{ OR } [a_i \, S^+ \, a_j \text{ AND } a_i \, I^- \, a_j] \text{ OR } [a_i \, I^+ \, a_j \text{ AND } a_i \, S^- \, a_j].$$

- The action a_i and the action a_j are indifferent if

$$a_i \, I \, a_j \Leftrightarrow [a_i \, I^+ \, a_j \text{ AND } a_i \, I^- \, a_j].$$

- The action a_i is incomaprable to the action a_j if

$$a_i \, J \, a_j \Leftrightarrow [a_i \, S^+ \, a_j \text{ AND } a_j \, S^- \, a_i] \text{ OR } [a_j \, S^+ \, a_i \text{ AND } a_i \, S^- \, a_j].$$

Here P stands for *global preference*, I for *global indifference* and J for *global incomparability*.

The PROMETHEE I ranking is a partial ranking; an incomparability exists if there is no preference or indifference relation between two actions. The partial ranking of Example 6.1 is depicted in Figure 6.24, where '→' means 'is preferred to'. Note here that no preference relation can be stated between A4 and A2 or between A3 and A2.

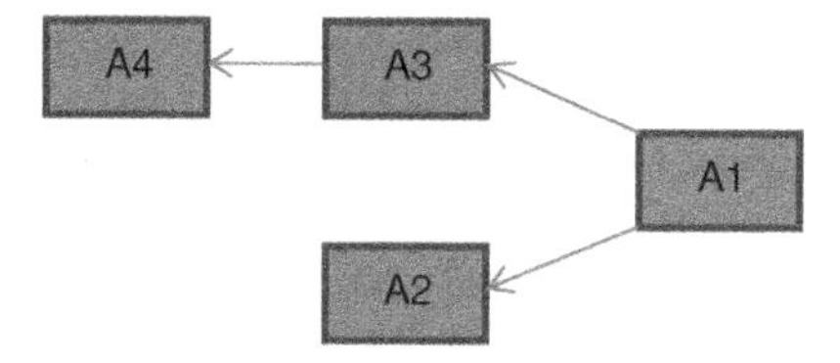

Figure 6.24 Representation of PROMETHEE I ranking.

The PROMETHEE II ranking is a complete ranking where a preference relation exists between any pair of actions. This is due to the fact that the PROMETHEE II ranking is based solely on the net flows, which are necessarily transitive. Based on Figure 6.23, the PROMETHEE II ranking is as follows: A1 is ranked first, followed by A4, A3 and finally A2.

6.4.5 The Gaia plane

The Gaia method permits a visual representation of a decision problem and therefore uses the unicriterion net flows computed by the PROMETHEE method. Let us consider the matrix $\mathbf{\Phi}$ containing the unicriterion net flows of all the actions of the decision problem (Mareschal and Brans 1988):

$$\phi = \begin{pmatrix} \phi_1(a_1) & \phi_2(a_1) & \cdots & \phi_j(a_1) & \cdots & \phi_q(a_1) \\ \phi_1(a_2) & \phi_2(a_2) & \cdots & \phi_j(a_2) & \cdots & \phi_q(a_2) \\ \cdots & \cdots & \cdots & \cdots & \cdots & \cdots \\ \phi_1(a_i) & \phi_2(a_i) & \cdots & \phi_j(a_i) & \cdots & \phi_q(a_i) \\ \cdots & \cdots & \cdots & \cdots & \cdots & \cdots \\ \phi_1(a_n) & \phi_2(a_n) & \cdots & \phi_j(a_n) & \cdots & \phi_q(a_n) \end{pmatrix} \tag{6.7}$$

This matrix is similar to the performance table since each row of the matrix represents an action and each column represents a criterion. However, the matrix contains some preference information given by the decision maker since it incorporates the preference functions and their parameters. It thus measures the intra-criteria information.

In order to represent the inter-criteria information, the decision maker can use the Gaia method (Brans and Mareschal 1994). The actions will be represented by points, denoted by $\boldsymbol{\alpha}_i$ with the coordinates $\boldsymbol{\alpha}_i$: $(\varphi_1(a_i), \ldots, \varphi_j(a_i), \ldots, \varphi_q(a_i))$ in q-dimensional space. The vector $\boldsymbol{\alpha}_i$ is a line in the matrix $\mathbf{\Phi}$.

To graphically represent this matrix in two dimensions, the GAIA method uses the statistical technique of principal component analysis (PCA). For this purpose, the variance–covariance matrix of our decision problem, denoted by $\mathbf{C}$, is first computed. This matrix can be obtained by using the relation

$$n\mathbf{C} = \Phi^T \cdot \Phi, \tag{6.8}$$

where $\mathbf{\Phi}^T$ denotes the transpose of $\mathbf{\Phi}$, and n is a positive integer. Then, two eigenvectors, denoted by $\mathbf{u}$ and $\mathbf{v}$, are calculated such that they have the greatest eigenvalues λ_1 and λ_2. These two eigenvectors are orthogonal ($\mathbf{u}\perp\mathbf{v}$) and define the best plane, called the Gaia plane, to use for the projection of the $\boldsymbol{\alpha}_i$ points while minimizing the loss of information (Brans and Mareschal 1994; Mareschal and Mertens 1990, 2003). By definition, this plane is the one that holds the maximum amount of information after the projection has been realized. The amount of information preserved can be calculated as follows:

$$\delta = \frac{\lambda_1 + \lambda_2}{\sum_{j=1}^{q} \lambda_j}. \tag{6.9}$$

The coordinates of $\mathbf{A}_i$, the projections of the $\boldsymbol{\alpha}_i$ points, on the Gaia plane will be:

$$\begin{cases} u_i = \boldsymbol{\alpha}_i^T \cdot \mathbf{u} \\ v_i = \boldsymbol{\alpha}_i^T \cdot \mathbf{v} \end{cases} \tag{6.10}$$

where $\boldsymbol{\alpha}_i^T$ denotes the transpose of $\boldsymbol{\alpha}_i$, $\boldsymbol{\alpha}_i$ the ith row of $\mathbf{\Phi}$.

In order to represent the intra-criteria information, each criterion f_j will be projected on the Gaia plane by considering each criterion axis e_j as follows:

$$e_j : (0, 0, \ldots, 1, 0, \ldots, 0), \quad j = 1, 2, \ldots, q \tag{6.11}$$

The angle between the projections of two criteria is a measure of similarity or conflict between two criteria. The smaller the angle, the more similar the two criteria. In contrast, the angle will be greater when criteria are conflicting, given the data set.

Finally, the information on the weights chosen by the decision maker can be added by finding the projection of the weights vector $\mathbf{w} = (w_1, \ldots, w_q)$.

The vector $\mathbf{D} = (\mathbf{wu}, \mathbf{wv})$ obtained is called a decision stick, as it represents the decision maker's priorities. An illustration of the Gaia plane is given in Figure 6.8.

We note here that Nemery et al. (2011) recently showed the importance of taking the weights into account in the GAIA projections.

6.4.6 Influence of pairwise comparisons

Several authors have studied the theoretical properties of the PROMETHEE I and PROMETHEE II ranking methods (Bouyssou 1992; Bouyssou and Perny 1992; Keyser and Peeters). We now describe the major characteristics of these ranking methods.

First, let us suppose that the preference degree π_{ij} does not result from an aggregation of several criteria, but corresponds to a percentage of voters considering that action a_i is preferred or indifferent with respect to a_j. We may note here that the PROMETHEE II ranking corresponds to the well-known method of Borda (Fishburn 1973) where the actions are ordered according to their sums of votes (Bouyssou 1992; Bouyssou and Perny 1992). Moreover, if we suppose that $\forall a_i, a_j \in A$: $\pi_{ij} \in \{0,1\}$, i.e.

binary values, then the PROMETHEE II ranking method amounts to the Copeland ranking method (Bouyssou 1992; Bouyssou and Perny 1992).

Furthermore, we can obtain the preference degrees π_{ij} by other means as given in previous sections. The preference degrees can be obtained while using other preference functions or can even be provided directly by the decision maker (when comparing a_i to a_j). Nevertheless, regardless of how these preference degrees are obtained, we may still compute the positive, negative and net flows. These ranking methods are called ranking methods based on leaving, entering and net flows (Bouyssou 1992; Bouyssou and Perny 1992). The PROMETHEE I and PROMETHEE II methods are thus particular ranking methods based on flows.

The ranking methods based on net flows and on the difference of the entering and leaving flows make use of the 'cardinal properties' of the valued preference degrees. In fact, if we transform the preference degrees by a strict increasing transformation $T(x)$ on the real line and such that $T(0) = 0$ and $T(1) = 1$, it may happen that the initial flow ranking is not preserved. As a consequence, it does not seem appropriate when the comparisons of the valuations (preference degrees) only have an ordinal meaning in term of credibility (Bouyssou 1992; Bouyssou and Perny 1992). The PROMETHEE methods may only be applied if the decision maker is able to express a preference between two actions, either on a certain criterion or on a ratio scale – and not on an ordinal scale (Keyser and Peeters 1996).

The ranking methods based on net flows and on the difference of the entering and leaving flows are said to be *neutral*. In particular, PROMETHEE I and PROMETHEE II do not discriminate actions in their ranking on the basis of their label or their given name (Bouyssou 1992; Bouyssou and Perny 1992). Assume that alternatives are numbered $a_1, a_2, a_3, \ldots$ and then renamed (while keeping the same order) to a_j, a_n, $a_3, \ldots$. The ranking obtained after renaming the alternatives will remain coherent with the initial ranking.

The PROMETHEE I and PROMETHEE II ranking methods are *strongly monotonic* since the rankings respond *in the right direction* to a modification of the preference degrees. This property excludes, in particular, the use of any threshold in the treatment of the valuations (Bouyssou 1992; Bouyssou and Perny 1992). When two actions are compared similarly to any other action of the set A (in terms of preference degrees: $\pi(a_i, x) = \pi(a_j, x)$, $\pi(x, a_i) = \pi(x, a_i)$ and $\pi(a_i, a_j) = \pi(a_j, a_i), \forall x \in A$), they will be considered as globally indifferent. This is often called the non-discriminatory property of a ranking method. Nevertheless, it has as a drawback that a situation of indifference or incomparability between two actions may be similarly treated in the final ranking.

When comparing actions by means of pairwise comparisons, cycles may occur: a_i is preferred to a_j which a_j is preferred to a_l which is preferred to a_i, etc. This is the so-called 'Condorcet paradox'. A consequence of aggregating the comparisons into a global complete ranking is that the order in the final ranking may not correspond to these pairwise comparisons (Mareschal et al. 2008). It may suffer from *pairwise rank reversal* since we may have that $\pi(a_i, a_j) > \pi(a_j, a_i)$ with $\Phi(a_i) < \Phi(a_j)$.

Since PROMETHEE II is not pairwise rank reversal free, it may suffer from the rank reversal phenomenon: the addition to the set A or the suppression of an action of the set A may lead to a rank reversal between two actions since the computed flows can vary when the initial set A is altered. Some examples of pairwise rank reversals and rank reversals are given in Keyser and Peeters (1996). Nevertheless, the rank reversal phenomenon occurs when the difference of flows between two actions is 'small'. The interested reader may find more information on situations where there is no rank reversal in Mareschal et al. (2008), Nemery (2008) and Verly and De Smet (in press).

Exercise 6.3

In this exercise, you will learn the step-by-step calculation of the preference matrix, and then have the opportunity to compute the positive, negative and net flows.

Learning Outcomes

- Understand the calculation of the performance matrix in *Microsoft Excel*
- Understand the calculation of the positive, negative and net flows in *Microsoft Excel*

Tasks

Open the file Car Selection.xls.

Answer the following questions:

a) Describe the meaning of each calculation cell and its formula. (Only read the comments in the red squares in the case of difficulty.)

b) Analyze Figure 6.22 and verify the complementaries and uniqueness of both approaches. Are there some local and global incompatibilities?

6.5 Extensions of PROMETHEE

Several extensions of the PROMETHEE method have been proposed in recent decades. Amongst others, let us mention the extension of the PROMETHEE method to uncertainty and imprecision (Teno and Mareschal 1998; D'Avignon and Vincke 1998; Drechsler 2004; Oberschmidt 2010; Saidi Mehrabad and Anvari 2010), group decision support (Macharis et al. 1998; Brans and Mareschal 2005; Silva et al. 2010) and sorting problems (Araz and Ozkarahan 2007; Nemery and Lamboray 2008). In this section we briefly describe the latter two. Nemery et al. (2010) propose a unification of problem formulations with the PROMETHEE methods.

6.5.1 PROMETHEE GDSS

Sometimes decisions have to be made by a group of people rather than a single person. Rarely do all the stakeholders of the problem requiring a decision have completely identical perceptions and preferences with regard to the problem. Moreover, it might happen that not all the decision makers have the same weight of importance in the final decision. To tackle this situation, the group can use a group decision support system (GDSS). In what follows, we assume that the group has agreed upon a set of actions and a set of criteria. These will not subsequently be altered by any decision maker.

The PROMETHEE GDSS method enables the tackling of a ranking problem involving several decision makers as follows (Macharis et al. 1998; Brans and Mareschal 2005; Ishizaka and Nemery 2012). Every person acts, at first, as if he were the only decision maker and uses the PROMETHEE II method to score and rank each action of the decision problem. In a second phase, the individual rankings are aggregated into a group ranking. This approach is recommended when the evaluations of the actions on the different criteria are not required to be identical for all decision makers (i.e. no consensus about the evaluations is specified).

To aggregate the individual rankings, each individual net flow ranking will be considered as a criterion of the group problem. In other words, in the second phase, each decision maker will act as a criterion of the group performance matrix. The evaluation of an action on a group criterion is equal to the net flows of that action in the individual ranking. This is illustrated in Figure 6.25.

Each decision maker can define his own evaluations and preference parameters (preference functions, thresholds, weights, etc.) in the first step. In the second step, the weights can be equal or not depending upon whether all decision makers have the same weight in the decision. The group of decision makers need to define whether the preference functions of the second phase are all identical or not.

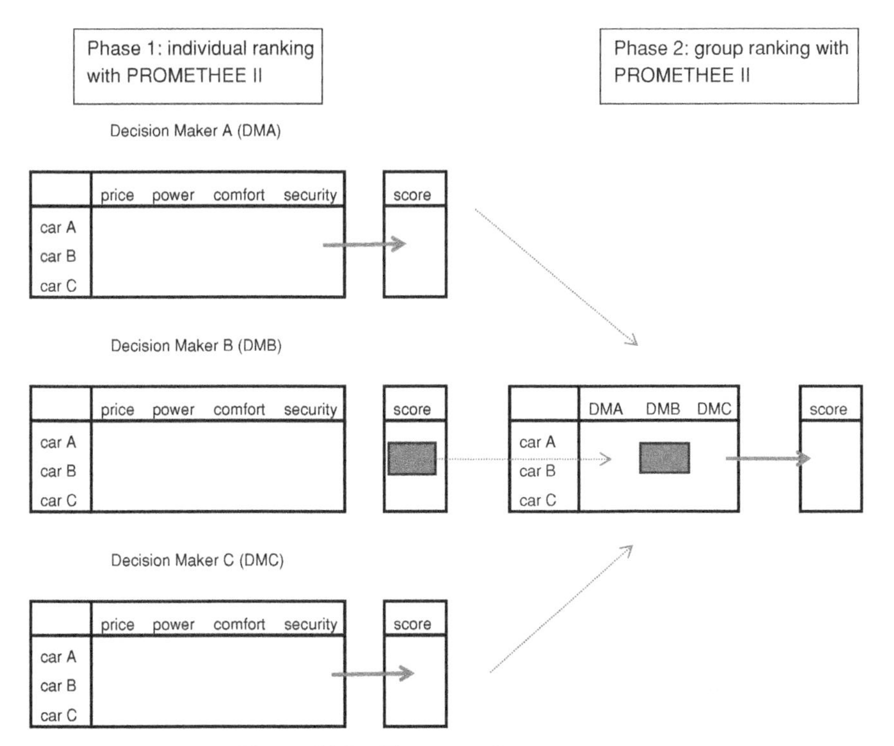

Figure 6.25 The group decision process.

The preference function will determine the role played by the difference between the net flows of the actions. Small differences can be considered as negligible (use of a linear preference function with a high indifference threshold), while sometimes a small difference can be significant (use of a linear preference function with the indifference and preference threshold equal to 0).

Finally, the criterion arrows of the Gaia representation of the second phase represent the individual rankings of the decision makers. Thus this means that if two arrows are close, then the corresponding decision makers are consensual and have similar rankings. On the other hand, two opposite arrows indicate opposite rankings and thus a situation of conflict between two decision makers. For detailed examples we refer the reader to Ishizaka and Nemery (2011, 2012).

Exercise 6.4

In this group exercise, you will be able to solve a group decision with PROMETHEE GDSS.

Prerequisites: Exercise 6.2

Learning Outcomes

- Structure a group hierarchy in *Smart Picker Pro*
- Understand the aggregation of individual rankings
- Understand the final group ranking

Tasks

a) Form a group of three or four people.

b) Using *Smart Picker Pro*, each person enters the performances of Table 6.1 and his own preference parameters.

c) Collect the net flows of each decision maker and open a new problem in *Smart Picker Pro*. Define the problem as follows: the actions are the cars and the criteria correspond to the different persons. The evaluation of the car for each criterion corresponds to the net flow of the corresponding person.

d) Compute the final ranking while choosing equal weights. Discuss the final ranking. Is everybody satisfied with the outcome? Identify the persons with similar and opposite rankings in the Gaia plane.

6.5.2 FlowSort: A sorting or supervised classification method

In a sorting problem, the decision maker wants to assign a set of actions $A = \{a_1, a_2, \ldots, a_n\}$ to K predefined categories $C_1, \ldots, C_K$ (Figure 6.26). The categories

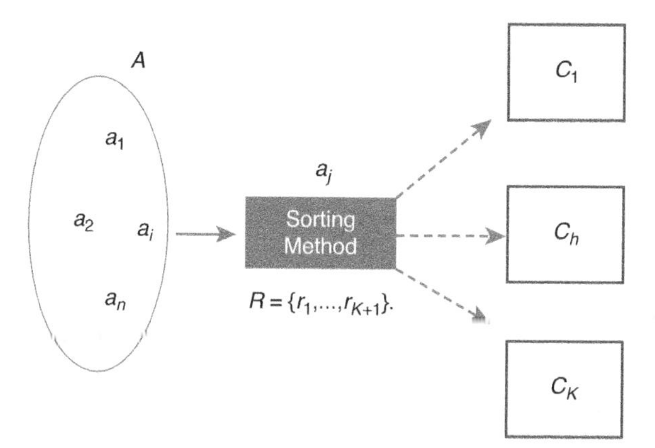

Figure 6.26 The sorting process.

are predefined in the sense that the decision maker knows their exact meaning. In a sorting context, there is a complete order on the categories. This means thus that the decision maker is able to order the categories from best to worst. Category C_1 is the best category and C_K the worst category.

An example of a sorting problem is the assignment of projects into two categories: the accepted or rejected projects. Several real-world decision problems have been addressed through sorting models, including financial decision-making problems, environmental decisions, marketing decisions, and even medical decisions (diagnosis) (Zopounidis and Doumpos, 2002).

An important requirement in a sorting problem is that the assignment of the actions is independent: the assignment of an action to a category must not depend on the assignment of another action (i.e. the actions are not pairwise compared). This constitutes a fundamental difference in dealing with the sorting problem compared to the ranking problem. This explains why we have chosen to describe the FlowSort method (instead of for instance the PromSort method (Araz and Ozkarahan 2007) or the sorting method proposed by Doumpos and Zopounidis (2004) where the assignment of an action is not independent of the other assignments).

The FlowSort method (Nemery 2008; Nemery and Lamboray 2008) is a direct extension of the PROMETHEE method and is based on the following idea: in order to assign an action to a category, compare the action to the profiles by means of the PROMETHEE method and deduce its category based on its rank.

In the FlowSort method, the K categories are predefined by either *limiting profiles* or *central profiles*. In the first case, each category is defined by an upper and a lower boundary. The category C_j is thus defined by the upper limiting profile r_j (maximum value in order to be categorized in the C_j category) and the lower profile r_{j+1} (minimum value in order to be categorized in the C_j category) of the limiting profiles set denoted by $R=\{r_1, \ldots, r_{K+1}\}$. The first category and the worst category can be defined as open classes or as closed classes. In the former case, the two extremes categories will be defined by an upper and lower limit respectively. In the latter case, the extreme

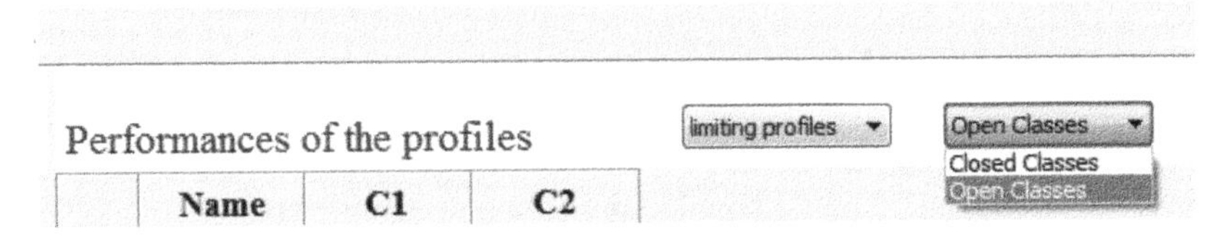

Figure 6.27 Limiting profiles and central profiles; open and closed categories. Reproduced by permission of Smart Picker.

categories will not be defined by these extreme limits. This can be chosen as the type of profile in the *Smart Picker Pro* software, as illustrated in Figure 6.27. Defining closed classes can be used to eliminate actions that are out of the boundaries defined by the two extreme profiles.

In the second case, the categories are defined by a central element. The category C_j is thus defined by the central profile r_j^*.

For simplicity, only the case of limiting profiles defining closed classes is presented in this chapter.

So if an action a_i is compared to the limiting profiles and if it is ranked between two successive limiting profiles r_h and r_{h+1} then it will be assigned to category C_h. Formally, we can define the assignment rule as follows:

$$C(a_i) = C_h, \quad \text{if} \phi^{R_i}(r_h) > \phi^{R_i}(a_i) \geq \phi^{R_i}(r_{h+1}),$$

where

$$\phi^{R_i}(a_i) = \sum_{l=1}^{q} w_l \cdot \phi_l^{R_i}(a_i) = \sum_{l=1}^{q} w_l \cdot \left[\frac{1}{|R_i| - 1} \sum_{j=1}^{K+1} \left[P_l(a_i, r_j) - P_l(a_j, r_i) \right] \right].$$

This therefore requires that the limiting profiles are always ranked with respect to the category they define (i.e. avoiding situations where, for instance, r_{h+1} has a better rank than r_h). This condition is met if a limiting profile dominates (i.e. is better on all criteria than) all the successive limiting profiles. Formally, if the criteria have to be maximized, we need the following condition:

$$\forall h = 1, \ldots, K; \ l = 1, \ldots q : f_l(r_h) \geq f_l(r_{h+1}) \text{ and } \exists j : f_j(r_h) > f_j(r_{h+1}).$$

Figure 6.28 illustrates this condition, where limiting profiles r_h and r_{h+1} define category C_h. In the left figure r_h and r_{h+1} respect the condition but don't respect it on the right since $f_j(r_h) < f_j(r_{h+1})$.

Let us stress that the ranking method is applied on the data set consisting of one action to be assigned and the reference profiles, $R_i = \{r_1, \ldots, r_{k+1}\} \cup \{a_i\}$. This means thus that if n actions have to be sorted, we will perform n rankings of the sets $R_1, R_2, \ldots, R_n$.

The FSGaia plane (illustrated in Figure 6.29) makes it easy to detect whether actions (represented by circles) assigned to a same category are incomparable (e.g.,

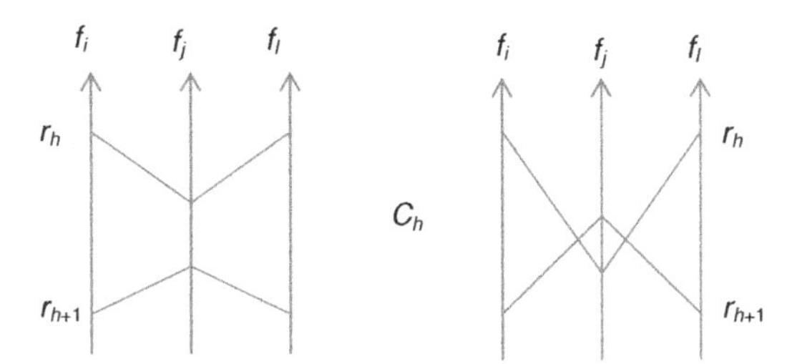

Figure 6.28 Limiting profiles which respect the dominance condition (left) and which do not respect it (right).

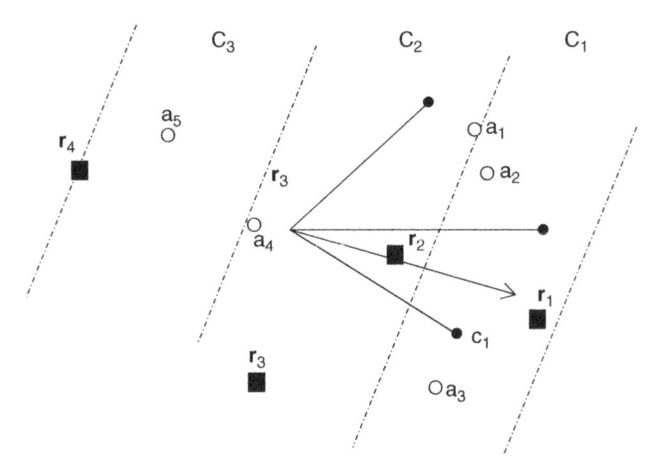

Figure 6.29 Plane representing the actions, reference profiles and criteria.

a_1 and a_3) or indifferent (e.g., a_1 and a_2). Moreover, it enables comparison of the actions in a global view of the profiles (represented by the rectangles). It is thus a descriptive approach, and is provided in *Smart Picker Pro*. The first application of the FSGaia plane in a real case study can be found in Nemery et al. (2012).

Exercise 6.5

In this exercise, you will learn how to use *Smart Picker Pro* in a sorting problem.

Learning Outcomes

- Structure a sorting problem in *Smart Picker Pro*
- Understand the steps of the FlowSort sorting method

Action	Price	Consumption	Power
Limiting profile 1	23000	8	80
Economic	15000	7.5	50
Sport	29000	9	110
Luxury	38000	8.5	90
Touring A	26000	9	75
Touring B	25500	7	85

Figure 6.30 Performance of the limiting profile and the actions to be sorted.

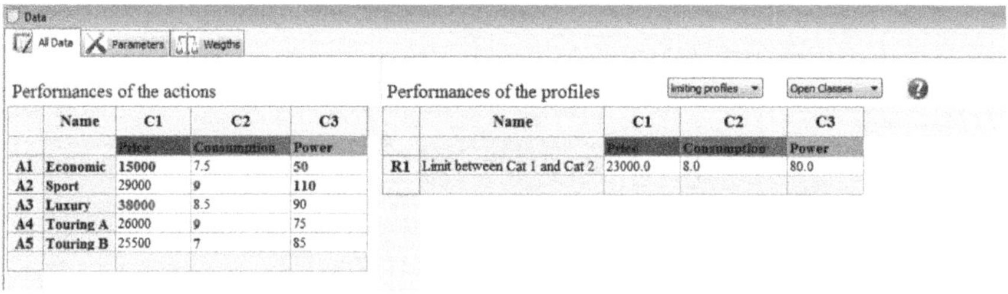

Data

All Data | Parameters | Weigths

Performances of the actions

	Name	C1	C2	C3
		Price	Consumption	Power
A1	Economic	15000	7.5	50
A2	Sport	29000	9	110
A3	Luxury	38000	8.5	90
A4	Touring A	26000	9	75
A5	Touring B	25500	7	85

Performances of the profiles — limiting profiles — Open Classes

	Name	C1	C2	C3
		Price	Consumption	Power
R1	Limit between Cat 1 and Cat 2	23000.0	8.0	80.0

Figure 6.31 Data in Smart Picker Pro *for Exercise 6.5. Reproduced by permission of Smart Picker.*

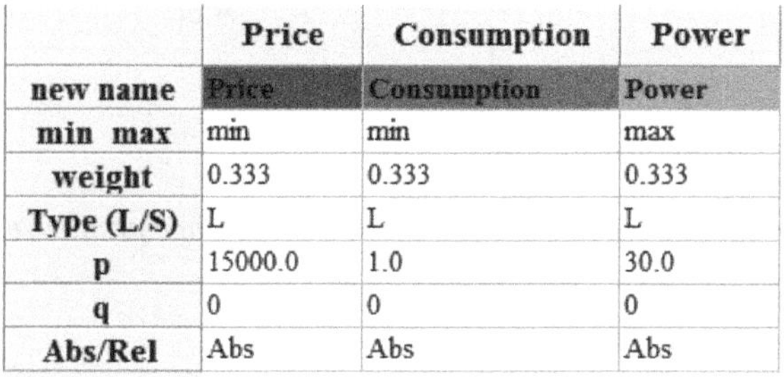

	Price	Consumption	Power
new name	Price	Consumption	Power
min max	min	min	max
weight	0.333	0.333	0.333
Type (L/S)	L	L	L
p	15000.0	1.0	30.0
q	0	0	0
Abs/Rel	Abs	Abs	Abs

Figure 6.32 Sorting parameters. Reproduced by permission of Smart Picker.

Tasks

a) Consider the actions and the limiting profiles given in Figure 6.30 and enter their performance in the left and right panels as shown in Figure 6.31.

b) Enter the sorting parameters as given in Figure 6.32.

c) Determine the category to which each action is assigned and explain the assignment of each action regarding the limiting profile.

d) Can you deduce anything from a comparison between the actions themselves?

References

Andreopoulou, Z. K. (2009). Assessment and optimization of e-commerce websites of fish culture sector. *Operational Research*, *9*(3), 293–309.

Araz, C. and Ozkarahan, I. (2007). Supplier evaluation and management system for strategic sourcing based on a new multicriteria sorting procedure. *International Journal of Production Economics*, *106*(2), 585–606.

Behzadian, M., Kazemzadeh, R., Albadvi, A., and Aghdasi, M. (2010). PROMETHEE: A comprehensive literature review on methodologies and applications. *European Journal of Operational Research*, *200*(1), 198–215.

Bouyssou, D. (1992). Ranking methods based on valued preference relations: A characterization of the net flow method. *European Journal of Operational Research*, *60*, 61–67.

Bouyssou, D., and Perny, P. (1992). Ranking methods for valued preference relations: A characterization of a method based on leaving and entering flows. *European Journal of Operational Research*, *61*, 186–194.

Brans, J., and Mareschal, B. (1994). The PROMCALC & GAIA decision support system for multicriteria decision aid. *Decision Support Systems*, *12*(4.5), 297–310.

Brans, J., and Mareschal, B. (2005). Promethee methods. In J. Figeira, S. Greco, and M. Ehrgott, *Multiple Criteria Decision Analysis: State of the Art Surveys*. New York: Springer.

Brans, J., Mareschal, B., and Vincke, P. (1986). How to select and how to rank projects: the Promethee method. *European Journal of Operational Research*, *24*, 228–338.

Brans, J., and Vincke, P. (1985). A preference ranking organization method: The Promethee method for multiple criteria decision making. *Management Science*, *31*, 647–656.

Castillo-Manzano, J. C.-N.-V.-Q. (2009). Low-cost port competitiveness index: Implementation in the Spanish port system. *Marine Policy*, *33*(4), 591–598.

Cornelissen, T. J. (2010). Flash co-pyrolysis of biomass: The infuence of biopolymers. *Journal of Analytical and Applied Pyrolysis*, *85*(1–2), 87–97.

D'Avignon, G., and Vincke, P. (1988). An outranking method under uncertainty. *European Journal of Operational Research*, *36*, 311–321.

Doumpos, M., and Zopounidis, C. (2004). Developing sorting models using preference disaggregation analysis: An experimental investigation. *European Journal of Operational Research*, *154*(3), 585–598.

Doumpos, M. and Zopounidis, C. (2010). A multicriteria decision support system for bank rating. *Decision Support Systems*, *50*(1), 55–63.

Dreschler, M. (2004). Model-based conservation decision aiding in the presence of goal conflicts and uncertainty. *Biodiversity and Conversation*, *13*, 141–161.

Fishburn, P. (1973). *The Theory of Social Choice*. Princeton, NJ: Princeton University Press.

Frenette, C. B.-Z. (2010). Multicriteria decision analysis applied to the design of light-frame wood wall assemblies. *Journal of Building Performance Simulation*, *3*(1), 33–52.

Ghafghazi, S. S. (2010). A multicriteria approach to evaluate district heating system options. *Applied Energy*, *87*(4), 1134–1140.

Ghazinoory, S. D. (2009). A new defnition and framework for the development of a national technology strategy: The case of nanotechnology for Iran. *Technological Forecasting and Social Change*, *76*(6), 835–848.

Halouani, N. C. (2009). PROMETHEE-MD-2T method for project selection. *European Journal of Operational Research*, *195*(3), 841–849.

Ishizaka, A., and Nemery, P. (2011). Selecting the best statistical distribution with PROMETHEE and GAIA. *Computers & Industrial Engineering*, *61*(4), 958–969.

Ishizaka, A., and Nemery, P. (2012). A multi-criteria group decision framework for partner grouping when sharing facilities. *Group Decision and Negotiation*, doi: 10.1007/s10726-012-9292-8, advance online.

Ishizaka, A., Nemery, P., and Lidouh, K. (2013). Location selection for the construction of a casino in the greater London region: A triple multi-criteria approach. *Tourism Management*, *34*(1), 211–220.

Juan, Y.-K. R.-L. (2010). Optimal decision making on urban renewal projects. *Management Decision* *48*(2), 207–224.

Keyser, W., and Peeters, P. (1996). A note on the use of PROMETHEE multicriteria methods. *European Journal of Operational Research*, *89*, 457–461.

Kodikara, P. P. (2010). Stakeholder preference elicitation and modelling in multi-criteria decision analysis – A case study on urban water supply. *European Journal of Operational Research*, *206*(1), 209–220.

Kwak, C., and Kim, C.O. (2009). A multicriteria approach to timeout collaboration protocol. *International Journal of Production Research*, *47*(22), 6417–6432.

Lanza, G., and Ude, J. (2010). Multidimensional evaluation of value added networks. *CIRP Annals – Manufacturing Technology*, *59*(1), 489–492.

Luk, J. F. (2010). A conceptual framework for siting biorefineries in the Canadian Prairies. *Biofuels, Bioproducts and Biorefning*, *4*(4), 408–422.

Macharis, C., Brans, J., and Mareschal, B. (1998). The GDSS PROMETHEE procedure: A PROMETHEE-GAIA based procedure for group decision support. *Journal of Decision Systems*, *7*, 283–307.

Mareschal, B., and Brans, J. (1988). Geometrical representations for MCDA. *European Journal of Operational Research*, *34*(1), 69–77.

Mareschal, B., De Smet, Y., and Nemery, P. (2008). Rank reversal in the PROMETHEE II method: Some new results. *Proceedings of the IEEE International Conference on Industrial Engineering and Engeneering Management*, Singapore, 959–963.

Marschal, B., and Mertens, D. (1990). Evaluation financière par la méthode Gaia: application au secteur bancaire. *Revue de la Banque*, *6/90*, 317–329.

Mareschal, B., and Mertens, D. (2003). A multiple criteria decision support system for financial evaluation in the international banking sector. *Journal of Decision Systems*, *1*, 175–189.

Nemery, P. (2008). On the use of multicriteria ranking methods in sorting problems. PhD thesis, Université Libre de Bruxelles.

Nemery, P., Ishizaka, A., Camargo, M., and Morel, L. (2012). Enriching descriptive information in ranking and sorting problems with visualizations techniques. *Journal of Modelling in Management*, *7*(2), 130–147.

Nemery, P., and Lamboray, C. (2008). FlowSort: a flow-based sorting method with limiting or central profiles. *TOP*, *16*(1), 90–113.

Nemery, P., Lidouh, K., and Mareschal, B. (2011). On the use of taking weights into account into account in the GAIA map. *International Journal of Information and Decision Sciences*, *3*(3), 228–251.

Nemery, P., Mareschal, B., and Ishizaka, A. (2010). Unification of problem formulation with PROMETHEE, Keynote Paper. *The 52th Operational Research Society Conference* (pp. 60–72). London.

Ni, Y. L. (2009). Multi-wavelength HPLC fingerprints from complex substances: An exploratory chemometrics study of the Cassia seed example. *Analytica Chimica Acta, 647*(2), 149–158.

Nikolic, D. J. (2010). Multi-criteria ranking of copper concentrates according to their quality – An element of environmental management in the vicinity of copper smelting complex in Bor, Serbia. *Journal of Environmental Management, 91*(2), 509–515.

Oberschmidt, J. G. (2010). Modified PROMETHEE approach for assessing energy technologies. *International Journal of Energy Sector Management, 4*(2), 183–212.

Ramzan, N. N. (2009). Multicriteria decision analysis for safety and economic achievement using PROMETHEE: A case study. *Journal of Analytical and Applied Pyrolysis, 85*(1–2), 87–97.

Safaei Mohamadabadi, H. T. (2009). Development of a multi-criteria assessment model for ranking of renewable and non-renewable transportation fuel vehicles. *Energy, 34*(1), 112–125.

Saidi Mehrabad, M. and Anvari, M. (2010). Provident decision making by considering dynamic and fuzzy environment for FMS evaluation. *International Journal of Production Research, 48*(15), 4555–4584.

Semaan, N. and. (2010). A stochastic diagnostic model for subway stations. *Tunnelling and Underground Space Technology, 25*(1), 32–41.

Silva, V. B. S., Morais, D. C., and Almeida, A. T. (2010). A multicriteria group decision model to support watershed committees in Brazil. *Water Resources Management, 24*(14), 4075–4091.

Soltanmohammadi, H., Osanloo, M., and Aghajani Bazzazi, A. (2009). Deriving preference order of post-mining land-uses through MLSA framework: Application of an outranking technique. *Environmental Geology, 58*(4), 877–888.

Teno, J., and Mareschal, B. (1998). An interval version of Promethee for the comparison of building products' design with ill-defined data on environmental quality. *European Journal of Operational Research, 109*, 522–529.

Tuzkaya, G. G. (2010). An integrated fuzzy multicriteria decision making methodology for material handling equipment selection problem and an application. *Expert Systems with Applications, 37*(4), 2853–2863.

Venkata Rao, R. and Patel, B. K. (2010). Decision making in the manufacturing environment using an improved PROMETHEE method. *International Journal of Production Research, 48*(16), 4665–4682.

Verly, C., and De Smet, Y. (in press). Some results about rank reversal instances in the PROMETHEE methods. *International Journal of Multicriteria Decision Making*.

Vincke, P. (1992). *Multicriteria Decision Aid*. Chichester: John Wiley & Sons, Ltd.

Zhang, K. K. (2009). A comparative approach for ranking contaminated sites based on the risk assessment paradigm using fuzzy PROMETHEE. *Environmental Management, 44*(5), 952–967.

Zhu, Z. X. (2010). Optimization on tribological properties of aramid fibre and $CaSO_4$ whisker reinforced non-metallic friction material with analytic hierarchy process and preference ranking organization method for enrichment evaluations. *Materials & Design, 31*(1), 551–555.

Zopounidis, C., and Doumpos, M. (2002). Multicriteria classification and sorting methods: A literature review. *European Journal of Operational Research, 138*, 229–246.

7

ELECTRE

7.1 Introduction

This chapter describes the theory and practical uses of the ELECTRE methods. You will learn how to use the *Electre III-IV* software package which permits the (partial) ranking of options. Section 7.3 is designed for readers interested in the methodological background of the ELECTRE methods. Section 7.4 is devoted to the extensions of ELECTRE in group decision and sorting problems.

The companion website provides illustrative examples with *Microsoft Excel*, and case studies and examples with the ELECTRE software package.

7.2 Essentials of the ELECTRE methods

The *ELimination Et Choix Traduisant la REalité* (elimination and choice expressing reality) methods, referred to as ELECTRE, belong to the outranking methods. They constitute one of the main branches of this family despite their relative complexity (due to many technical parameters and a complex algorithm).

The outranking methods are based on pairwise comparisons of the options. This means that every option is compared to all other options. As we will see, this will be computed for the user by the *Electre III-IV software*. Based on these pairwise comparisons, final recommendations can be drawn.

The main characteristic and advantage of the ELECTRE methods is that they avoid compensation between criteria and any normalization process, which distorts the original data.

B. Roy, the father of the outranking methods, presented ELECTRE I for the first time at a conference in 1965 and published the first paper on this topic in 1968 (Roy 1968). This initiated a long series of improvements, research and developments

Multi-Criteria Decision Analysis: Methods and Software, First Edition. Alessio Ishizaka and Philippe Nemery.

Table 7.1 Overview of the different ELECTRE methods.

Decision Problem	Method	Software
Choice problem	ELECTRE I	–
	ELECTRE Iv	–
	ELECTRE Is	*Electre Is*
Ranking problem	ELECTRE II	–
	ELECTRE III	*Electre III–Electre IV*
	ELECTRE IV	*Electre III–Electre IV*
Sorting problem	ELECTRE-Tri-B	*Electre-Tri*
	ELECTRE-Tri-C	*IRIS*
Elicitation problem	Elicitation of the weights in ELECTRE	*SRF*
		IRIS
	Elicitation for ELECTRE-Tri:	*Electre Tri Assistant*
	• IRIS method	
	• other elicitation methods	

of the ELECTRE methods in order to tackle new decision problems. They can be subdivided according to the type of problem they solve (cf. Table 7.1).

ELECTRE methods are relevant when facing decision problems with more than two criteria and if at least one the following conditions is satisfied (Figueira et al. 2005):

- The performances of the criteria are expressed in different units (e.g. duration, weight, price, colour, etc.) and the decision maker wants to avoid defining a common scale, which is difficult and complex.
- The problem does not tolerate a compensation effect (e.g. the weak performance of the time delay can not be compensated by good quality).
- There is a need to use indifference and preference thresholds, such that small differences may be insignificant although the sum of small differences is decisive (e.g. we are indifferent to an additional grain of sugar in a cup of tea but not to an additional 100 grains of sugar).
- The options are evaluated on a scale presenting an order or on a 'weak' interval scale (temperature and calendar dates are examples of interval scales), where it is difficult to compare differences (e.g. a temperature of 60°F is 30°F more than 30°F, but it cannot be said to be twice as warm as 30°F, because interval variables do not have a true zero point).

The first ELECTRE method, ELECTRE I, and its variants ELECTRE Iv and ELECTRE Is (cf. Table 7.1) were developed to solve choice problems. In a choice problem the decision maker will select, amongst a given set of options, the smallest subset containing the best options. The only difference between ELECTRE I and ELECTRE Iv is the introduction of the veto concept: if an option performs badly on a single criterion compared to another option, the option will then be considered as outranked, irrespective of its performance on the other criteria. The novelty of ELECTRE Is is the use of pseudo-criteria. Pseudo-criteria are introduced to model the fact that a decision maker might not have a preference between two options of a criterion, if the difference in their performance is smaller than the indifference threshold. On the other hand, it is also used to reflect a situation where the preference might be strong if the difference is higher than a preference threshold. Such thresholds permit situations to be handled where data are imprecise or uncertain. Today, choice problems are mostly tackled with the ELECTRE Is method.

ELECTRE II, ELECTRE III and ELECTRE IV (cf. Table 7.1) are ranking methods, which may lead to a partial order on a set of options (i.e. the ranking accepts that two options are incomparable) but without assigning a score to the alternatives. The preference order amongst the options is the output of the methods. ELECTRE III is distinguished from ELECTRE II by the use of pseudo-criteria and outranking degrees (instead of binary outranking relations). ELECTRE IV, on the other hand, does not require the relative importance of criteria (i.e. the weights). ELECTRE III is the most used ranking method in the ELECTRE family and is implemented, along with ELECTRE IV, in the *Electre III and IV* software.

ELECTRE-Tri-B (more commonly known as ELECTRE-Tri) and ELECTRE-Tri-C are sorting methods that enable the independent assignment of a set of options to one or several predefined categories. These methods are thus supervised classification methods, but with the particularity of a preference relation amongst the categories, that is, they can be ordered from best to worst. The difference between the two methods lies in the definition of the categories: either by limiting profiles or Boundaries (hence ELECTRE-Tri-B), or by typical or Central profiles (ELECTRE-Tri-C). A detailed description is provided in Section 7.5.

The drawback of the ELECTRE methods is that they require various (difficult) technical parameters, which means that it is not always easy to fully understand them. As a result, researchers have made some significant progress in the *automatic elicitation* of those parameters. This requires that the decision maker rank (real or fictitious) options that have a clear ranking in order to infer parameters such as the weights of the criteria, and the thresholds. These methods cannot, however, always be considered as a complete panacea for fixing the parameters. They may point to some of the decision maker's inconsistencies or contradictions, which means re-evaluating the judgements. This might form the basis for a discussion to set the value of the parameters.

The ELECTRE methods have been successfully applied in many areas such as environmental management, agriculture and forest, energy, water management, finance, calls for tender, transportation and military (Figueira et al. 2005). In particular, ELECTRE III is a well-established partial ranking method with successful real-world applications such as environmental and energy management (Parent and Schnabele 1988; Hokkanen and Salminen 1996; Karagiannidis and Moussiopoulos

1997; Rogers and Bruen 1998a, 1998b; Karagiannidis and Papadopoulos 2008; Figueira et al. 2005), and strategic planning (Kangas and Pykäläinen 2001).

The next section describes the ELECTRE III ranking method along with the corresponding *Electre III-IV* software, given its success with the ELECTRE methods. Section 7.5 is devoted to the ELECTRE-Tri sorting method.

7.2.1 ELECTRE III

ELECTRE III is divided into two phases. First, the outranking relationship between the options is constructed and then exploited, although most of the information from the decision maker is required in the first phase: the weight of the criteria, the indifference, the preference and the veto thresholds. The meaning of those parameters is explained in this section as well as the final results of the ELECTRE III method. The intermediate steps, which require a more advanced understanding, are explained in Section 7.4.

In what follows, the term 'alternative' instead of 'option' will be used, in line with the ELECTRE methods and software terminology. Without loss of generality, the preference directions of all criteria are taken to be increasing: in other words, all the criteria have to be maximized.

ELECTRE III makes use of outranking relations. An outranking relation, where a outranks b (denoted by a S b),[1] expresses the fact that there are sufficient arguments to decide whether a is at least as good as b and there are no essential reasons to refute this (Roy 1974). An outranking degree $S(a,b)$ between a and b will be computed in order to 'measure' or to 'evaluate' this assertion.

Case Study 7.1 illustrates this concept.

Case Study 7.1

Governmental organizations and companies are often faced with the task of recruiting new employees, promoting promising staff and awarding grants, etc. In today's meeting, a committee has a travel grant to award to one of six research students. In order to evaluate the candidates, the committee has agreed on five criteria:

1. Number of years of university study (to be maximized)
2. Their professional experience expressed in years (to be maximized)
3. The requested grant amount expressed in pounds (to be minimized)
4. The evaluation of the application letter (to be maximized)
5. The potential return of the allocated grant (to be maximized).

For the last two criteria, the committee will allocate a score between 0 and 10.

[1] S stands for 'surclasse' in French, which means 'outranks'.

Suppose that the committee has agreed on the evaluations and the preference parameters. Table 7.2 suggests that there is no ideal candidate as none of the applicants have the highest score for each of the five criteria: none of the candidates are *efficient*. The committee will have to compromise.

Table 7.2 Performance table of the six applications evaluated on five criteria.

	Education (years) f_1	Experience (years) f_2	Grant (× £100) f_3	Letter (score) f_4	Potential return (score) f_5
Candidate 1	6	5	28	5	5
Candidate 2	4	2	25	10	9
Candidate 3	5	7	35	9	6
Candidate 4	6	1	27	6	7
Candidate 5	6	8	30	7	9
Candidate 6	5	6	26	4	8

The ELECTRE III method will help the decision maker in the decision process and deduce the (final) partial order. It will compare the candidates pairwise by calculating the outranking degrees.

The strength of the assertion *a outranks b* is given by the credibility or outranking degree $S(a,b)$. It is a score between 0 and 1, where the closer $S(a,b)$ is to 1, the stronger the assertion. This outranking degree $S(a,b)$ considers two perspectives: the concordance and the discordance of the statement that *a* outranks *b*. The concordance and discordance are measured respectively while incorporating the decision maker's preference on various (often conflicting) criteria. The user is required to provide the indifference and preference thresholds for calculating the concordance degree, while the veto threshold is needed for the discordance degree.

7.2.1.1 Concordance

A partial concordance degree $c_i(a,b)$ measures the assertion '*a* outranks *b*' or '*a* is at least as good as *b*' on the specific criterion f_i. Table 7.2 concludes that candidate 1 is as good as candidate 2 on the *experience* criterion. As experience has to be maximized, this assertion is strong because the *experience* of candidate 1 is higher than that of candidate 2 and as a result as good: the partial concordance degree will be equal to 1.

Consider now the opposite comparison: candidate 2 is at least as good as candidate 1 on the experience criterion (where the performances are respectively 4 versus 6). This can be true or false depending on the perception of the decision maker: one person can consider a difference of 2 as negligible whereas another may feel this

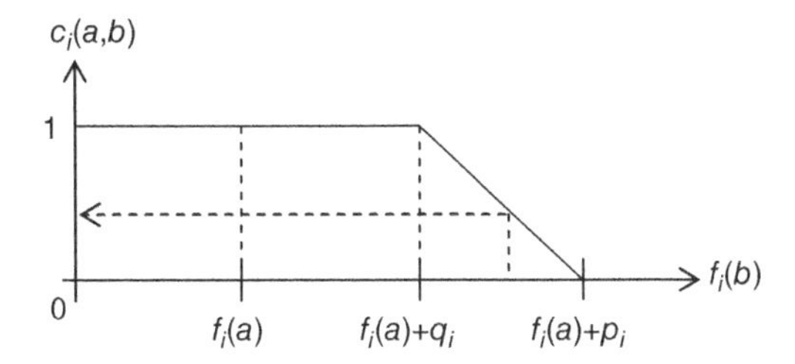

Figure 7.1 The partial concordance index $c_i(a,b)$.

is decisive. Therefore, the decision maker needs to specify the indifference (q_i) and preference (p_i) threshold in order to measure the difference in performance.

The *indifference threshold* indicates the largest difference between the performances of the alternatives on the criterion considered such that they remain indifferent for the decision maker.

The *preference threshold* indicates the largest difference between the performances of the alternatives such that one is preferred over the other on the considered criterion. Between these two thresholds, the partial concordance degree is computed on the basis of a linear interpolation, represented in Figure 7.1:

- If the performance of alternative *b* on f_i is higher than *a* augmented with the preference threshold p_i, there is a strict preference for *b* over *a*. The concordance degree stating that *a is as good as* b *on* f_i is thus zero. Formally, if $f_i(b)$ is higher than $f_i(a)+p_i$ then $c_i(a,b) = 0$.
- If the performance of *b* is between the performance of *a* augmented with the indifference threshold and the performance of *a* augmented with the preference threshold, then *b* is weakly preferred to *a*. The concordance degree is deduced by linear interpolation: if $f_i(b)$ is between $f_i(a)+q_i$ and $f_i(a)+p_i$ then $c_i(a,b)$ is between 0 and 1.
- If the performance of *b* is smaller than the performance of *a* augmented with the indifference threshold, *a* and *b* are indifferent. The concordance degree stating that *a is at least as good as* b *on* f_i is 1: if $f_i(b)$ is smaller or equal to $f_i(a)+q_i$ then $c_i(a,b) = 1$.

Table 7.3 gives some examples of the variation of the concordance index for criterion 1 between candidate 2 and candidate 1 for different values of q_1 and p_1. The reader can easily check that c_1(candidate 1,candidate 2) is always 1 for any value of q_1 and p_1.

A global concordance degree $C(a, b)$ aggregates all the partial concordance indices on the different criteria by taking into account their corresponding criteria weight. This global index is thus the weighted sum of all the partial concordance indices and measures how concordant the assertion '*a is at least as good b*' is regarding all the criteria. The *Electre III-IV* software computes these global concordance degrees.

Table 7.3 Some examples of $c_1(a,b)$ in Case Study 7.1 for different values of q and p.

q_1	p_1	c_1(candidate 2,candidate 1)
0	0	0
1	1.5	0
1	3	0.5
1	4	0.66
≥2	$\geq q_1$	1

7.2.1.2 Discordance

On the other hand, the partial discordance degree $d_j(a,b)$ measures the decision maker's discordance with the assertion '*a* is at least as good as *b*' on criterion f_j. If the decision maker, when considering criterion f_j, strongly disagrees with the assertion, the discordance degree reaches its maximum value 1 and reflects the fact that f_j sets its veto. This is the case if the difference in performances (i.e. $f_j(b) - f_j(a)$) is higher than a so-called veto threshold, denoted by v_i. The discordance degree has minimum value 0, when there is no reason to refute the assertion. As in the case of the partial concordance degree, between these two extremes, $d_j(a,b)$ will vary linearly between the preference and veto thresholds as a function of the difference $f_j(b) - f_j(a)$, as shown in Figure 7.2:

- If $f_i(b)$ is higher than $f_i(a)+v_i$, the difference between *b* and *a* exceeds the veto threshold which means a total discordance with the assertion: $d(a,b) = 1$.
- If the performance of *b* is between the performance of *a* augmented with the preference threshold and the performance of *a* augmented with the veto threshold, *b* is slightly preferred to *a*. The concordance degree is deduced by linear interpolation: if $f_i(b)$ is between $f_i(a)+p_i$ and $f_i(a)+v_i$ then, $c_i(a,b)$ is between 0 and 1.

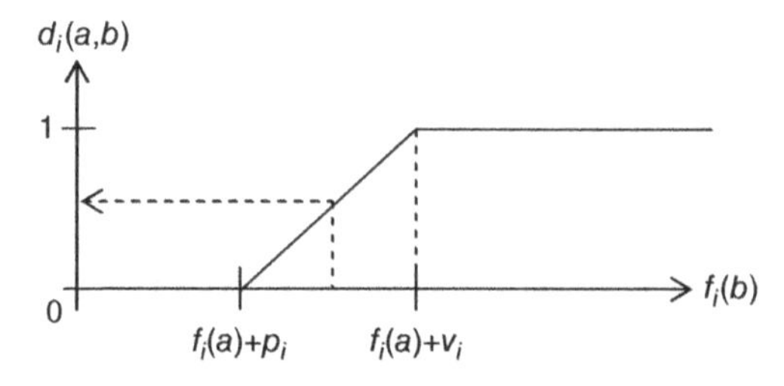

Figure 7.2 The partial discordance degree $d_j(a,b)$.

Table 7.4 Some examples of $d_1(a,b)$ in Case Study 7.1 for different values of p and v.

p_1	v_1	d_1(candidate 2, candidate 1)
0	0	1
1	1.5	1
1	2	1
1	3	0.5
≥ 2	$> p_1$	0

- If $f_i(b)$ is smaller than or equal to $f_i(a)+p_i$, the assertion is correct. There is no discordance hence $d_i(a,b) = 0$.

Table 7.4 gives some examples of the variation of the partial discordance index for criterion 1, between candidate 2 and candidate 1, for different values of p_1 and v_1. It is easy to see that d_1(candidate 1,candidate 2) is always 0 for any value of p_1 and v_1.

7.2.1.3 Outranking degree

Finally, a global outranking degree $S(a,b)$ summarizes the concordance and discordance degrees into one measure of the assertion '*a* outranks *b*' using a rather complicated formula, shown in Section 7.4. One of the intermediary outputs of *Electre III-IV* software is the value of the ordered outranking degrees of the alternatives. This is the reason for temporarily ignoring the exact formula.

7.2.1.4 Distillation

The second phase consists of exploiting these pairwise outranking degrees: the *ascending* and *descending* distillation procedures lead each to a complete (i.e. transitive) pre-order. Each pre-order takes into account respectively the outranking and outranked behaviour of an alternative with regard to the others. Since these procedures may lead to two different procedures, a final ranking is generated as the intersection of the two pre-orders. The final ranking, as illustrated in Figure 7.3, is a partial ranking resulting from the preference parameters given in Table 7.5.

In Figure 7.3, we can see that the best alternative in both distillations is A5 and thus also in the final ranking. Alternatives A4 and A6 are incomparable in the final graph: there is no arrow (and thus no preference relation) between the two alternatives. This is the result of the fact that A6 has a different ranking (compared to A4) in the descending distillation than in the ascending distillation. A4 is ranked third in both distillations, whereas A6 respectively fourth in the descending distillation and second in the ascending distillation.

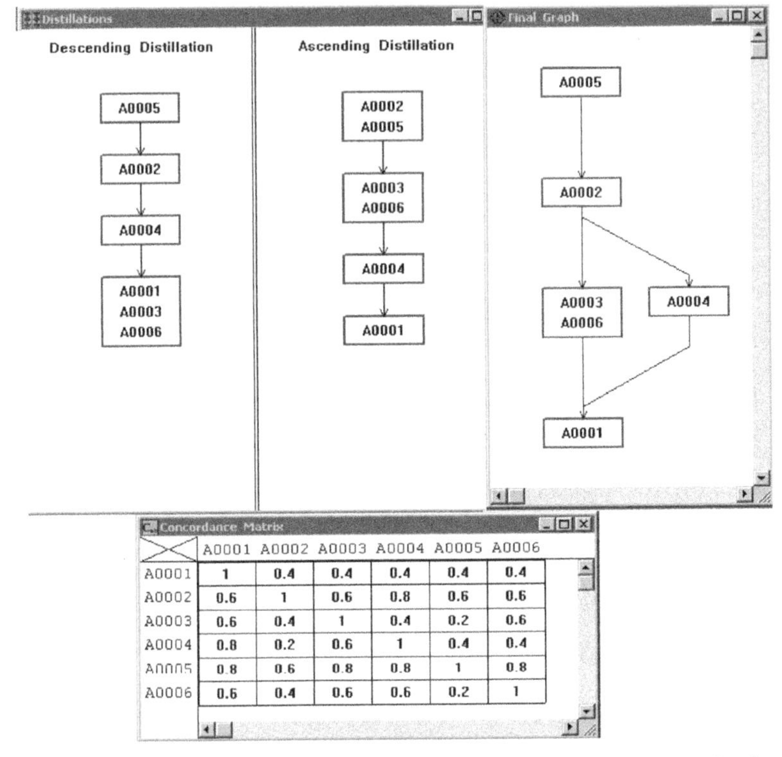

Figure 7.3 Descending and ascending distillation graphs anf the final graph obtained with the Electre III-IV *software for Case Study 7.1. Reproduced by permission of LAMSADE.*

Table 7.5 Preference parameters for Case Study 7.1.

	f_1	f_2	f_3	f_4	f_5
w_i	0.2	0.2	0.2	0.2	0.2
q_i	0	0	0	0	0
p_i	1	1	1	1	1
v_i	0	0	0	0	0

Exercise 7.1

The following multiple-choice questions test your knowledge on the basics of ELECTRE. Only one answer is correct. Answers can be found on the companion website.

1. What does ELECTRE stand for?

   a) *ELimination Et Choix Traduisant la REalité*

   b) *ELicit, Evaluate Criteria Through REferences*

c) ELECitation and TRained Evaluation

d) Evidence Limitée Et Confidence Transcripte de la Réalité

2. Which statement is incorrect?

 a) ELECTRE can be used in a wide range of applications

 b) Every decision maker will find the same ranking

 c) The ELECTRE method requires a lot of input parameters

 d) Results can be explained

3. What is the main purpose of ELECTRE III?

 a) ELECTRE III prioritizes alternatives based on criteria and constraints

 b) ELECTRE III assigns goals to alternatives

 c) ELECTRE III ranks alternatives based on criteria

 d) ELECTRE III sorts alternatives

4. On what scale are pairwise comparisons based?

 a) Ratio scale

 b) Interval scale

 c) Ordinal scale

 d) Nominal scale

5. How many input parameters does a decision maker need to specify for each criterion in ELECTRE III?

 a) 5

 b) 4

 c) 3

 d) 2

7.3 The Electre III-IV software

It is possible to implement all the ELECTRE III computation steps in a spreadsheet but it is not a simple task. Only one software package supports the ELECTRE III and ELECTRE IV methods: the *Electre III-IV* software. It is an old software package, which runs on Windows 3.1, 95, 98, 2000, XP and Vista. As the software is no longer maintained, it is not guaranteed to run on more recent operating systems. A free version is available at Lamsade at Université Paris-Dauphine with no time expiration and with no limitation on the number of alternatives or criteria. Later in this chapter, software supporting other ELECTRE methods (see Table 7.1) such as ELECTRE-Tri is described. *Electre III-IV* computes the concordance matrix, discordance matrix and the different intermediate rankings. Compared to other existing software, it lacks the functionality to enable the user to model the decision problem. It is difficult to understand what is happening behind the scenes and there are not many graphical representations explaining the results. Moreover, it does not perform a thorough sensitivity analysis: the user is unable to ask questions such as 'What if I change this parameter?'. Nevertheless, entering the data is relatively simple and the final results are easy to interpret.

As an illustration, Case Study 7.1 will be solved according to the set of preference parameters set out in Table 7.5. The basic steps of the ELECTRE III calculations can be found on the companion website in a spreadsheet format, as well as the *Electre III-IV* input files.

7.3.1 Data entry

From the menu bar, select *File/New Project* to create a new decision problem. Specify the name of the owner, a small description of the decision problem and the desired decision support method (ELECTRE III in this case).

The next consecutive steps are: the definition of the criteria, the alternatives, performance of alternatives and threshold of criteria. This is done via *Edit* from the menu bar.

In the Edit Criteria Table window (Figure 7.4, left), click on *Insert* which opens the Edit Criterion window (Figure 7.4, right). A name for the criterion, its associated weight, direction of preference (increasing if the criterion is to be maximized or decreasing if it is to be minimized) and optionally a code or abbreviation for the criterion, are specified.

Once the Edit Criterion window is completed, click on *OK* to add the criterion to the list of criteria (displayed in the Edit Criteria Table). To add a new criterion, click on the *Insert* button again, which opens a blank Edit Criterion window. To avoid going back to Edit Criteria Table, the Auto Insert Mode in the Edit Criteria Table

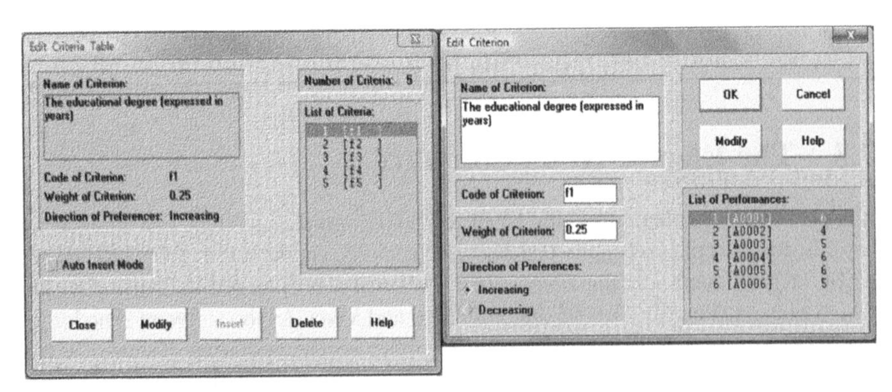

Figure 7.4 The Edit Criteria Table *(left) and* Edit Criterion *(right) windows. Reproduced by permission of LAMSADE.*

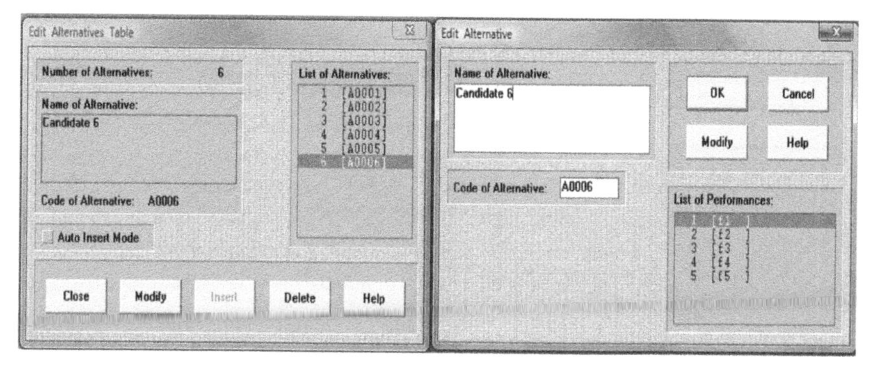

Figure 7.5 The Edit Alternatives Table (left) and Edit Alternative (right) windows. Reproduced by permission of LAMSADE.

window can be activated: the Edit Criterion window will continuously appear after clicking OK until Cancel is selected. In order to modify the parameters of an already defined criterion, double-click on the name of the criterion from the list.

The definition of the alternatives is done in an analogous way: from the Edit Alternatives Table (via *Edit/Alternatives*) open the Edit Alternative window (see Figure 7.5).

Once the criteria and alternatives have been defined, enter the performance of the alternatives in the Edit Performances Table dialogue box (see Figure 7.6) by selecting *Edit/Performances* from the menu bar. We remark that only numerical values can be entered in the software.

At this stage, the criteria, alternatives and their performance have been entered. Following this, the decision maker needs to introduce the preference settings.

7.3.2 Entering preference parameters

The user needs to define the preference parameters for each criterion via the Edit Thresholds Table (see Figure 7.7). Analogously, the user can change the settings by

Edit Performances Table

	f1	f2	f3	f4	f5
A0001	6	5	28	5	5
A0002	4	2	25	10	9
A0003	5	7	35	9	6
A0004	6	1	27	6	7
A0005	6	8	30	7	9
A0006	5	6	26	4	8

Number of Criteria: 5
Number of Alternatives: 6
8
Close
Help

Figure 7.6 The Edit Performances Table. Reproduced by permission of LAMSADE.

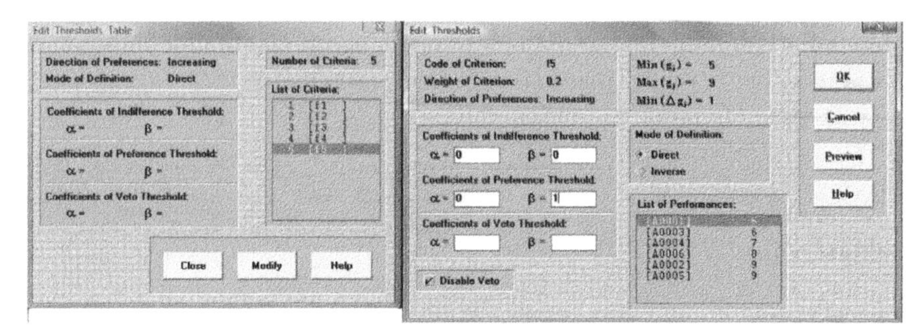

Figure 7.7 Input boxes for preference information. Reproduced by permission of LAMSADE.

choosing a criterion in the Edit Thresholds Table window (Figure 7.7, left), which leads to the Edit Thresholds window (Figure 7.7, right).

Various parameters are required. The user has to define the indifference and preference threshold for each criterion. The veto threshold is optional (by checking the *Disable Veto* box).

The ELECTRE III preference thresholds can be constant values or proportionate to the performance of the alternative. If the user wants to use a fixed threshold, the α-coefficient must be equal to 0. The β-coefficient takes the value of the fixed threshold. The *Mode of Definition* does not play any part in this situation.

In Case Study 7.1, the indifference and preference thresholds for the criterion f_5 were set to 0 and 1, respectively (see Table 7.5). These parameters are entered into *Electre III-IV* (see Figure 7.7, right):

- α-indifference coefficient = 0,
- β-indifference coefficient = 0,
- α-preference coefficient = 0;
- β-preference coefficient = 1.

Moreover, we need to disable the veto.

7.3.2.1 Advanced settings

Let us suppose that with regard to the potential return criterion of Case Study 7.1, the decision maker considers that two alternatives are indifferent if their score difference is less than 10%. In this case, the indifference threshold is not absolute, but relative to the performance of the alternatives.

This relative proportion is introduced via the coefficient α of the indifference and preference thresholds in the software. The user needs to specify which performance the proportion is taken from: either 10% from the highest performance or 10% from the lowest one. This can be specified by the ‘mode of definition’ (see Figure 7.7, left): if the direct mode is chosen, the worst performance between the two will be used to compute the relative threshold. The term ‘worst’ means ‘the least preferred’, which might be the lowest value for criteria to be maximized or highest for criteria to be minimized. If the mode of definition is ‘inverse’, the best performance is chosen.

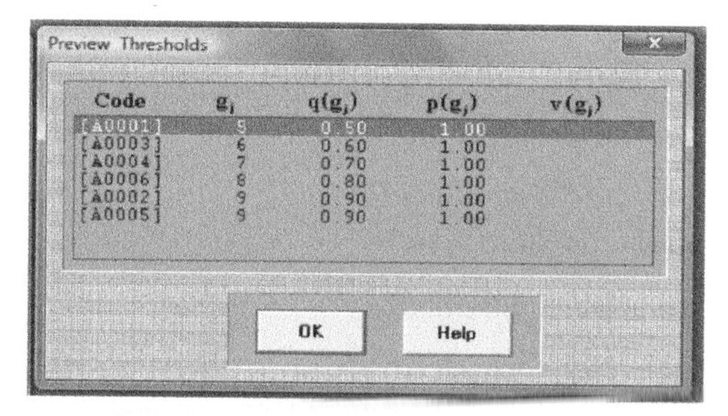

Code	g_i	$q(g_i)$	$p(g_i)$	$v(g_i)$
[A0001]	5	0.50	1.00	
[A0003]	6	0.60	1.00	
[A0004]	7	0.70	1.00	
[A0006]	8	0.80	1.00	
[A0002]	9	0.90	1.00	
[A0005]	9	0.90	1.00	

Figure 7.8 Display of the relative indifference threshold and absolute preference threshold for f_5. *Reproduced by permission of LAMSADE.*

The absolute threshold is defined via the coefficient β. Relative and absolute thresholds are added if neither is zero.

There is a menu in *Electre III-IV* to display the thresholds as a function of the alternatives (Figure 7.8). The defined threshold also depends on the mode of definition (direct or inverse).

Throughout the rest of this section, all thresholds are in absolute terms as defined in Table 7.5.

7.3.3 Results

The most useful results are presented in Figure 7.3 (accessible via: *Menu/Results*): the final (partial) ranking, distillation results, concordance and credibility matrix. Figure 7.9 gives additional information such as the final rank of the alternatives. We can thus see that candidate A5 is ranked first, while candidates 4 and 6 are ranked the same but are incomparable. This information is deduced from the ranking matrix, which includes the global reference relations (indifference, preference or incomparability, denoted respectively $\equiv$, $\prec$ and $\square$) amongst the alternatives. If the decision maker wants a complete pre-order, which suppresses incomparabilities, they might want to display the median pre-order. The median pre-order takes into account the relative rank of the two alternatives in the two partial rankings. The difference in the two partial rankings for candidate 4 is –1 (i.e. 3 – 4) and for candidate 6 is 1 (4 – 3). Therefore, candidate 4 has a better rank than candidate 6.

Ranks in Final Preorder

Rank	Alternative
1	A0005
2	A0002
3	A0003
4	A0004 A0006
5	A0001

Ranking Matrix

	A0001	A0002	A0003	A0004	A0005	A0006
A0001	≡	≺	≺	≺	≺	≺
A0002	≻	≡	≻	≻	≺	≻
A0003	≻	≺	≡	≻	≺	≻
A0004	≻	≺	≺	≡	≺	■
A0005	≻	≻	≻	≻	≡	≻
A0006	≻	≺	≺	■	≺	≡

Median Preorder

Rank	Alternative
1	A0005
2	A0002
3	A0003
4	A0004
5	A0006
6	A0001

Figure 7.9 Ranks, ranking matrix and median pre-order for Case Study 7.1. Reproduced by permission of LAMSADE.

Exercise 7.2

In this exercise, you will solve the problem set out in Case Study 7.1 with *Electre III-IV*.

Learning Outcomes

- ➢ Structure a problem in *Electre III-IV*
- ➢ Enter the preference parameters
- ➢ Understand the results

Consider Case Study 7.1 where the performance of the alternatives is given in Table 7.2 and the preference parameters in Table 7.5. Enter the data in *Electre III-IV* and cross-reference your results with those set out in Figure 7.3 and Figure 7.9.

Tasks

a) Read the description of Case Study 7.1 on page 183 and open the *Electre III-IV* software. Choose *New Project* from the *File* menu; specify the name of the owner and choose ELECTRE III as the ranking method.

b) From *Edit* select *Criteria,* which displays the Edit Criteria Table where a new criterion can be inserted. For each criterion specify its name, weight and direction of preference.

c) From *Edit* select *Alternatives,* which displays the Edit Alternative Table where new alternatives can be inserted. For each alternative specify the name and code name.

d) From *Edit* select *Performances,* which presents the Edit Performances Table where the performance of the alternatives can be edited.

e) From *Edit* select *Thresholds,* which displays the Edit Threshold Table where the preference parameters of the criteria can be edited.

g) Check the results against those in Figure 7.3 and Figure 7.9.

g) Check that the concordance matrix corresponds to Table 7.6.

h) Are there any incomparable actions?

7.4 In the black box of ELECTRE III

7.4.1 Outranking relations

ELECTRE III makes use of outranking relations. An outranking relation on a set A is a binary relation: a collection of ordered pairs of elements of A and thus a subset of the Cartesian product $A \times A$ (where $A \times A = \{(a,b)|a\in A \text{ and } b\in A\}$; this means that it is

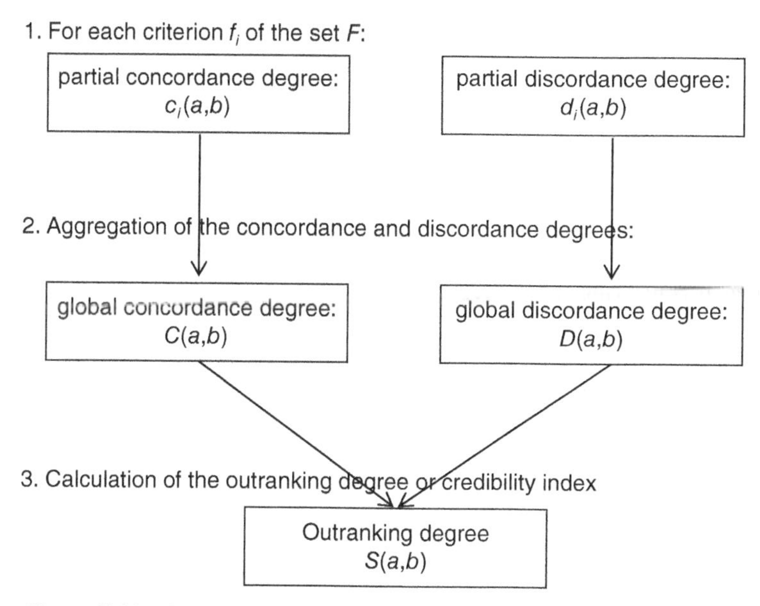

Figure 7.10 Steps in the calculation of the outranking degree S(a,b).

the set of all possible ordered pairs). According to (Roy 1974), an outranking relation is such that for two elements *a* and *b* of A, *a* outranks *b* (written *a* S *b*), if, given what is known about the decision maker's preferences, the quality of the valuations of the alternatives and the nature of the problem, there are enough arguments to decide that *a* is at least as good as *b*, where there is no essential reason to refute this statement.

An outranking degree $S(a,b)$ between *a* and *b* will be computed to 'measure' or 'evaluate' this assertion. The score will be between 0 and 1: the stronger the assertion, the closer $S(a,b)$ to 1. The outranking relation is a non-symmetric relation:$\exists a, b \in A : S(a, b) \neq S(b, a)$.

The outranking degree is calculated in three steps as shown in Figure 7.10:

1. The partial concordance and partial discordance degrees are computed for each criterion of the set F.
2. The criteria are aggregated to the global concordance and global discordance degree.
3. The aggregation of the concordance and discordance degree leads to the outranking degree or credibility index.

7.4.2 Partial concordance degree

For each criterion f_j from F the assertion '*b* is at least as good as *a*' or '*b* outranks *a*' is measured by the partial concordance index noted $c_j(b,a)$. This degree is obtained as follows (Roy and Bouyssou 1993):

$$c_j(b,a) = \begin{cases} 1 & \text{if } f_j(b) + p_j < f_j(a) \\ \dfrac{f_j(b) + p_j - f_j(a)}{p_j - q_j} & \text{if } f_j(b) + q_j < f_j(a) < f_j(b) + p_j \\ 0 & \text{otherwise,} \end{cases} \tag{7.1}$$

where q_j, p_j (satisfying $p_j > q_j$) represent respectively the indifference and preference thresholds as illustrated in Figure 7.1.

These thresholds may be absolute or dependent (i.e. relative) on the performances of a or b: $p_j = p_j(f_j(b))$ and $q_j = q_j(f_j(b))$. The stronger the confidence of the decision maker with the outranking assertion, the higher the concordance index. Its value is always between 0 and 1. A concordance degree of 0 means that b does not outrank a. A score of 1 means that b is as least as good as a (on this particular criterion).

7.4.3 Global concordance degree

The global concordance degree $C(a, b)$ aggregates all the partial concordance indices on the different criteria by taking into account their corresponding weight, denoted by w_j for all f_j with $j = 1, \ldots q$. It is the weighted sum of all the partial concordance indices, which measures how concordant the assertion 'a is at least as good b' is, regarding all the criteria. We have

$$C(b,a) = \sum_{j=1,\ldots,q} w_j \cdot c_j(b,a). \tag{7.2}$$

We remark that the weight of the criteria cannot be considered as substitution rates as in compensatory methods such as AHP (Chapter 2), MAUT (Chapter 4) and MACBETH (Chapter 5) (Figueira et al. 2005). The weights depend neither on the range nor the scale of the criteria (Figueira et al. 2005).

To illustrate, consider the performance matrix of Case Study 7.1 given in Table 7.2 and the preference parameters given in Table 7.5. This leads to the concordance matrix in Table 7.6.

7.4.4 Partial discordance degree

For each criterion f_j from F, the measure of the discordance with the assertion 'b is at least as good as a' is given by the partial discordance index $d_j(a,b)$. This index is calculated as follows:

$$d_j(b,a) = \begin{cases} 1 & \text{if } f_j(b) + v_j < f_j(a) \\ 0 & \text{if } f_j(a) < f_j(b) + p_j \\ \dfrac{f_j(a) - p_j - f_j(b)}{v_j - p_j} & \text{otherwise,} \end{cases} \tag{7.3}$$

Table 7.6 Concordance matrix of Case Study 7.1 with the preference parameters in Table 7.5 obtained with the *Electre-III* software.

	A1	A2	A3	A4	A5	A6
A1	1	0.5	0.35	0.5	0.35	0.45
A2	0.5	1	0.5	0.75	0.5	0.5
A3	0.65	0.5	1	0.45	0.2	0.7
A4	0.75	0.25	0.55	1	0.35	0.45
A5	0.9	0.7	0.8	0.9	1	0.9
A6	0.55	0.5	0.55	0.55	0.1	1

where v_j (satisfying $v_j > p_j$) represents the veto threshold for criterion f_j as illustrated in Figure 7.2. The veto threshold can be absolute or relative: $v_j = v_j(f_j(b))$.

7.4.5 Outranking degree

The (global) outranking degree *S(a,b)* summarizes the concordance and discordance degree into one measure of the assertion '*a* outranks *b*' using the following formula:

$$S(b,a) = C(b,a) \cdot \prod_V \left[\frac{1 - d_j(b,a)}{1 - C(b,a)} \right], \tag{7.4}$$

where *V* is the set of criteria for which $d_j(b,a) > C(b,a)$.

If the concordance index *C(b,a)* is greater than or equal to the partial discordance indexes (i.e. $C(b,a) \geq d_j(b,a)$, for all *j*), then the outranking degree is equal to the concordance index.

The outranking degree *S(a,b)* always lies between 0 and 1 and is not symmetrical. The information contained in both *S(a,b)* and *S(b,a)* can be combined to express whether *a* is preferred over *b* (*a* P *b*) while considering the whole set of actions of *A* (Giannoulis and Ishizaka 2010):

$$a \text{ P } b \Leftrightarrow S(a,b) > \lambda_2 \quad \text{and} \quad S(a,b) - S(b,a) > s(\lambda_0), \tag{7.5}$$

where λ_2 is the largest credibility index, which is just below the cut-off level λ_1. Here

$$\lambda_2 = \max_{\{S(a,b) \leq \lambda_1\}} S(a,b) \quad \forall a, b \in A; \tag{7.6}$$

λ_1 is the cut-off level

$$\lambda_1 = \lambda_0 - s(\lambda_0); \tag{7.7}$$

λ_0 is the highest degree of credibility in the credibility matrix,

$$\lambda_0 = \max_{a,b \in A} S(a, b); \tag{7.8}$$

and $s(\lambda_0)$ is the discrimination threshold,

$$s(\lambda_0) = \alpha + \beta\lambda_0. \tag{7.9}$$

In this obscure formula, α and β are technical parameters that Roy and Bouyssou (1993) suggest setting to $\alpha = -0.15$ and $\beta = 0.3$.

It is worth noting that the preference relation between two actions, in ELECTRE III, is dependent on the outranking degrees between the other actions given the definitions of λ_2 and λ_0. This is not the case with the preference relation, defined in ELECTRE-Tri (Section 7.5).

To illustrate, consider Case Study 7.1, where the preference relations of the alternatives are computed based on the concordance matrix given in Table 7.6. The parameters are given in Table 7.7. Based on these parameter values, the following preference relation between two alternatives of A exist:

$$a \text{ P } b \Leftrightarrow S(a, b) > 0.8 \text{ and } S(a, b) - S(b, a) > 0.15. \tag{7.10}$$

Equation (7.11) leads to the following conclusions: alternative A1 is preferred to no other alternative as its outranking relation to the other alternatives is always lower than 0.8. Alternative A5 is the only alternative preferred over A1: $S(\text{A5,A1}) = 0.9 > 0.8$ and $S(\text{A5,A1}) - S(\text{A1,A5}) = 0.55 > 0.15$ and as a result A5 $\succ$ A1.

Considering two outranking degrees $S(a,b)$ and $S(c,d)$ with $S(a,b) > S(c,d)$, it is tempting to say that the assertion 'a outranks b' is more credible (or more likely) than

Table 7.7 Values of the parameters for Case Study 7.1.

Parameters	Values
λ_0	1
α, β	α=–0.15 and β=0.3
$s(\lambda_0)$	0.15
λ_1	0.85
λ_2	0.8

the assertion 'c outranks d'. Due to the arbitrary nature of the computation of these indices, this should be avoided (Tervoren et al. 2004). However, if $\lambda \in [0,1]$ and if $[S(a,b) = \lambda$ and $S(c,d) = \lambda - \eta$ with $\eta > s(\lambda_0)]$, then we can say that a S b is strictly more credible than c S d.

7.4.6 Partial ranking: Exploitation of the outranking relations

The second phase consists of exploiting the outranking degrees by the *ascending* and *descending* procedures, which give two complete (i.e. transitive) pre-orders O_1 and O_2. The final ELECTRE III partial pre-order, O, is obtained by calculating the intersection of O_1 and O_2.

The distillation procedures are based on the qualification of the alternatives. The qualification score of an alternative is a score which characterizes its global behaviour with regard to the other alternatives. Each time one action is preferred to another, the score is incremented by 1 (strength), whereas if it is preferred by another, this score is reduced by 1 (weakness). The qualification of an alternative is thus the balance of its strengths and weaknesses.

For instance, in Case Study 7.1, the strength of candidate 1 is 0 (candidate 1 is not preferred to any other alternative) and its weakness is 1 (one other candidate (candidate 5) is preferred to candidate 1). Therefore, the qualification of candidate 1 will be –1.

The descending distillation procedure leading to the complete pre-order O_1 can be explained as follows:

- We start with the complete set of alternatives. From this set, the alternative(s) from A with the highest qualification is extracted. This constitutes the first group (denoted by C_1). This means that it is not possible to decide between the remaining alternatives in the subset, and therefore they are declared indifferent and belong to C_1.
- From the remaining *set* of alternatives (i.e. $A \backslash C_1$), the best alternative is again extracted to obtain the second group C_2. On the successive distillations, the cut-off level λ_1 (defined in (7.8)) is progressively reduced, which makes the condition weaker and easier for an alternative to be preferred over another.
- This procedure is repeated until A has been distilled completely (i.e. all alternatives of A belong to a subgroup).

The complete *descending* pre-order corresponds to the order C_1, C_2, etc. Figure 7.11 illustrates the descending distillation procedure, while Figure 7.3 represents the descending pre-order of Case Study 7.1.

The ascending distillation procedure is comparable in some respects; however, instead of starting with the best subset, it starts with the worst.

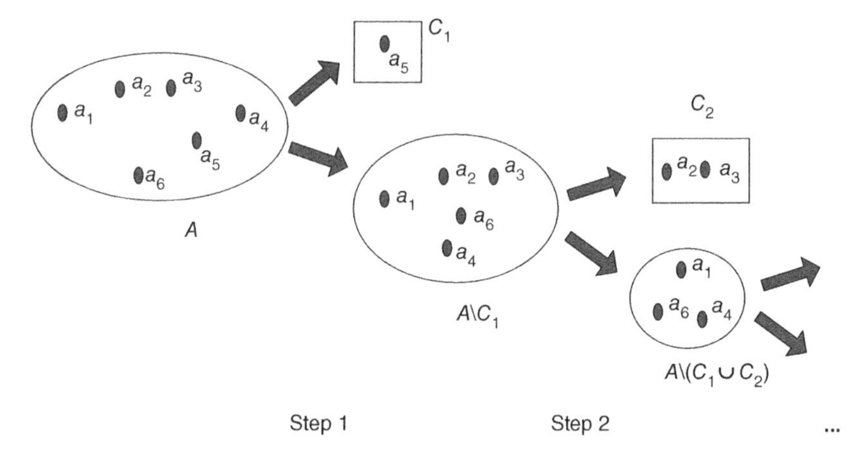

Figure 7.11 The descending distillation procedure.

The final partial pre-order O is defined as the intersection of O_1 and O_2. The global relations are defined as follows:

- a is globally better than b, written $a \succ b$, if and only if:
  - a is better than b in O_1 and in O_2, or
  - a is indifferent with respect to b in O_1 but better than b in O_2, or
  - a is better than b in O_2 and indifferent with respect to b in O_2.
- a and b are globally indifferent, written $a \equiv b$, if and only if a and b are indifferent in O_1 and O_2.
- a is globally incomparable to b, written $a \,\square\, b$, if and only if:
  - a is better than b in O_1 but b is better than a in O_2,
  - b is better than a in O_1 but a is better than b in O_2.
- a is globally worse than b, written $a \prec b$, if and only if:
  - b is better than a in O_1 and in O_2, or
  - a is indifferent with respect to to b in O_1 but b is better than a in O_2, or
  - b is better than a in O_2 and indifferent with respect to a in O_2.

Example 7.1 Consider the following example, where four alternatives are to be ranked with the ELECTRE III partial ranking method according to two criteria. These criteria are to be maximized; suppose that they are true criteria (i.e. for all $j = 1$, 2: $q_j = p_j = 0$ and with $v_j = 0$). The performance matrix of A is given in Table 7.8. The binary relations (based on the values of the credibility matrix S) are given in Table 7.10 and represented in Figure 7.12 when fixing $\lambda > 0.5$.

Table 7.8 The performance matrix of the alternatives in Example 7.1.

	f_1	f_2
a_1	1	1
a_2	0	0.5
a_3	0.5	0
a_4	0	0

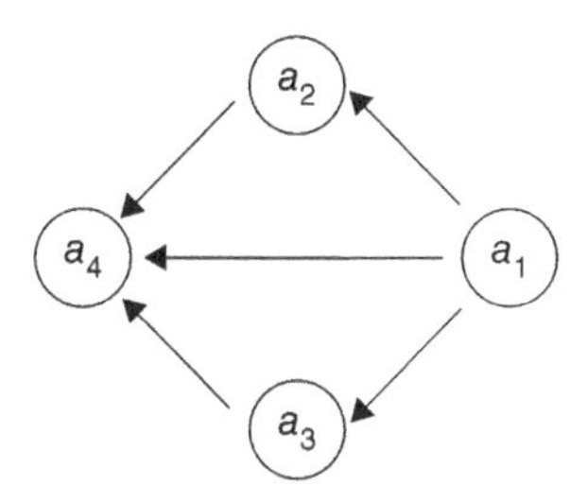

Figure 7.12 Outranking graph of A.

Table 7.9 The outranking degrees for Example 7.1.

S	a_1	a_2	a_3	a_4
a_1	1	1	1	1
a_2	0	1	0.5	1
a_3	0	0.5	1	1
a_4	0	0	0	1

Based on the outranking matrix given in Table 7.9, the outranking graph given in Figure 7.12 can be drawn, where an arrow between two alternatives represents a (binary) preference relation; thus a_3 and a_2 are incomparable.

The ascending and descending distillation procedures lead to the complete ranking represented in Figure 7.13 with the ranking relations given in Table 7.10. The results were acquired from *Electre III-IV*.

$O = O_1 = O_2$:

a_1 → a_2, a_3 → a_4

Figure 7.13 The partial pre-order for Example 7.1 obtained with ELECTRE III.

From Figure 7.13, ELECTRE III leads to a complete pre-order, although a more intuitive partial ranking could be expected on the basis of the outranking graph in Figure 7.12. If only the transitive outranking relations are represented, the partial

Table 7.10 The binary relation between the alternatives in the global ranking O.

S	a_1	a_2	a_3	a_4
a_1	$\equiv$	$\succ$	$\succ$	$\succ$
a_2	$\prec$	$\equiv$	$\equiv$	$\succ$
a_3	$\prec$	$\equiv$	$\equiv$	$\succ$
a_4	$\prec$	$\prec$	$\prec$	$\equiv$

pre-order given in Figure 7.14 (what we will call the reduced outranking graph) is obtained. The main difference lies in the incomparability between a_2 and a_3 being preserved.

A potential drawback of the ELECTRE III method is that it cannot always clearly differentiate between indifference and incomparability between two alternatives in the final ranking (Roy and Bouyssou, 1993, p. 423). As one might notice, the preference relations given in Table 7.10 are based on a 'global' level. However, based on these outranking degrees and a cut-off level λ defined by the decision maker, four possible pairwise (local) comparisons can be defined when comparing alternative a to alternative b:

- $a\ P^+\ b$: a is preferred to b if and only if $S(a, b) \geq \lambda$ *and* $S(b, a) < \lambda$, that is, we have that a is at least as good as b but b is not at least as good as a.
- $a\ P^-\ b$: b is preferred to a if and only if $S(a, b) < \lambda$ *and* $S(b, a) \geq \lambda$, that is, we have that a is not at least as good as b but b is at least as good as a.
- $a\ I\ b$: a and b are indifferent if and only if $S(a, b) \geq \lambda$ *and* $S(b, a) \geq \lambda$, that is, we have that a is at least as good as b and b is at least as good as a.
- $a\ R\ b$: a and b are incomparable if and only if $S(a, b) < \lambda$ *and* $S(b, a) < \lambda$, that is, we have that a is not at least as good as b but neither is b compared to a.

If we define a value higher than 0.5 for the cut-off level λ, we obtain the pairwise preference relations given in Table 7.11.

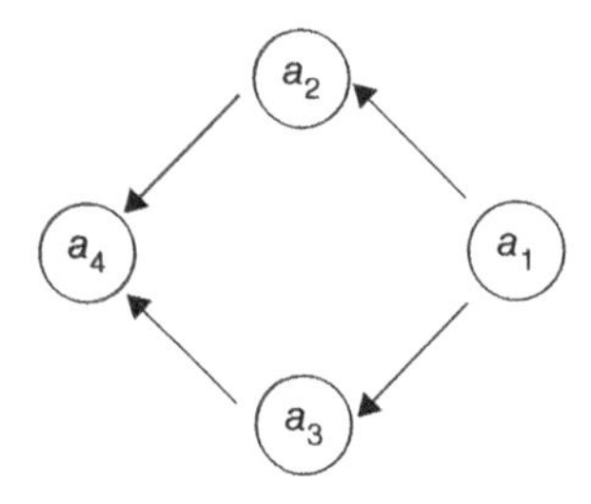

Figure 7.14 The partial pre-order of the set of alternatives A *obtained by 'reducing' the outranking graph.*

Table 7.11 The pairwise preference relations between the alternatives.

	a_1	a_2	a_3	a_4
a_1	I	P^+	P^+	P^+
a_2	P^-	I	R	P^+
a_3	P^-	R	I	P^+
a_4	P^-	P^-	P^-	I

As we can see from Table 7.11, the alternatives a_2 and a_3 are (locally) incomparable but are considered as indifferent in the global ranking O (see Table 7.10). The decision maker can nevertheless accept this as a_2 and a_3 behave similarly with respect to the alternatives a_1 and a_4. An analogous result would be obtained with PROMETHEE. This is a direct consequence of the aggregation of the pairwise comparisons. To avoid this situation, a modified version of ELECTRE III has been proposed and can be found in Roy and Bouyssou (1993).

7.4.7 Some properties

When comparing alternatives by means of pairwise outranking degrees, cycles may occur, for example, a_i outranks a_j outranks a_k outranks a_i. This is the 'Condorcet paradox'. The consequence of aggregating the comparisons into a global complete ranking is that the order in the final ranking may not correspond to these pairwise comparisons (Mareschal et al. 2008). It may suffer from *pairwise rank reversal* since a S b with b having a better rank in the final ranking than a. In O_1 of Case Study 7.1, notice that a_6 has a lower rank than a_4 (see Figure 7.14) even though $S(a_4, a_6) < S(a_6, a_4)$.

A consequence of pairwise rank reversal, as described by Mareschal et al. (2008), is that the addition or the suppression of an alternative to set A may lead to a rank reversal phenomenon in the final ranking. The pre-orders O_1 and O_2 can be modified. Wand and Triantaphyllou (2006) conducted some computational experiments on randomly generated and real-life decision problems to test the rank reversal phenomenon with the ELECTRE III method. They observed that the rates of ranking irregularities were significant in both the simulated and real-life decision problems.

ELECTRE III is *neutral* to the name or label given to the alternatives, as it does not discriminate alternatives in their ranking on the basis of their label or their given name (Bouyssou and Perny 1992; Bouyssou 1992). Assume that alternatives are numbered $a_1, a_2, a_3, \ldots$, and then renamed (while keeping the same order) to a_j, a_n, $a_3, \ldots$. The ranking obtained after renaming the alternatives will remain coherent with the initial ranking.

When two alternatives, a and b, are compared similarly to any other alternative of set A (i.e. $\forall x \in A$: $S(a, x) = S(b, x)$; $S(x, a) = S(x, b)$ and $S(a, b) = S(b, a)$), they will be considered globally indifferent. This is often called the non-discriminatory property of a ranking method.

Finally, Gabrel (1990) and Perny (1992) have pointed out that ELECTRE III does not fulfill the property of *monotonicity* since the rankings do not respond 'in the right direction' to a modification of performances of the alternatives: the amelioration of the performances of an alternative may lead to deterioration in its final ranking.

Exercise 7.3

You will learn the calculation of the outranking matrix step by step, and then have the opportunity to compute the rank of the alternatives based on the distillation procedures.

Learning Outcomes

- Understand the calculation of the outranking matrix in *Microsoft Excel*
- Understand the calculation of the ascending and descending distillation procedures
- Understand the calculation of the final ranking

Tasks

Open the file Grant.xls. The spreadsheet contains the steps of the ELECTRE III procedure. Answer the following questions:

a) Describe the meaning of each calculation cell and its formula. (Read the comments in the red square in case of difficulty.)

b) The spreadsheets are incomplete because they calculate only one local alternative. Complete them in order to calculate the other local alternatives.

7.5 ELECTRE-Tri

7.5.1 Introduction

ELECTRE-Tri is a multi-criteria sorting method used for the assignment of a set of alternatives A into K completely ordered categories $C_1, \ldots, C_K$ where category C_1 is the *best* category and C_K the *worst*. For example, the prioritization of projects, which are categorized as low, medium or high priority.

In ELECTRE-Tri, the categories can be defined either by limiting profiles (or boundaries (Yu 1992)) or by central profiles (or centroids (Dias et al. 2010)). In the first case, the method is named ELECTRE-Tri-B and in the second ELECTRE-Tri-C. In this section, ELECTRE-Tri-B, which has been named ELECTRE-Tri, will be described.

A limiting profile r_h is the upper reference profile for category C_h and the lower reference profile for category C_{h-1}. The best profile will be denoted by r_1 and the

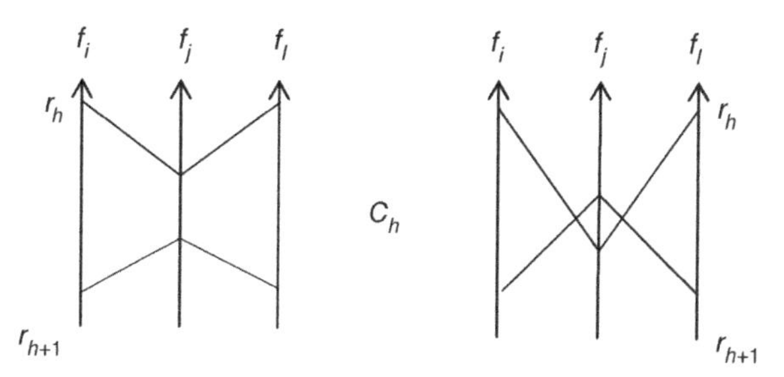

Figure 7.15 Limiting profiles respecting (left) and infringing (right) the dominance condition.

worst r_{K+1} . This convention has been chosen because the best category is ranked 1 and the worst has is ranked K.

Since the limiting profiles define ordered categories, they need to respect the *condition of dominance*: a limiting profile dominates, written $\succ^D$, all the successive limiting profiles of worse categories. Formally, if all the criteria f_l ($l = 1, \ldots, q$) are to be maximized, the following condition has to be satisfied:

$$\begin{aligned}\forall h = 1, \ldots, K; r_h \succ^D r_{h+1} \Leftrightarrow \forall l = 1, \ldots, q : f_l(r_h) \geq f_l(r_{h+1}) \\ \text{and } \exists j : f_j(r_h) > f_j(r_{h+1}).\end{aligned} \quad (7.11)$$

This condition implies that the performance of the upper limit of a class must be at least as good as the performance of its lower limit and at least better on one criterion.

Figure 7.15 illustrates the condition of dominance, where limiting profiles r_h and r_{h+1} define category C_h. On the left, r_h and r_{h+1} respect the condition dominance. The condition of dominance is not respected in the figure on the right, because on criterion f_j, profile r_h is lower than r_{h+1}.

7.5.2 Preference relations

To assign alternative a to one of the categories, the outranking relations between a and the limiting profiles are built based on a set of coherent criteria F: $S(a,r_h)$ and $S(r_h,a)$, $\forall h = 1, \ldots, K + 1$. These outranking degrees measure the strength of the assertion that 'a is at least as good as r_h' (and vice versa) and are calculated in the same way as ELECTRE III (Section 7.4.5).

On the basis of these outranking degrees and a cut-off level λ defined by the decision maker, four possible situations may occur when comparing alternative a to the limiting profile r_h:[2]

- $a \succ r_h$: a is preferred to r_h if and only if $S(a, r_h) \geq \lambda$ and $S(r_h, a) < \lambda$, that is, a is at least as good as r_h but r_h is not at least as good as a.

[2] We remark that we use the symbols of a pairwise (local) preference relation (and not a global one as in Section 7.4.6) in order to keep consistent with the symbols of the *Electre-Tri* software.

- $a \prec r_h$: r_h is preferred to a if and only if $S(a, r_h) < \lambda$ and $S(r_h, a) \geq \lambda$, that is, a is not at least as good as r_h but r_h is at least as good as a.
- $a \, \mathrm{I} \, r_h$: a and r_h are indifferent if and only if $S(a, r_h) \geq \lambda$ and $S(r_h, a) \geq \lambda$, that is, a is at least as good as r_h and r_h is at least as good as a.
- $a \, \mathrm{R} \, r_h$: a and r_h are incomparable if and only if $S(a, r_h) < \lambda$ and $S(r_h, a) < \lambda$, that is, a is not at least as good as r_h but neither is r_h compared to a.

Although the outranking relations S are calculated the same way in ELECTRE III and ELECTRE-Tri, the preference relations are not identical. ELECTRE-Tri has the advantage of a clear definition of preference, indifference and incomparability relationships based on a fixed cut-off level. In ELECTRE III, this cut-off level varies across different distillation steps.

Based on these preference relations, an additional condition needs to be introduced on the limiting profiles, which translates the fact that the categories are completely ordered:

$$\forall h = 1, \ldots, K : r_h \succ r_{h+1}. \tag{7.12}$$

Condition (7.13) is stronger than the dominance condition (7.12) as it imposes a preference relation between successive profiles (instead of the dominance relation). This condition implies that the categories are not 'too close' to each other: unlike the dominance relation, the preference relation requires that the performance of successive limiting profiles are such that there is a preference relation (and avoids an indifference relation between limiting profiles).

When comparing alternative a with regard to the reference profiles, three different situations may occur (Roy and Bouyssou 1993, p. 392):

1. Alternative a is, in the sense of the preference relation, 'in between' two consecutive limiting profiles: $r_1 \succ a, \ldots, r_j \prec a, a \succ r_{j+1}, a \succ r_{j+2}, \ldots, a \succ r_{K+1}$.
2. Alternative a is indifferent with respect to one or several (i.e. $k+1$) consecutive limiting profiles: $r_1 \succ a, \ldots, r_{j-1} \succ a, a\mathrm{I}r_j, a\mathrm{I}r_{j+1}, \ldots, a\mathrm{I}r_{j+k}, a \succ r_{j+k+1}, \ldots, a \succ r_{K+1}$.
3. Alternative a is incomparable to one or several (i.e. $k+1$) consecutive limiting profiles: $r_1 \succ a, \ldots, r_{j-1} \succ a, a\mathrm{R}r_j, a\mathrm{R}r_{j+1}, \ldots, a\mathrm{R}r_{j+k}, a \succ r_{j+k+1}, \ldots, a \succ r_{K+1}$.

If alternative a behaves in a similar way (i.e. indifference or incomparability) to several limiting profiles, these profiles must be consecutive. In other words, there cannot be a 'hole' in the sequence of similar profiles. This monotone behaviour is due to the dominance condition (7.12) imposed on the limiting profiles and the way outranking degrees are calculated.

7.5.3 Assignment rules

ELECTRE-Tri proposes two different assignment rules: the optimistic and pessimistic assignment rule. Both rules use different preference relations to compare the limiting profiles with the alternative to be assigned. The optimistic assignment rule uses the preference relation ($\succ$) and the pessimistic assignments rule uses the outranking relation (S). Both procedures handle the situation in a different way.

Optimistic assignment rule. Alternative a will be assigned to category C_h if the upper limiting profile r_h is the worst (lowest) profile which is preferred to a. Formally:

- Compare successively a and r_h, with h from $K+1$ to 1, where $K+1$ is the worst profile.
- If r_h is the first reference profile such that $r_h \succ a$, then a is assigned to C_h.

Pessimistic assignment rule. Alternative a will be assigned to category C_h if the lower limiting profile r_{h+1} is the best (highest) profile, which is outranked by a or with which a is at least as good. Formally:

- Compare successively a and r_h with h from 1 to $K+1$.
- If r_{h+1} is the first reference profiles such that a S r_{h+1}, then a is assigned to C_h.

Let us consider situations 1, 2 and 3 defined in Section 7.5.2. The assignment rules lead to the assignments given in Table 7.12. Note that the two rules lead to different assignments in situation 3, when an alternative is incomparable to one limiting profile. In the optimistic case, it is assigned to the 'best' category or to the category whose lower limiting profile is the best profile to which the alternative is incomparable. In the pessimistic case, the alternative is assigned to the 'worst' category or the category whose upper limiting profile is the worst one to which the alternative is incomparable (i.e. r_{j+k}).

Table 7.12 Summary of the assignment results when using the ELECTRE-Tri rules.

	Optimistic rule	Pessimistic rule
Situation 1	C_j	C_j
Situation 2	C_{j-1}	C_{j-1}
Situation 3	C_{j-1}	C_{j+k}

7.5.4 Properties

ELECTRE-Tri has the following properties (Yu, 1992; Roy and Bouyssou 1993):

- Every alternative is assigned to one category according to one of the procedures ('uniqueness property'). However, optimistic and pessimistic procedures may assign an alternative to a different category.

- The assignment of an alternative does not depend on the assignment of the other alternatives of A. ('independence property').
- When two identical alternatives are compared to the reference profiles (i.e. the outranking relations between the alternatives and the profiles are the same), they are assigned to the same categories ('strong homogeneity property').
- If alternative a dominates alternative b, then a will be assigned to the category which is at least as good as the category to which b will be assigned ('monotonicity property').
- The fusion of two successive categories or the separation of a category into two new categories does not affect the assignment of the alternatives in the other categories ('stability property').
- If the performance of alternative a is 'between' the performance of two consecutive limiting profiles, it will unequivocally be assigned to the category delimited by these profiles ('conformity property').

Exercise 7.4

In this exercise, you will learn how to use the *Electre-Tri* software in a sorting problem.

Learning Outcomes

- ➢ Structure a sorting problem in the *Electre-Tri* software (download from http://www.lamsade.dauphine.fr/spip.php?rubrique64)
- ➢ Understand the steps of the sorting method ELECTRE-Tri

Tasks

a) Consider the two limiting profiles in Table 7.13, evaluated on five criteria defining three ordered categories. In the software the best profile is denoted by r_{K+1} (r_2 in our example) and the worst by r_1. Enter the limiting profiles in the software.

b) Consider the alternatives to be sorted, whose performance is given in the top left matrix of Figure 7.16. Enter their performances in the software.

Table 7.13 Evaluation of the performances of the limiting profiles.

	f_1	f_2	f_3	f_4	f_5
r_2	15	15	15	15	15
r_1	10	10	10	10	10

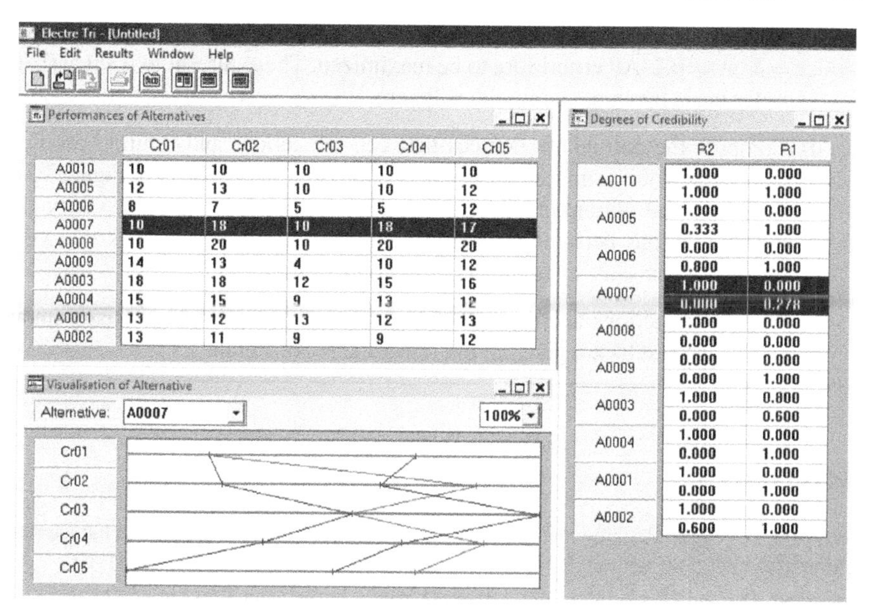

Performances of Alternatives

	Cr01	Cr02	Cr03	Cr04	Cr05
A0010	10	10	10	10	10
A0005	12	13	10	10	12
A0006	8	7	5	5	12
A0007	10	18	10	18	17
A0008	10	20	10	20	20
A0009	14	13	4	10	12
A0003	18	18	12	15	16
A0004	15	15	9	13	12
A0001	13	12	13	12	13
A0002	13	11	9	9	12

Degrees of Credibility

	R2	R1
A0010	1.000	0.000
	1.000	1.000
A0005	1.000	0.000
	0.333	1.000
A0006	0.000	0.000
	0.800	1.000
A0007	1.000	0.000
	0.000	0.278
A0008	1.000	0.000
	0.000	0.000
A0009	0.000	0.000
	0.000	1.000
A0003	1.000	0.800
	0.000	0.600
A0004	1.000	0.000
	0.000	1.000
A0001	1.000	0.000
	0.000	1.000
A0002	1.000	0.000
	0.600	1.000

Figure 7.16 The performances of the alternatives, a visualization of the performances of alternative 7 compared to the profiles and the outranking degrees between the alternatives and the profiles in the Electre-Tri *software. Reproduced by permission of LAMSADE.*

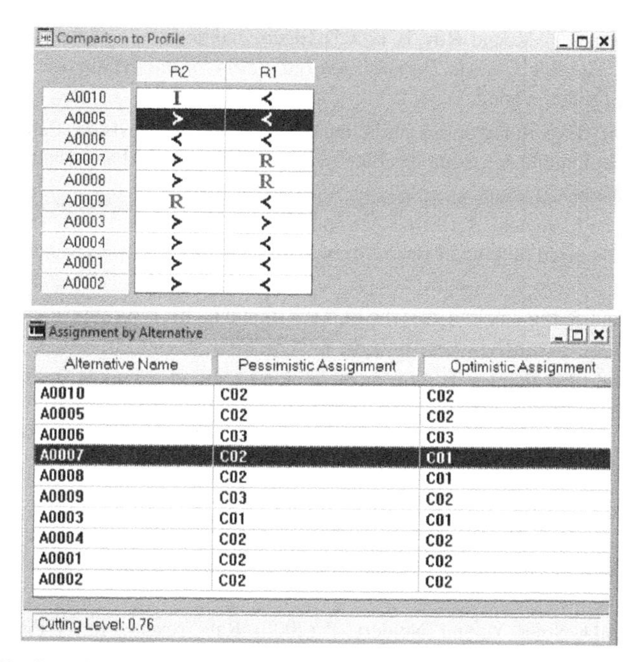

Comparison to Profile

	R2	R1
A0010	I	<
A0005	>	<
A0006	<	<
A0007	>	R
A0008	>	R
A0009	R	<
A0003	>	>
A0004	>	<
A0001	>	<
A0002	>	<

Assignment by Alternative

Alternative Name	Pessimistic Assignment	Optimistic Assignment
A0010	C02	C02
A0005	C02	C02
A0006	C03	C03
A0007	C02	C01
A0008	C02	C01
A0009	C03	C02
A0003	C01	C01
A0004	C02	C02
A0001	C02	C02
A0002	C02	C02

Figure 7.17 Results in the Electre Tri *software: the preference relations between alternatives to be assigned and the profiles as well the pessimistic and optimistic assignments.*

c) The parameters associated to each criterion are identical: $q_j = 1$, $p_j = 2$, $v_j = 4$, $w_j = 0.2$. All criteria are to be maximized. The λ-threshold is set at 0.76. Enter these parameters into the software.

d) Compute the outranking degrees between the actions and limiting profiles. Check that your results are the same as in the right-hand matrix of Figure 7.16. Based on the outranking degrees, and the λ-threshold of 0.76, verify the binary relations between the actions and limiting profiles. Check that your results are the same as in the upper matrix of Figure 7.17.

e) Assign the actions according to the pessimistic and optimistic assignment rules. The results are given in the lower matrix of Figure 7.17.

References

Bouyssou, D. (1992). Ranking methods based on valued preference relations: A characterization of the net flow method. *European Journal of Operational Research*, *60*, 61–67.

Bouyssou, D., and Perny, P. (1992). Ranking methods for valued preference relations: A characterization of a method based on leaving and entering flows. *European Journal of Operational Research*, *61*, 186–194.

Dias, J., Figueira, J., and Roy, B. (2010). Electre Tri-C: a multiple criteria sorting method based on central reference actions. *European Journal of Operational Research*, *204*(3), 565–580.

Figueira, J., Mousseau, V., and Roy, B. (2005). Electre methods. In J. Figueira, S. Greco, and M. Ehrgott, *Multiple Criteria Decision Analysis: State of the Art Surveys* (pp. 133–162). New York: Springer-Verlag.

Gabrel, V. (1990). Experimentations sur la non-indépendance vis-à-vis des tierces alternatives et la non-monotonicité des méthodes Electre III et IV. Université Paris-Dauphine, Mémoire de DEA 103.

Giannoulis, C., and Ishizaka, A. (2010). A web-based decision support system with ELECTRE III for a personalised ranking of British universities. *Decision Support Systems*, *48*(3), 488–497.

Hokkanen, J., and Salminen, P. (1996). ELECTRE III and IV decision aids in an environmental problem. *Multicriteria Decision Analysis*, *4*, 215–226.

Kangas, A., and Pykäläinen, J. (2001). Outranking methods as tools in strategic natural resources planning. *Silva Fennica*, *35*(2), 215–277.

Karagiannidis, A., and Moussiopoulos, N. (1997). Application of ELECTRE III for the integrated management of municipal solid wastes in the Greater Athens area. *European Journal of Operational Research*, *97*(11), 439–449.

Karagiannidis, A., and Papadopoulos, A. (2008). Application of the multi-criteria analysis method ELECTRE III for the optimisation of decentralised energy systems. *Omega*, *5*, 766–776.

Mareschal, B., De Smet, Y., and Nemery, P. (2008). Rank reversal in the PROMETHEE II method: Some new results. *Procedeeings of the IEEE International Conference on Industrial Engineering and Engeneering Management*, Singapore, 959–963, 2008.

Parent, E., and Schnabele, P. (1988). Le choix d'un aménagement aquacole: Exemple d'utilisation de la méthode Electre III et comparaison avec d'autres methodes multicritères d'aide à la décision. Cahier du LAMSADE 47, Université Paris-Dauphine.

Perny, P. (1992). Modélisation, agrégation et exploitation des préférences floues dans une problématique de rangement: bases axiomatiques, procedures et logiciels. Université Paris-Dauphine.

Rogers, M., and Bruen, M. (1998a). Choosing realistic values of indifference, preference and veto thresholds for use with environmental criteria within ELECTRE. *European Journal of Operational Research*, *107*(3), 542–551.

Rogers, M., and Bruen, M. (1998b). A new system for weighting environmental criteria for use within Electre III. *European Journal of Operational Research*, *107*(3), 552–563.

Roy, B. (1968). Classement et choix en présence de points de vue multiples (la méthode ELECTRE). *Revue d'Informatique et de Recherche Opérationnelle*, *2*(8), 57–75.

Roy, B. (1974). Critères multiples et modélisation des préférences: l'apport des relations de surclassment. *Revue d'Economie Politique*, *1*, 1–44.

Roy, B., and Bouyssou, D. (1993). *Aide multicritère à la decision: Méthodes et cas.* Paris: Economica.

Tervoren, T., Figueira, J., Lahdelma, R., and Salminem, R. (2004). An inverse approach for ELECTRE III. Technical report.

Wand, X., and Triantaphyllou, E. (2006). Ranking irregularities when evaluating alternatives by using some Electre methods. *International Journal of Management Science*, *36*, 45–63.

Yu, W. (1992). Aide multicritère à la décision dans le cadère de la problématique du tri: concepts, méthodes et applications. PhD thesis, UER Sciences de l'organisation, Universite de Paris Dauphine.

Part III

GOAL, ASPIRATION OR REFERENCE-LEVEL APPROACH

8

TOPSIS

8.1 Introduction

This chapter explains the theory and practical uses of TOPSIS, which stands for 'Technique of Order Preference Similarity to the Ideal Solution'. In this chapter *Microsoft Excel* is used to illustrate problem solving with TOPSIS, while *DECERNS*, one of the few available software packages supporting TOPSIS, is described in Chapter 11. TOPSIS is not based on a complex algorithm and therefore a 'black box' section is unnecessary.

The companion website provides an illustrative example with *Microsoft Excel*.

8.2 Essentials of TOPSIS

The TOPSIS method requires only a minimal number of inputs from the user and its output is easy to understand. The only subjective parameters are the weights associated with the criteria. Several applications can be found in Behzadian et al. (2012). The fundamental idea of TOPSIS is that the best solution is the one which has the *shortest distance* to the ideal solution and the *furthest distance* from the anti-ideal solution (Hwang and Yoon 1981; Lai et al. 1994; Yoon 1980). For example, in Figure 8.1, where both criteria are to be maximized, alternative A is closer to the ideal solution than B and further from the anti-ideal solution if the criteria weights are equivalent. As a result, TOPSIS presents alternative A as a better solution than alternative B.

The TOPSIS method is illustrated by Case Study 8.1.

Multi-Criteria Decision Analysis: Methods and Software, First Edition. Alessio Ishizaka and Philippe Nemery.

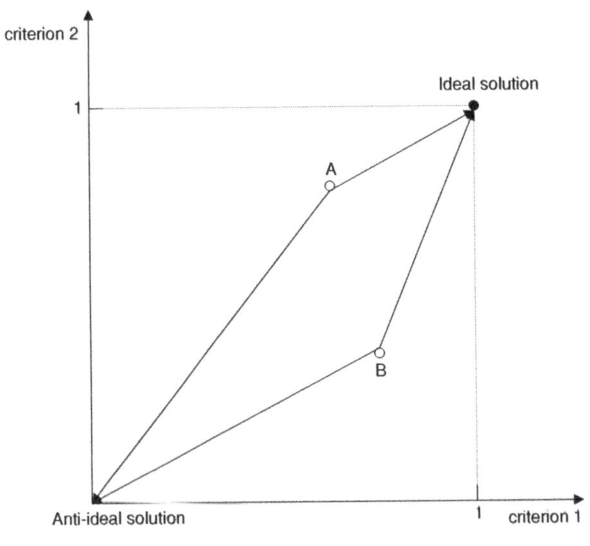

Figure 8.1 TOPSIS method.

Case Study 8.1

A company wants to recruit a new principal assistant for its international market. Four candidates have been shortlisted: Anna, Tom, Jack and Emma. Four criteria have been selected to make the decision. As the post requires intensive contact with various customers, it is necessary for the principal assistant to have strong interpersonal skills, with the ability to interact effectively with diverse client styles within different working environments. The role involves dealing with the international market, and as a result, extensive experience of living abroad would be advantageous. Similar work experience would be beneficial. Each candidate is required to sit a written exam to assess their knowledge of international culture. The performances of each candidate against the four criteria are shown in Table 8.1.

Table 8.1 Weights of the criteria and performances of the alternatives.

	Interpersonal skills (score out of 10)	Living abroad (years)	Written test (score out of 10)	Work experience (years)
Weight	0.1	0.4	0.3	0.2
Anna	7	9	9	8
Tom	8	7	8	7
Jack	9	6	7	12
Emma	6	11	8	6

The TOPSIS method is based on five computation steps. The first step is the gathering of the performances of the alternatives on the different criteria. These performances need to be normalized in the second step. The normalized scores are then weighted and the distances to an ideal and anti-ideal point are calculated. Finally, the closeness is given by the ratio of these distances. These five steps are explained in more detail below.

The performances of n alternatives a with respect to m criteria i are collected in a decision matrix $\mathbf{X} = (x_{ia})$ as in Table 8.1 where $i = 1, \ldots, m$ and $a = 1, \ldots, n$.

1. The performances of the different criteria are normalized in order to be able to compare the measure on different units (e.g. pounds, years,...). Several normalization methods can be found for this purpose:

   (a) The *distributive normalization* requires that the performances are divided by the square root of the sum of each squared element in a column.

$$r_{ia} = \frac{x_{ia}}{\sqrt{\sum_{a=1}^{n} x_{ia}^2}} \quad \text{for } a = 1, \ldots, n \quad \text{and} \quad i = 1, \ldots, m. \tag{8.1}$$

   If we consider the performances of Table 8.1, the distributive normalization method gives the scores shown in Table 8.2.

   (b) The *ideal normalization* requires dividing each performance by the highest value in each column if the criterion has to be maximized. If the criterion has to be minimized, each performance is divided by the lowest score in each column.

$$r_{ai} = \frac{x_{ai}}{u_a^+} \quad \text{for } a = 1, \ldots, n \quad \text{and} \quad i = 1, \ldots, m, \tag{8.2}$$

   where $u_a^+ = \max(x_{ai})$ for all $a = 1, \ldots, n$;

$$r_{ai} = \frac{x_{ai}}{u_a^-} \quad \text{for } a = 1, \ldots, n \quad \text{and} \quad i = 1, \ldots, m, \tag{8.3}$$

   where $u_a^- = \min(x_{ai})$ for all $a = 1, \ldots, n$.

   For the performances of Table 8.1, the ideal normalization method gives the scores shown in Table 8.3.

Table 8.2 Distributive normalization.

	Interpersonal skills	Living abroad	Written test	Work experience
Anna	0.46	0.53	0.56	0.47
Tom	0.53	0.41	0.50	0.41
Jack	0.59	0.35	0.44	0.70
Emma	0.40	0.65	0.50	0.35

Table 8.3 Ideal normalization.

	Interpersonal skills	Living abroad	Written test	Work experience
Anna	0.78	0.82	1.00	0.67
Tom	0.89	0.64	0.89	0.58
Jack	1.00	0.55	0.78	1.00
Emma	0.67	1.00	0.89	0.50

2. Now the weights are taken into account: A weighted normalized decision matrix is constructed by multiplying the normalized scores r_{ai} by their corresponding weights w_i:

$$v_{ai} = w_i \cdot r_{ai}. \tag{8.4}$$

   For the distributive normalized scores, we obtain the weighted scores shown in Table 8.4.

3. The weighted scores will be used to compare each action to an ideal (zenith) and anti-ideal (or nadir or negative ideal) virtual action. There are three different ways of defining these virtual actions.

   (a) By collecting the best and worst performance on each criterion of the normalized decision matrix. For the ideal action we have

$$A^+ = \left(v_1^+, \ldots, v_m^+\right), \tag{8.5}$$

   and for the anti-ideal action

$$A^- = \left(v_1^-, \ldots, v_m^-\right), \tag{8.6}$$

   where $v_i^+ = \max_a(v_{ai})$ if criterion i is to be maximized and $v_i^- = \min_a(v_{ai})$ if criterion i is to be minimized.

   (b) Assuming an absolute ideal and anti-ideal point, which are defined without considering the actions of the decision problem, $A^+ = (1, \ldots, 1)$ and $A^- = (0, \ldots, 0)$

   (c) The ideal and anti-ideal points are defined by the decision maker. These points must be between the ideal and anti-ideal points calculated with the two other methods explained above. This method is not often used as it requires an input from the user, which is often difficult to elicit.

4. Calculate the distance for each action to the ideal action,

$$d_a^+ = \sqrt{\sum_i (v_i^* - v_{ai})^2}, \quad a = 1, \ldots, m, \tag{8.7}$$

Table 8.4 Weighted normalized scores.

	Interpersonal skills	Living abroad	Written test	Work experience
Anna	0.046	0.213	0.168	0.093
Tom	0.053	0.165	0.149	0.082
Jack	0.059	0.142	0.131	0.140
Emma	0.040	0.260	0.149	0.070

and the anti-ideal action,

$$d_a^- = \sqrt{\sum_i (v_i^- - v_{ai})^2}, \quad a = 1, \ldots, m. \tag{8.8}$$

In (8.7) and (8.8), we use a Euclidean distance (L^2), which is the most popular, but another metric could be adopted (e.g. L^1, the Manhattan metric).

5. Calculate the relative closeness coefficient of each action:

$$C_a = \frac{d_a^-}{d_a^+ + d_a^-}. \tag{8.9}$$

The closeness coefficient is always between 0 and 1, where 1 is the preferred action. If an action is closer to the ideal than the anti-ideal, then C_a approaches 1, whereas if an action is closer to the anti-ideal than to the ideal, C_a approaches 0.

Table 8.5 contains the ideal and anti-ideal action with the ideal and distributive normalization. Table 8.6 contains the details of the closeness calculation. Emma is the selected action independently of the normalization adopted.

TOPSIS has been criticized because it sometimes gives *illogical* results. Opricovic and Tzeng (2004) presented a simple example where an extreme action (A1), evaluated on two similarly weighted criteria, is preferred over a superior compromise (A2). In a slightly modified example (Table 8.7), Figure 8.2, produced with DECERNS, shows that compromise A2 will never be ranked first indifferently of the assigned weights.

Table 8.5 Ideal and anti-ideal action.

	Ideal normalization				Distributive normalization			
	Interpersonal skills	Living abroad	Written test	Work experience	Interpersonal skills	Living abroad	Written test	Work experience
A^+	0.100	0.400	0.300	0.200	0.059	0.260	0.168	0.140
A^-	0.067	0.218	0.233	0.100	0.040	0.142	0.131	0.070

Table 8.6 Closeness calculation.

	Ideal normalization				Distributive normalization			
	Anna	Tom	Jack	Emma	Anna	Tom	Jack	Emma
d_a^+	0.101	0.171	0.194	0.111	0.068	0.113	0.124	0.075
d_a^-	0.133	0.057	0.105	0.185	0.084	0.035	0.073	0.120
C_a	0.507	0.250	0.350	0.630	0.550	0.240	0.370	0.610

Table 8.7 Table of scores.

	A1	A2	A3
Criterion 1 (to be maximized)	3000	3750	4500
Criterion 2 (to be minimized)	1	2	5

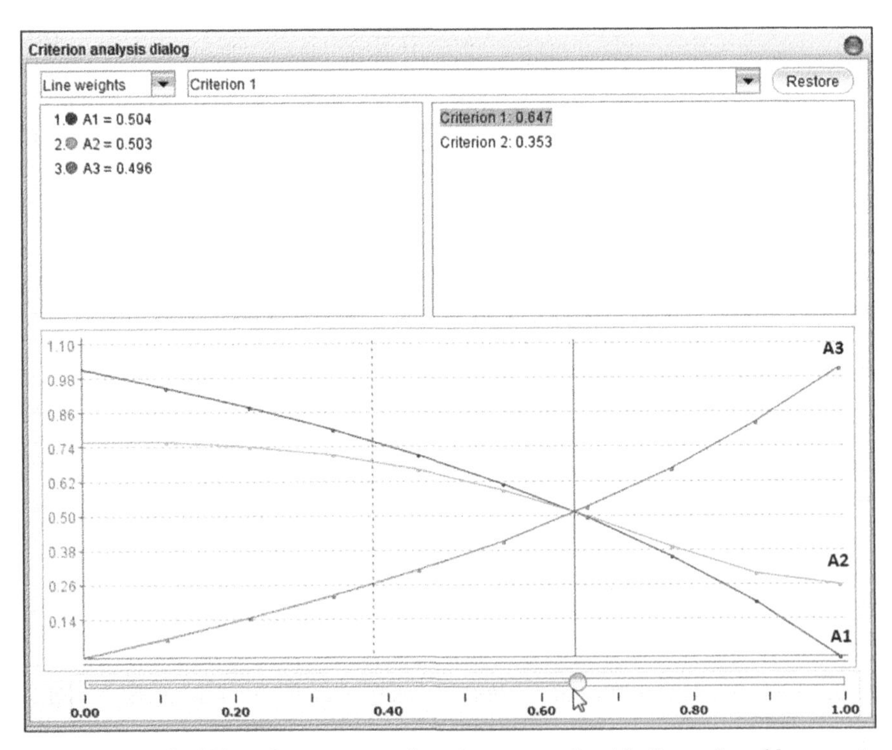

Figure 8.2 TOPSIS prefers never prefers the comprmise A2. Reproduced by permission of Boris Yatsalo.

Finally, we remark that using the Euclidean distance, as in (8.7) and (8.8), which magnifies large distances (Lai et al. 1994), may lead to different results than methods based on Manhattan distances (L_1).

Exercise 8.1

First you will familiarize yourself with the step-by-step TOPSIS calculations in *Microsoft Excel* and then have an opportunity to complete the spreadsheet for the other criteria.

Learning Outcomes

- Understand the calculation of the closeness with a distributive normalization
- Understand the calculation of the closeness with an ideal normalization
- Understand the calculation of the closeness with an ideal and absolute ideal solution

Tasks

Open the file Recruitment.xls. It contains four spreadsheets with variants of TOPSIS.

Answer the following questions:

a) Describe the meaning of each calculation cell and its formula. Compare the four variants.

b) Customize the spreadsheets for a problem of your choice.

References

Behzadian, M., Otaghsara, K., Yazdani, M., and Ignatius, J. (2012). A state-of the-art survey of TOPSIS applications. *Expert Systems with Applications*, *39*(17), 13051–13069.

Hwang, C.-L., and Yoon, K. (1981). *Multiple Attribute Decision Making: Methods and Applications*. New York: Springer-Verlag.

Lai, Y.-J., Liu, T.-Y., and Hwang, C.-L. (1994). TOPSIS for MODM. *European Journal of Operational Research*, *76*(3), 486–500.

Opricovic, S., and Tzeng, G.-H. (2004). Compromise solution by MCDM methods: A comparative analysis of VIKOR and TOPSIS. *European Journal of Operational Research*, *156*(2), 445–455.

Yoon, K. (1980). Systems selection by multiple attribute decision making. Kansas State University.

9

Goal programming

9.1 Introduction

This chapter explains the theory and practical use of the goal programming methods. After a theoretical section, Section 9.3 explains how to solve goal programming problems with *Microsoft Excel*. Three variants of goal programming are presented. This chapter does not include a black box section as the algorithm behind the goal programming is the well-known simplex method (see Ignizio and Cavalier 1994; Schniederjans 1984). Since goal programming is an extension of linear programming to handle multiple conflicting objectives, the reader is advised to first read the Appendix on linear programming to ensure a better understanding.

The companion website provides illustrative examples with *Microsoft Excel*.

9.2 Essential concepts of goal programming

The idea of goal programming is that there is an ideal goal to be achieved while also satisfying hard constraints. This goal is composed of several objectives that may be conflicting. The main difficulty is the modelling of the problem: to find the goal and the soft and hard constraints. Case Study 9.1 will be used to illustrate this point.

Case Study 9.1

A company produces two types of product: A and B. To manufacture, product A requires 5 parts of type I and 3 parts of type II; B requires 4 parts of type I and 2 parts of type II. The profit is £20 for A and £30 for B. The company aims to achieve a weekly profit of £2000. The production time per unit of A is 7 man-hours and per unit of B is 3 man-hours. The company employs seven people in the production department and would like to keep within the 250 available hours of work each

Multi-Criteria Decision Analysis: Methods and Software, First Edition. Alessio Ishizaka and Philippe Nemery.

week. A contract has been signed with a supplier to deliver up to 80 parts of type I and 60 parts of type II. It is possible to order more parts but they would be much more expensive. The manufacturing capacity of the machine for both products combined is limited to a maximum of 80 products per week. The company has a strategic target to produce at least 50 units of each product per week.

The modelling of the problem requires first the identification of the decision variables, goals and constraints.

Identification of the decision variables. The decision variables are independent variables that are changed until the desired quantity is obtained. In Case Study 9.1, they are given by x_1, the number of units of product A manufactured per week, and x_2, the number of units of product B manufactured per week.

Identification of the goals and soft and hard constraints. A *hard constraint* is an inequality that describes a threshold that cannot be exceeded as it represents an unfeasible region of solution. All solutions below the threshold have the same preference. A *goal* with a *soft constraint* has a threshold which is an ideal point, but can be exceeded because solutions over this point are feasible even if they are not attractive. In this case, both deviational variables should be added. Why should unattractive solutions be accepted? As there are often several goals, not all can be achieved simultaneously; therefore, some good solutions may be unattractive to some goals. A *goal* with a *hard constraint* has a threshold which is an ideal point and cannot be exceeded. The nearest solutions to the ideal point are preferred. In this case, only the deviational variable to be minimized should be introduced into the equation.

In Case Study 9.1, there are seven constraints:

- The profit target of the company of £2000 sets the first constraint:

$$30x_1 + 20x_2 \geq 2000.$$

Is this a soft or hard constraint? The company will still survive if the benefit is lower; therefore it is a desirable goal. The inequality can be transformed into an equality with two adjustment variables: n_1 for a negative deviation of the goal and p_1 for a positive deviation:

$$30x_1 + 20x_2 + \boldsymbol{n}_1 - p_1 = 2000.$$

One direction of the deviation will be preferred over the other. In this case, it is preferable to have a greater than a lesser benefit. The deviational variable n_1 is thus to be minimized.

- The second constraint is set by the total working hours in a week:

$$7x_1 + 3x_2 \leq 250.$$

Again, is it a soft or hard constraint? As employees can do overtime or temporary staff can be hired, this inequality can be considered a goal and transformed into:

$$7x_1 + 3x_2 + n_2 - \boldsymbol{p}_2 = 250.$$

In this case, the *extra hours* variable (p_2) is the deviational variable to minimize.

- The next two constraints are set by the capacity of the suppliers: for parts of type I,

$$5x_1 + 4x_2 \leq 80;$$

For parts of type II,

$$3x_1 + 2x_2 \leq 60.$$

As the suppliers are able to deliver additional parts, these are soft constraints and can be transformed into:

$$5x_1 + 4x_2 + n_3 - \boldsymbol{p}_3 = 80,$$
$$3x_1 + 2x_2 + n_4 - \boldsymbol{p}_4 = 60.$$

Additional parts are more expensive and as a result, the positives deviations p_3 and p_4 are to be minimized.

- The following two constraints are set by the strategic aims of the company, which is to produce a minimum of 50 units of product A and B. The constraints can be written as:

$$x_1 \geq 50,$$
$$x_2 \geq 50.$$

It is possible to produce less, therefore the constraints above are transformed into goals:

$$x_1 + \boldsymbol{n}_5 - p_5 = 50,$$
$$x_2 + \boldsymbol{n}_6 - p_6 = 50.$$

- The last constraint is set by the capacity of the machine:

$$x_1 + x_2 \leq 80.$$

This is a hard constraint because the machine cannot produce more than its capacity. Therefore, it cannot be transformed into a goal.

Putting all goals and constraints together, we have the following goal program:

$$\min z = n_1 + p_2 + p_3 + p_4 + n_5 + n_6$$

subject to:

$$\begin{aligned}
&30x_1 + 20x_2 + \boldsymbol{n_1} - p_1 = 2000\\
&7x_1 + 3x_2 + n_2 - \boldsymbol{p_2} = 250\\
&5x_1 + 4x_2 + n_3 - \boldsymbol{p_3} = 80\\
&3x_1 + 2x_2 + n_4 - \boldsymbol{p_4} = 60\\
&x_1 + \boldsymbol{n_5} - p_5 = 50\\
&x_2 + \boldsymbol{n_6} - p_6 = 50\\
&x_1 + x_2 \leq 80\\
&x_1, x_2 \leq 0\\
&n_i, p_i \geq 0, i = 1, \ldots, 6.
\end{aligned}$$

Note that the goal program does not contain any weight. They will be introduced in the section 9.4.1. In the program modelling, supplementary constraints are added to ensure that all decisional and deviational variables are positives.

It is important to transform hard constraints into goals where possible. This allows the examination of a larger solution space, which may contain potentially good or optimal solutions. It is also important to incorporate both deviational variables n and p, when possible, for the same reason.

The goal programming method cannot detect a Pareto front. A Pareto front contains the set of solutions that, when compared to other solutions, are at least as good with respect to all objectives and strictly better with respect to at least one objective. This is problematic if the goals are set too low because a better solution may exist but will not be proposed as it is too far from the goal set by decision maker. For example, if the Pareto optimum is at profit of £10 000, it will not be detected if the decision maker sets a low goal as £2000. Techniques have been developed to detect and restore the Pareto inefficiency (Hannan 1980; Romero 1991).

Exercise 9.1

The following multiple-choice questions allow you to test your knowledge on the basics of goal programming. Only one answer is correct. Answers can be found on the companion website.

1. How many deviational variables are in the objective function of goal programming?
   a) Equivalent to the number of goals
   b) Twice the number of goals
   c) Equivalent to the number of soft and hard constraints
   d) Twice the number of soft and hard constraints
2. What is a soft constraint?
   a) An inequality
   b) A constraint with a threshold indicating unfeasible solutions
   c) A goal
   d) A constraint that is not needed in the modelling of the problem
3. Which statement is incorrect?
   a) All deviational variables should be included in the equation
   b) A hard constraint is not a goal
   c) Several solutions may exist
   d) All goals are always satisfied
4. What type of problems can goal programming solve?
   a) Problems with a discrete solution space
   b) Problems with continuous solution space
   c) Problems with a binary solution space
   d) Sorting problems
5. Goal programming is a generalization of which method?
   a) MACBETH
   b) Linear programming
   c) TOPSIS
   d) DEA

	A–C	D	E	F	G	H	I	J	K	L	M	N	O	P	Q	R	S	T	U	V
1																				
2																				
3	Variables	x1	x2	n1	n2	n3	n4	n5	n6	p1	p2	p3	p4	p5	p6					
4	Deviations to be minimised			1				1	1		1	1	1							Achieved
5	Profit	30	20	1						-1						2000	=	2000		2000
6	Time	7	3		1						-1					250	=	250		425
7	Parts I	5	4			1						-1				80	=	80		350
8	Parts II	2	3				1						-1			60	=	60		175
9	Min parts I	1						1						-1		50	=	50		50
10	Min parts II		1						1						-1	50	=	50		25
11	Machine capacity	1	1													75	≤	80		75
12																				
13	Variables	50	25	0	0	0	0	0	25	0	175	270	115	0	0					
14																				
15	Min unwanted deviations		585																	

Figure 9.1 Microsoft Excel *spreadsheet for the Case Study 9.1 goal program.*

9.3 Software description

Linear goal programs can be solved using linear programming software, for example the *Microsoft Excel Solver* add-in or *LINGO* package. The widespread use and knowledge of *Microsoft Excel* is the main reason to opt for the *Microsoft Excel Solver*. This section describes how to use it.

9.3.1 Microsoft Excel Solver

To explain the *Microsoft Excel Solver*, Case Study 9.1 will be used. The essential first step to solving a goal programming problem is correct modelling. A goal program must be written by hand as in Section 9.2 as the software is unable to do this. When it has been defined, it must be rewritten in a *Microsoft Excel* spreadsheet. There is no strict rule on how to enter it, but it is helpful to have a clear structure.

In Figure 9.1:

- Line 4 determines the deviational variables to be minimized. In this case, they all have the same weight. Section 9.4 describes how to weight them differently.
- Lines 5–10 contain the goals, and line 11 the hard constraint.
- Line 13 contains the decision and deviational variables that need to be found by the solver.
- Cell E15 represents the objective to minimized, which is the sum of n_1, n_5, n_6, p_2, p_3 and p_4. This information is entered in the Solver parameters shown in Figure 9.2.

Exercise 9.2

You will learn the use of the *Microsoft Excel Solver* for solving goal programs.

Learning Outcomes

- ➢ Understand the modelling of a goal programme
- ➢ Understand the parameters in *Microsoft Excel Solver*

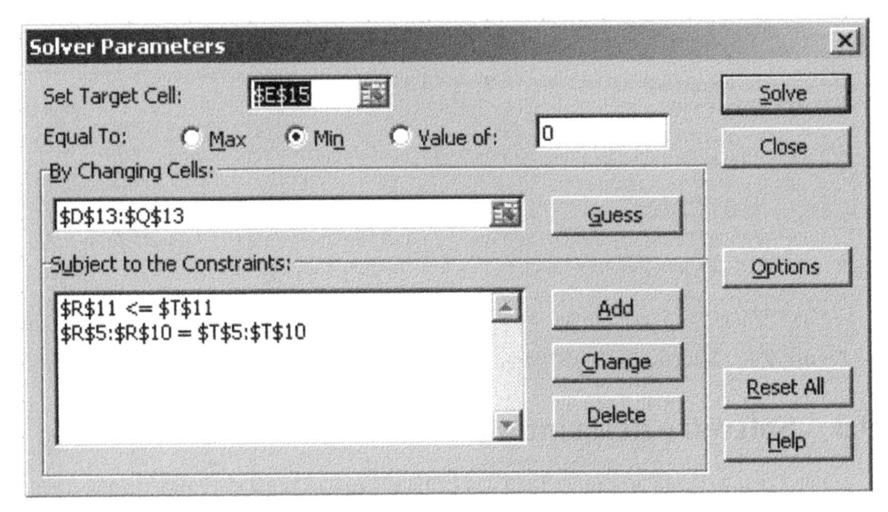

Figure 9.2 Solver parameters.

Tasks

Open the file Production.xls on the un-weighted GP tab. It contains a spreadsheet with the goal programme of the problem of the Case Study 9.1.

Answer the following questions:

a) In the spreadsheet, find the objective of the problem, the decision variables, the goals and hard constraints. (Read the comments in the red square in case of difficulty.)

b) Open the *Solver*. What is entered in the set target cell? What is entered in the box 'by changing cells'? What is entered in the box 'Subject to constraints?'

9.4 Extensions of the goal programming

As not all goals have the same importance, several variants have been conceived to weight goals differently (Jones and Mehrdad 2010).

9.4.1 Weighted goal programming

In weighted goal programming, penalty weights are attached to the unwanted deviational variables. These weights are composed of two parts: a conversion factor in order to assure a commensurability of the deviational variables, and the importance of the penalization given at each deviation.

There are different ways to normalize deviational variables. A practical way is to use percentage normalization, which indicates the percentage away from the goal. For this purpose, the deviational variable is divided by the goal of the objective. The weights are set in the numerator part. For example, in Case Study 9.1, we consider 1% deviation from the target of 250h/week is five times more important than the other goals. The new weighted goal program is:

$$\min z = \frac{1}{2000} n_1 + \frac{5}{250} p_2 + \frac{1}{80} p_3 + \frac{1}{60} p_4 + \frac{1}{50} n_5 + \frac{1}{50} n_6$$

subject to:

$$\begin{aligned}
&30x_1 + 20x_2 + \mathbf{n_1} - p_1 = 2000 \\
&7x_1 + 3x_2 + n_2 - \mathbf{p_2} = 250 \\
&5x_1 + 4x_2 + n_3 - \mathbf{p_3} = 80 \\
&3x_1 + 2x_2 + n_4 - \mathbf{p_4} = 60 \\
&x_1 + \mathbf{n_5} - p_5 = 50 \\
&x_2 + \mathbf{n_6} - p_6 = 50 \\
&x_1 + x_2 \leq 80 \\
&x_1, x_2 \geq 0 \\
&n_i, p_i \geq 0, i = 1, \ldots, 6
\end{aligned}$$

(the deviational variables to be minimized are in bold).

Exercise 9.3

You will learn the use of the *Microsoft Excel Solver* for solving a weighted goal program.

Learning Outcomes

- Understand the modelling of a weighted goal program
- Understand the parameters in *Microsoft Excel Solver*

Tasks

Open the file Production.xls on the weighted GP tab. It contains a spreadsheet with the weighted goal program of the problem in Case Study 9.1.

Answer the following questions:

a) In the spreadsheet, find the objective of the problem, the decision variables, the goals, the hard constraints and weights. (Read the comments in the red square in case of difficulty.)

b) How has the optimal result changed?

9.4.2 Lexicographic goal programming

The lexicographic goal programming is used when the decision maker has a clear preference order for satisfying the goals. For example, in Case Study 9.1, the goal of the total working hours in a week is strictly preferable to the other goals because the staff could strike. The first step is to minimize this goal:

$$\min z = p_2$$

Subject to:

$$\begin{aligned}
&30x_1 + 20x_2 + \boldsymbol{n_1} - p_1 = 2000\\
&7x_1 + 3x_2 + n_2 - \boldsymbol{p_2} = 250\\
&5x_1 + 4x_2 + n_3 - \boldsymbol{p_3} = 80\\
&3x_1 + 2x_2 + n_4 - \boldsymbol{p_4} = 60\\
&x_1 + \boldsymbol{n_5} - p_5 = 50\\
&x_2 + \boldsymbol{n_6} - p_6 = 50\\
&x_1 + x_2 \leq 80\\
&x_1, x_2 \geq 0\\
&n_i, p_i \geq 0, i = 1, \ldots, 6
\end{aligned}$$

(the deviational variables to be minimized are in bold).

An optimum is found at $p_2 = 0$. Then the second priority level is optimized. This goal program has all the goals and constraints of the previous formulation plus the additional constraint $p_2 = 0$:

$$\min z = n_1 + p_3 + p_4 + n_5 + n_6$$

subject to:

$$\begin{aligned}
&30x_1 + 20x_2 + \boldsymbol{n_1} - p_1 = 2000\\
&7x_1 + 3x_2 + n_2 - \boldsymbol{p_2} = 250\\
&5x_1 + 4x_2 + n_3 - \boldsymbol{p_3} = 80\\
&3x_1 + 2x_2 + n_4 - \boldsymbol{p_4} = 60\\
&x_1 + \boldsymbol{n_5} - p_5 = 50\\
&x_2 + \boldsymbol{n_6} - p_6 = 50\\
&x_1 + x_2 \leq 80\\
&p_2 = 0\\
&x_1, x_2 \geq 0\\
&n_i, p_i \geq 0, i = 1, \ldots, 6.
\end{aligned}$$

	B	C	D	E	F	G	H	I	J	K	L	M	N	O	P	Q	R	S	T	U	V
1																					
2																					
3		Variables	x1	x2	n1	n2	n3	n4	n5	n6	p1	p2	p3	p4	p5	p6					
4		Weights			1				1	1		1	1	1							Achieved
5		Profit	30	20	1						-1						2000	=	2000		1428.57
6		Time	7	3		1						-1					250	=	250		250
7		Parts I	5	4			1						-1				80	=	80		271.429
8		Parts II	2	3				1						-1			60	=	60		178.571
9		Min parts I	1						1						-1		50	=	50		14.2857
10		Min parts II		1						1						-1	50	=	60		50
11	Machine capacity		1	1													64.3	≤	80		64.2857
12																					
13		Variables	14.29	50	571.4	0	0	0	35.7	0	0	0	191.4	118.6	0	0					
14																					
15																					
16																					
17			P1									1					0	=	0		
18																					
19																					
20																	P1 Value		0		

Figure 9.3 Lexographic goal program, first-level priority.

	B	C	D	E	F	G	H	I	J	K	L	M	N	O	P	Q	R	S	T	U	V
1																					
2																					
3		Variables	x1	x2	n1	n2	n3	n4	n5	n6	p1	p2	p3	p4	p5	p6					
4		Weights			1				1	1		1	1	1							Achieved
5		Profit	30	20	1						-1						2000	=	2000		1625
6		Time	7	3		1						-1					250	=	250		250
7		Parts I	5	4			1						-1				80	=	80		322.5
8		Parts II	2	3				1						-1			60	=	60		237.5
9		Min parts I	1						1						-1		50	=	50		2.5
10		Min parts II		1						1						-1	50	=	50		77.5
11	Machine capacity		1	1													80	≤	80		80
12																					
13		Variables	2.5	77.5	375	0	0	0	47.5	0	0	0	242.5	177.5	0	27.5					
14																					
15																					
16																					
17			P1									1					0	=	0		
18			P2		1				1	1			1	1			843				
19																					
20																	P1 Value		0		
21																	P2 Value		842.5		

Figure 9.4 Lexographic goal program, second-level priority.

In *Microsoft Excel*, each priority level is solved in a new spreadsheet (Figure 9.3 and Figure 9.4). The model is solved for each priority level at a time by adding a constraint in the spreadsheet (line 18 in Figure 9.4) and in the *Microsoft Excel Solver* parameters box. For a large number of priority levels, the process can be automated by programming a macro in Visual Basics for Applications.

Exercise 9.4

You will learn the use of the *Microsoft Excel Solver* for solving a lexicographic goal program.

Learning Outcomes

- ➢ Understand the modelling of a lexicographic goal program
- ➢ Understand the parameters in *Microsoft Excel Solver*

Tasks

Open the file Production.xls on the LGP1 and LGP12 tabs. It contains two spreadsheets with the lexicographic goal program of the problem in Case Study 9.1.

Answer the following questions:

a) In the two spreadsheets, find the objective of the problem, the decision variables, the goals, the hard constraints and weights. (Read the comments in the red square in case of difficulty.)

b) Compare the *Solver* box of the two spreadsheets. How do the parameters differ?

c) How has the optimal result changed from the traditional goal program?

9.4.3 Chebyshev goal programming

Unweighted, weighted and lexographic goal programming often find extreme points (intersections of goals, constraints and axes), which lead to an unbalanced solution: some goals are achieved and others are far from satisfactory. In order to counter this problem, Chebyshev goal programming has been developed (Flavell 1976). The idea of this method is to introduce additional constraints to the model to ensure a balance between the objectives.

If we assume equal preferential weights and a percentage normalization, the Chebyshev goal programme of Case Study 9.1 is given by:

$$\min z = \lambda$$

subject to:

$$\frac{1}{2000} n_1 \leq \lambda$$
$$\frac{5}{250} p_2 \leq \lambda$$
$$\frac{1}{80} p_3 \leq \lambda$$
$$\frac{1}{60} p_4 \leq \lambda$$
$$\frac{1}{50} n_5 \leq \lambda$$
$$\frac{1}{50} n_6 \leq \lambda$$
$$30x_1 + 20x_2 + n_1 - p_1 = 2000$$
$$7x_1 + 3x_2 + n_2 - p_2 = 250$$
$$5x_1 + 4x_2 + n_3 - p_3 = 80$$

	A	B	C	D	E	F	G	H	I	J	K	L	M	N	O	P	Q	R	S	T	U	V	W
1																							
2																							
3			Variables	x1	x2	n1	n2	n3	n4	n5	n6	p1	p2	p3	p4	p5	p6	Lambda					
4			Weights			0.0005				0.02	0.02		0.004	0.013	0.0167			1					Achieved
5			Profit	30	20	1						-1							2000	=	2000		20
6			Time	7	3		1						-1						250	=	250		3
7			Parts I	5	4			1						-1					80	=	80		4
8			Parts II	2	3				1						-1				60	=	60		3
9			Min parts I	1						1						-1			50	=	50		0
10			Min parts II		1						1						-1		50	=	50		1
11			Machine capacity	1	1														30.1886792	≤	80		1
12																							
13			Lambda 1			1						1						-1	-0.0754717	<=	0		
14			Lambda 2				1						1					-1	-0.6981132	<=	0		
15			Lambda 3					1						1				-1	4.1397E-12	<=	0		
16			Lambda 4						1						1			-1	-0.4402516	<=	0		
17			Lambda 5							1						1		-1	4.6692E-11	<=	0		
18			Lambda 6								1						1	-1	2.8525E-10	<=	0		
19																							
20			Variables	15.1	15.1	1245.3	99.1	0	0	34.9	34.9	0	0	55.85	15.472	0	0	0.7					
21																							
22			minimise lambda	0.7																			

Figure 9.5 Chebyshev goal program.

$$3x_1 + 2x_2 + n_4 - \boldsymbol{p_4} = 60$$
$$x_1 + \boldsymbol{n_5} - p_5 = 50$$
$$x_2 + \boldsymbol{n_6} - p_6 = 50$$
$$x_1 + x_2 \leq 80$$
$$p_2 = 0$$
$$x_1, x_2 \geq 0$$
$$n_i, p_i \geq 0, i = 1, \ldots, 6$$

(the deviational variables to be minimized are in bold). In the *Microsoft Excel* spreadsheet, the additional constraints are also added (lines 13–18 in Figure 9.5).

Exercise 9.5

You will learn the use of the *Microsoft Excel Solver* for solving a Chebyshev goal program.

Learning Outcomes

- Understand the modelling of a Chebyshev goal program
- Understand the parameters in *Microsoft Excel Solver*

Tasks

Open the file Production.xls on the Chebyshev tab. It contains a spreadsheet with the Chebyshev goal program of the problem in Case Study 9.1.

Answer the following questions:

a) In the spreadsheet, find the objective of the problem, the decision variables, the goals, the hard constraints and weights. (Read the comments in the red square in case of difficulty.)

b) Open the *Solver* in the spreadsheets. How have the parameters of the *Solver* boxes been changed compared to Exercise 9.2?

c) How has the optimal result changed from the traditional goal program?

References

Flavell, R. (1976). A new goal programming formulation. *Omega*, *4*(6), 731–732.

Hannan, E. (1980). Non-dominance in goal programming. *INFOR*, *18*(4), 300–309.

Ignizio, J., and Cavalier, T. (1994). *Linear programming*. Englewood Cliffs, NJ: Prentice Hall.

Jones, D., and Mehrdad, T. (2010). *Practical Goal Programming*. New York: Springer-Verlag.

Romero, C. (1991). *Handbook of Critical Issues in Goal Programming*. Oxford: Pergamon Press.

Schniederjans, M. (1984). *Linear Goal Programming*. Princeton, NJ: Petrocelli Books.

10

Data Envelopment Analysis

Jean-Marc Huguenin[1]

10.1 Introduction

This chapter introduces a performance measurement technique called DEA, which stands for 'Data Envelopment Analysis'. Firm efficiency is defined as the ratio of the sum of its weighted outputs to the sum of its weighted inputs (Thanassoulis et al. 2008, p. 264). As Giannoulis and Ishizaka (2010) point out, the analogy with other multi-criteria methods is striking: firms can be considered as alternatives, outputs as criteria to be maximized and inputs as criteria to be minimized. What distinguishes DEA is that the weights assigned to outputs and inputs are not allocated by users. Moreover, it does not rely on a common set of weights for all firms. Instead, a different set of weights is calculated by a linear optimization procedure in order to show each firm in its best possible light.

DEA helps decision makers in the following ways:

- By calculating an efficiency score, it indicates if a firm is completely efficient or has capacity for improvement.

[1]Jean-Marc Huguenin is a Senior Lecturer and a Project Manager at the Swiss Graduate School of Public Administration at the University of Lausanne, Switzerland. He also heads the Independent Economists, an economic intelligence unit founded in 1999. He was previously an economist at the Swiss National Bank, an economist for the Swiss in-service training centre for secondary school teachers and an economics professor at the College of Bienne, Switzerland. His areas of research and expertise are the evaluation of productivity and efficiency of organizations and the economics and management of education systems.

Multi-Criteria Decision Analysis: Methods and Software, First Edition. Alessio Ishizaka and Philippe Nemery.

- By setting target values for input and output, it calculates by how much input must be decreased or output increased in order to become efficient.
- By identifying the nature of returns to scale, it indicates if a firm has to decrease or increase its scale (or size) in order to minimize the average total cost.
- By identifying a set of benchmarks, it specifies which other firms' processes need to be analyzed in order to improve its own practices.

After this introduction, Section 10.2 presents the essentials of DEA, alongside a case study to give an intuitive understanding of its application. Section 10.3 introduces *Win4DEAP*, a software package that conducts efficiency analysis based on DEA methodology. Section 10.4 is designed for more demanding readers interested in the methodological background of DEA. Finally, four advanced DEA topics are presented in Section 10.5: adjustment to the environment, preferences, sensitivity analysis and time series data.

The companion website provides a case study using *Win4DEAP*, solutions to exercises, and an illustrative example using *Microsoft Excel Solver.*

10.2 Essential concepts of DEA

10.2.1 An efficiency measurement method

DEA is used to measure the performance of firms or entities (called *decision-making units*, DMUs) which convert multiple inputs into multiple outputs. It is suitable for the use of both private sector firms and public sector organizations (and even for entities such as regions or countries). DEA was formulated in Charnes et al. (1978, 1981) in order to evaluate a US federal government programme in the education system called 'Program Follow Through'. The use of DEA then spread to other public organizations (hospitals, elderly care facilities, social service units, unemployment offices, police forces, army units, prisons, waste management services, power plants, public transportation companies, forestry companies, libraries, museums, theatres, etc.) and to the private sector (banks, insurance companies, retail stores, etc.).

Each DMU's efficiency score is calculated relative to an efficiency frontier. DMUs located on the efficiency frontier have an efficiency score of 1 (or 100%). DMUs operating beneath the frontier have an efficiency score less than 1 and so have the capacity to improve future performance. Note that no DMU can be located above the efficiency frontier because they cannot have an efficiency score greater than 1. DMUs located on the frontier serve as benchmarks – or peers – for inefficient DMUs. These benchmarks (i.e. real DMUs with real data) are associated with best practices. DEA is therefore a powerful benchmarking technique.

10.2.2 A DEA case study

To better understand the mechanics behind DEA, this section presents a simple practical case study. It includes only one input and one output, although DEA can handle multiple inputs and multiple outputs.

Case Study 10.1

Five Register Offices (A to E) produce one output (total number of documents, such as marriage or birth certificates) with one input (number of full-time equivalent public servants).[2] The data are listed in Table 10.1. For example, first row of the table shows that two public servants work in Register Office A. They produce one document (during a certain period of time).

Table 10.1 Five Register Offices produce documents with public servants.

	Input	Output
Register Office	Public servants (x)	Documents (y)
A	2	1
B	3	4
C	5	5
D	4	3
E	6	7

10.2.2.1 Two basic DEA models

Two basic models are used in DEA, leading to the identification of two different frontiers:

- The first model assumes constant returns to scale (CRS model). This is appropriate when all DMUs are operating at an optimal scale. However, note that this is quite an ambitious assumption. To operate at an optimal scale, DMUs should evolve in a perfectly competitive environment, which is seldom the case. The CRS model calculates an efficiency score called constant returns to scale technical efficiency (CRSTE).
- The second model assumes variable returns to scale (VRS model). This is appropriate when DMUs are not operating at an optimal scale. This is usually

[2] In order to represent this example in a two-dimensional graph, we consider a total of two outputs and inputs (one output, one input; no variable representing the quality of the variables).

the case when DMUs face imperfect competition, government regulations, etc. The VRS model calculates an efficiency score called variable returns to scale technical efficiency (VRSTE).

Comparison between the two models reveals the source of inefficiency. CRSTE corresponds to the global measure of a DMU performance. It is composed of a 'pure' technical efficiency measure (captured by the VRSTE score) and a scale efficiency (SE) measure. Section 10.2.2.6) demonstrates how these three notions (CRSTE, VRSTE and SE) relate to each other.

10.2.2.2 Input or output orientation

A DEA model can be input- or output-oriented:

- In an input orientation, DEA minimizes input for a given level of output; in other words, it indicates how much a DMU can decrease its input for a given level of output.
- In an output orientation, DEA maximizes output for a given level of input; in other words, it indicates how much a DMU can increase its output for a given level of input.

The efficiency frontier will be different in a CRS or a VRS model (see Section 10.2.2.6). However, within each model, the frontier will not be affected by an input or an output orientation. DMUs located on the frontier in an input orientation will also be on the frontier in an output orientation.

In a CRS model, technical efficiency scores have the same values in an input or an output orientation. But these values will be different according to the model's orientation when VRS is assumed. However, Coelli and Perelman (1996, 1999) note that, in many instances, the choice of orientation has only a minor influence upon the technical efficiency scores calculated.

The model's orientation should be chosen according to which variables (inputs or outputs) the decision maker has most control over. For example, a school principal will probably have more control over his teaching staff (input) than over the number of pupils (output). An input orientation will be more appropriate in this case.

In the public sector, but sometimes also in the private sector, a given level of input can be assigned and secured to a DMU. In this case, the decision maker may want to maximize the output (and therefore choose an output orientation). Alternatively, if the decision maker's task is to produce a given level of output (e.g. a quota) with the minimum input, he will opt for an input orientation.

If the decision maker is not facing any constraints and has control of both input and output, the model's orientation will depend on his objectives. Does he need to cut costs (input orientation) or does he want to maximize production (output orientation)?

10.2.2.3 CRS efficient frontier

Figure 10.1 shows the efficient frontier for Case Study 10.1 assuming constant returns to scale (CRS efficient frontier). The CRS efficient frontier starts at the origin and runs through Register Office B. Register Office B happens to be the observation with the steepest slope, or the highest productivity ratio, among all Register Office (4/3 = 1.33, meaning that one public servant produces 1.33 document). Register Office B is on the frontier; it is 100% efficient. Register Offices A, C, D and E are below the frontier. Their respective efficiency scores are less than 100%. DEA assumes that the production possibility set is bounded by this frontier. This actually implies that DEA calculates relative and not absolute efficiency scores. Although DMUs on the efficient frontier are assigned a 100% efficiency score, it is likely that they could further improve their productivity.

Figure 10.1 also illustrates how DEA measures efficiency scores. The example of Register Office A is described below:

- In an input orientation, A's efficiency score is equal to the distance $SA_{CRS\text{-}I}$ divided by the distance SA. $A_{CRS\text{-}I}$ is the projection of point A on the efficient frontier (assuming constant returns to scale and an input orientation). Note that one can easily calculate efficiency scores using a ruler and measuring the distances on the graph. A's score is 37.5%. This means that Register Office A could reduce the number of public servants employed (input) by 62.5% (100 – 37.5) and still produce the same number of documents (1).
- In an output orientation, A's efficiency score is equal to the distance TA divided by the distance $TA_{CRS\text{-}O}$. $A_{CRS\text{-}O}$ is the projection of point A on the efficient

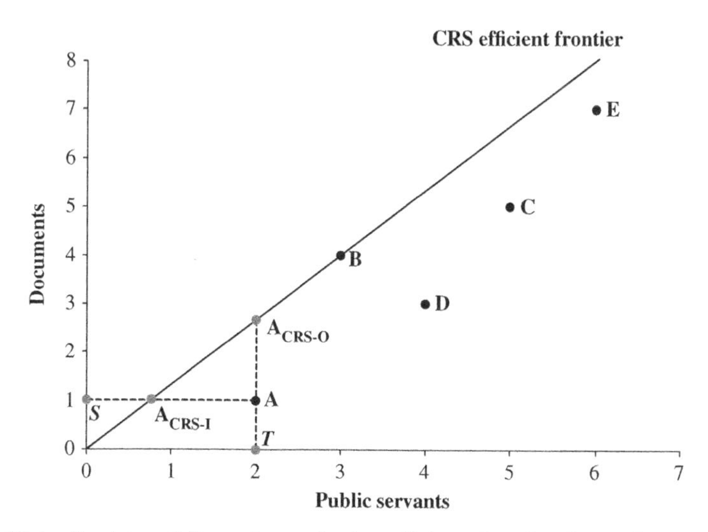

Figure 10.1 Register Offices beneath the efficient frontier have the capacity to improve performance.

frontier (assuming constant returns to scale –CRS– and an output orientation). A's score is 37.5%, as in an input orientation.[3] This means that Register Office A could increase its production of documents (output) by 62.5% (100 – 37.5) whilst holding the number of public servants constant at 2.

10.2.2.4 VRS efficient frontier

Figure 10.2 shows the efficient frontier for Case Study 10.1 assuming variable returns to scale technology (VRS efficient frontier). The VRS efficient frontier is formed by covering all the observations. Register Offices A, B and E are on the frontier. They are 100% efficient. Register Offices C and D are below the frontier. Their respective efficiency scores are less than 100%. DEA assumes that the production possibility set is bounded by this frontier. Again, this implies that DEA calculates relative and not absolute efficiency scores. Although DMUs on the efficient frontier are assigned a 100% efficiency score, it is likely that they could further improve their productivity.

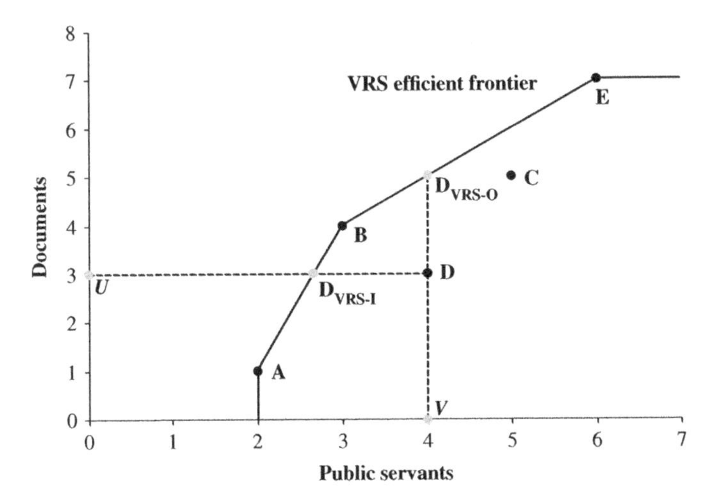

Figure 10.2 For the same level of input, Register Office D could improve its output up to the projected values of point $D_{VRS\text{-}O}$ (i.e. from 3 to 5 documents).

Figure 10.2 also illustrates how DEA measures efficiency scores. The example of Register Office D, one of the two inefficient offices, is described below:

- In an input orientation, D's efficiency score is equal to the distance $UD_{VRS\text{-}I}$ divided by the distance UD. $D_{VRS\text{-}I}$ is the projection of point D on the efficient frontier (assuming variable returns to scale and an input orientation). Note that one can easily calculate efficiency scores using a ruler and measuring the

[3] Note that the efficiency scores in a CRS model are always the same for an input or an output orientation.

distances on the graph. D's score is 66.7%. This means that Register Office D could reduce the number of public servants employed (input) by 33.3% (100 – 66.7) and still produce the same number of documents (3).

- In an output orientation, D's efficiency score is equal to the distance VD divided by the distance $VD_{VRS\text{-}O}$. $D_{VRS\text{-}O}$ is the projection of point D on the efficient frontier (assuming variable returns to scale and an output orientation). D's score is 60%.[4] This means that Register Office D could increase its production of documents (output) by 40% (100 – 60) whilst holding the number of public servants constant at 4.

10.2.2.5 Interpreting efficiency scores according to the DEA model's output or input orientation

Register Office C has an efficiency score of 75% in the CRS model. It will get the same efficiency score in an output or in an input-oriented model under the constant returns to scale assumption. However:

- In the input-oriented model, the capacity to improve input (i.e. a reduction) by 25% (100 – 75) is calculated using the original input value of 5 public servants. The DEA model calculates a projected value of 3.75. The 25% improvement is then calculated according to the *original* value: ((5 – 3.75) / 5) 100 = 25. From a practical point of view, the capacity to improve input by 25% means that the Register Office should reduce all of its inputs by 25% in order to become efficient.
- In the output-oriented model, the capacity to improve output (i.e. an augmentation) by 25% (100 – 75) is calculated using the projected output value. Register Office C has an original output value of 5 documents. The DEA model calculates a projected value of 6.67 documents. The 25% improvement is then calculated according to the *projected* value: ((6.67 – 5) / 6.67) 100 = 25. From a practical point of view, the capacity to improve output by 25% means that the Register Office should augment all of its outputs by 25% in order to become efficient.

10.2.2.6 CRS, VRS and scale efficiency

Figure 10.3 shows both the CRS and VRS efficient frontiers on the same graph. Register Office B is CRS and VRS efficient, as it is located on both frontiers. Register Offices A and E are efficient under the VRS assumption but inefficient under the CRS assumption. Finally, Register Office D and C are both CRS and VRS inefficient; they are located neither on the CRS nor on the VRS frontiers.

The gap observed between the CRS and the VRS frontiers is due to a problem of scale. For example, Register Office A is VRS efficient. To become CRS efficient, A

[4] Note that the efficiency scores in a VRS model are different for an input or an output orientation.

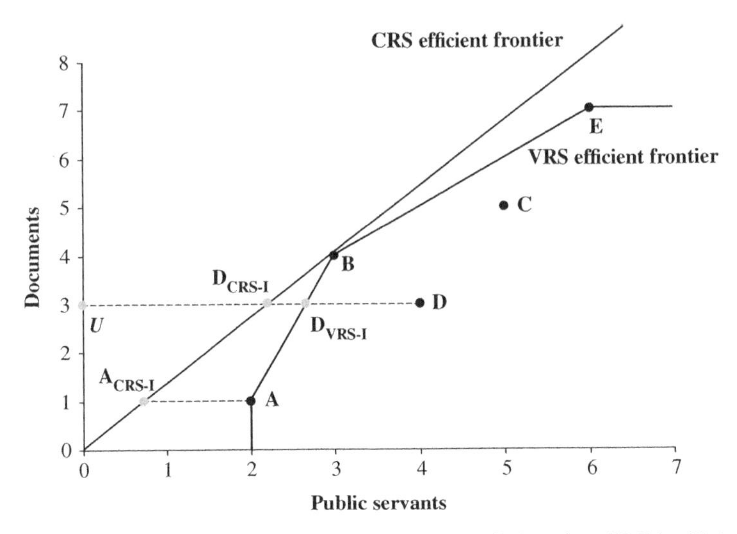

Figure 10.3 Register Offices A and E are VRS efficient but CRS inefficient.

should modify its scale (or size). Only by operating at point $A_{CRS\text{-}I}$ would A be as productive as B, which is the only CRS efficient Register Office (and thus the most productive one).

Some Register Office (D and C) are not even located on the VRS frontier. These Register Office not only have a scale problem but are also poorly managed. For example, D should move to point $D_{VRS\text{-}I}$ located on the VRS frontier in order to become VRS efficient (i.e. to eliminate the inefficiency attributable to poor management). Furthermore, D should move from point $D_{VRS\text{-}I}$ to point $D_{CRS\text{-}I}$ located on the CRS frontier in order to become CRS efficient (i.e. to eliminate the inefficiency attributable to a problem of scale).

As a result, the CRS efficiency (also called 'total' efficiency) can be decomposed into two components: the VRS efficiency (also called 'pure' efficiency) and the scale efficiency. The following ratios represent these three types of efficiency for Register Office D (input orientation):

Technical efficiency of D under CRS	Technical efficiency of D under VRS	Scale efficiency of D
$TE_{CRS} = \frac{UD_{CRS-I}}{UD} = 56.3\%;$	$TE_{VRS} = \frac{UD_{VRS-I}}{UD} = 66.7\%;$	$SE = \frac{UD_{CRS-I}}{UD_{VRS-I}} = 84.4\%;$

Knowing TE under CRS and TE under VRS, the scale efficiency is easily calculated. As $TE_{CRS} = TE_{VRS} \times SE_k$, the scale efficiency is obtained through the division of TE under CRS by TE under VRS: $SE = \frac{TE_{CRS}}{TE_{VRS}}$.

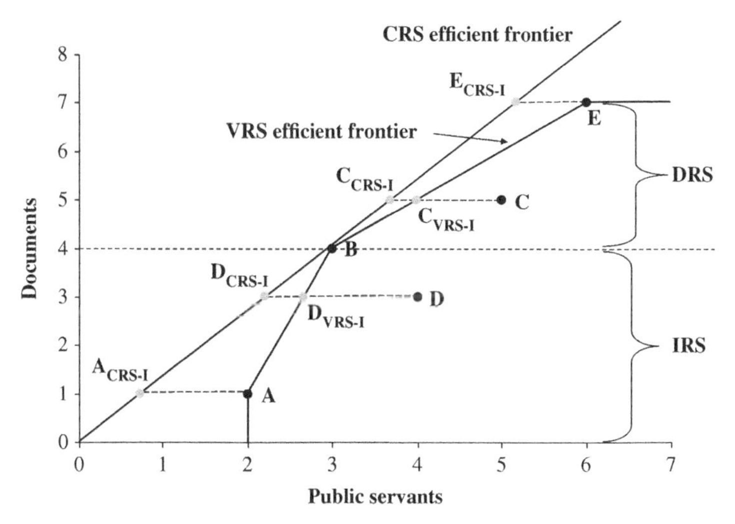

Figure 10.4 Register Offices A and D face increasing returns to scale (IRS, economies of scale), while C and E face decreasing returns to scale (DRS, dis-economies of scale).

10.2.2.7 Nature of returns to scale

The nature of returns to scale of Register Offices not located on the CRS frontier (in other words, scale inefficient) has to be identified. Figure 10.4 shows the CRS efficient points A_{CRS-I} and E_{CRS-I} of Register Offices A and E (which are CRS inefficient but VRS efficient). It also shows the CRS efficient points D_{CRS-I} and C_{CRS-I} and the VRS efficient points D_{VRS-I} and C_{VRS-I} of Register Offices D and C (which are CRS and VRS inefficient).

To identify the nature of returns to scale, one has to focus on the slope of the VRS efficient points A, D_{VRS-I}, B, C_{VRS-I} and E (or productivity). Three situations can occur:

- A Register Office is located both on the CRS and the VRS efficient frontiers (such as point B). Register Office B has the highest productivity of all VRS efficient points ($4/3 = 1.33$). It is facing constant returns to scale. Such a firm achieves its optimal size (or efficient scale).[5] It is operating at a point where the scale (or size) has no impact on productivity. This situation occurs when the average input consumption is minimized and does not vary with output. In

[5] In the economic context, a firm operates at the optimal size (or efficient scale) when it minimizes its average cost. In the context of DEA, we can measure efficiency in physical or monetary terms. Because cost and price information is not always available or appropriate, the use of technical efficiency is often preferred. As this latter measure is based on physical terms, we prefer to talk about average input consumption rather than average cost.

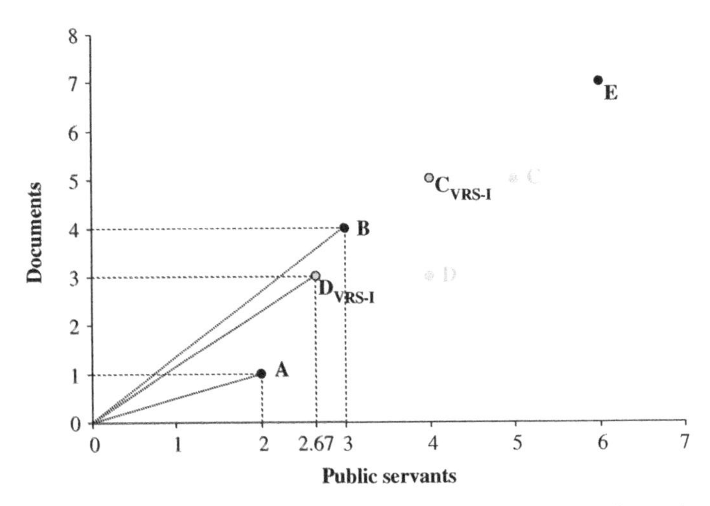

Figure 10.5 The ratio of productivity is increasing with the scale.

a situation of constant returns to scale, an increase in output of 1% requires a proportionate (i.e. 1%) increase in input.

- A Register Office (or the projected point of a Register Office) is located at a point where the scale (or size) has a positive impact on productivity. Points A and $D_{VRS\text{-}I}$ are in such a position (see Figure 10.5). The productivity of A ($1/2 = 0.5$) is less than the productivity of $D_{VRS\text{-}I}$ ($3/2.67 = 1.125$). The ratio of productivity is increasing with the scale. This situation occurs until point B, which has a productivity of 1.33. Register Offices A and D are therefore facing increasing returns to scale (or economies of scale). In this situation, the average inputs consumption declines whilst output rises. Register Offices A and D have not yet reached their optimal size (or efficient scale). To improve their scale efficiency, they have to expand their output. In a situation of economies of scale, a variation in output of 1% results in a variation in input of less than 1%. Hence, an increase in output results in a reduction of the average input consumption.

- A Register Office (or the projected point of a Register Office) is located at a point where the scale (or the size) has a negative impact on productivity. Points $C_{VRS\text{-}I}$ and E are in such a position (see Figure 10.6). The productivity of $C_{VRS\text{-}I}$ ($5/4 = 1.25$) is superior to the productivity of E ($7/6 = 1.17$). The ratio of productivity is decreasing with the scale. This situation occurs from point B, which has a productivity of 1.33. Register Offices C and E are therefore facing decreasing returns to scale (or diseconomies of scale). In this situation, the average input consumption rises whilst output rises. Register Offices C and E have exceeded their optimal size (or efficient scale). To improve their scale

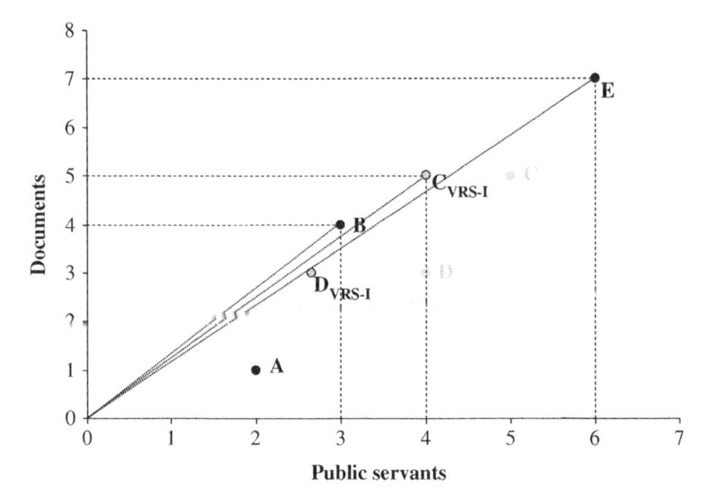

Figure 10.6 The ratio of productivity is decreasing with the scale.

efficiency, they have to reduce their output. In a situation of diseconomies of scale, a variation in output of 1% results in a variation in input of more than 1%. Hence, a decrease in output results in a reduction of the average input consumption.

The specific cases of the five Register Offices are described below (see Figure 10.4):

- A is located on the VRS frontier but not on the CRS frontier. Its inefficiency is due to an inappropriate scale. A is facing increasing returns to scale. A variation in output of 1% results in a variation in input of less than 1%.
- D is located neither on the CRS nor on the VRS frontier. Its inefficiency is due to an inappropriate scale and to poor management. D is facing increasing returns to scale. A variation in output of 1% results in a variation in input of less than 1%.
- B is located both on the CRS and on the VRS frontier. It has no inefficiency at all. B is facing constant returns to scale. A variation in output of 1% results in a variation in input of 1%.
- C is located neither on the CRS nor on the VRS frontier. Its inefficiency is due to an inappropriate scale and to poor management. C is facing decreasing returns to scale. A variation in output of 1% results in a variation in input of more than 1%.
- E is located on the VRS frontier (but not on the CRS frontier). Its inefficiency is due to an inappropriate scale. E is evolving in a situation of decreasing returns

to scale. A variation in output of 1% results in a variation in input of more than 1%.

10.2.2.8 Peers (or benchmarks)

DEA identifies, for each inefficient DMU, the closest efficient DMUs located on the frontier. These efficient DMUs are called peers or benchmarks. If inefficient DMUs want to improve their performance, they have to look at the best practices developed by their respective peers.

Under the CRS assumption, Register Office B is the only DMU located on the efficient frontier. Hence it is identified as the peer for all other inefficient Register Offices.

Figure 10.7 illustrates the peers under the VRS assumption. Three Register Offices (A, B and E) are located on the efficient frontier. Two (C and D) are inefficient. C has two assigned peers, B and E, because $C_{VRS\text{-}I}$, the projected point of C on the VRS frontier, lies between these two benchmarks. D has also two assigned peers, A and B, because $D_{VRS\text{-}I}$, the projected point of D on the VRS frontier, lies between these two benchmarks.

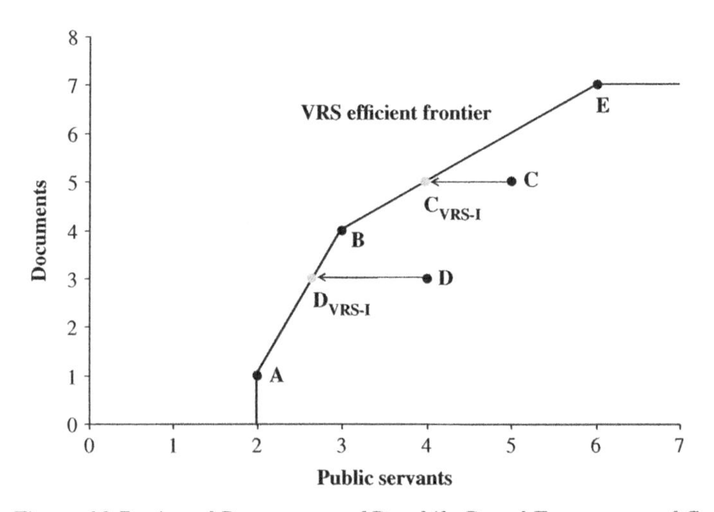

Figure 10.7 A and B are peers of D, while B and E are peers of C.

10.2.2.9 Slacks

Particular positions located on the frontier are inefficient. Assume there is an additional Register Office in our sample, F. It produces 0.5 documents with two public servants. Figure 10.8 illustrates the efficient frontier under VRS. F is not located on the frontier. In order to become efficient, it has first to move to point $F_{VRS\text{-}I \text{ without slacks}}$.

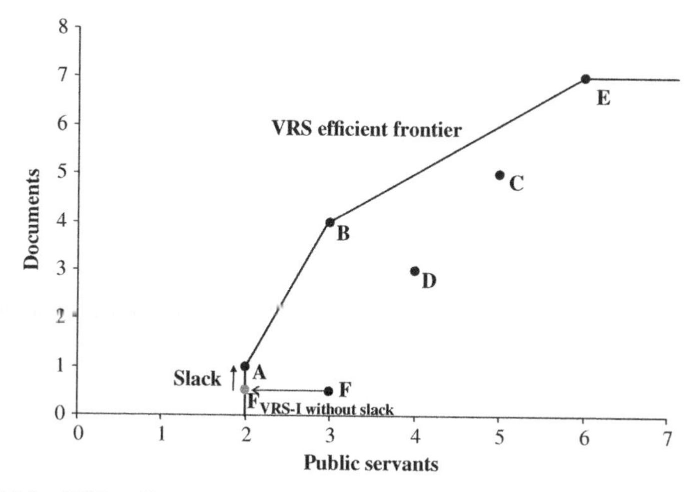

Figure 10.8 DEA adjusts the projected values of inefficient DMUs to take slacks into account.

At this location, F should have an efficiency score of 100%, as it is located on the frontier. But A, next to it on the frontier, is also 100% efficient. The difference between F and A is striking. With the same number of inputs (2 public servants), F produces 0.5 document and A produces 1 document (i.e. 0.5 more than F). Therefore point $F_{\text{VRS-I without slacks}}$ cannot be considered as 100% efficient, because it produces less output with the same amount of inputs than another Register Office (A). To get a 100% efficiency score, point $F_{\text{VRS-I without slacks}}$ has to move further up to point A. This additional improvement needed for a DMU to become efficient is called a *slack*.

Indeed, every point located on the sections of the frontier which run parallel to either the *x* or the *y* axes has to be adjusted for slacks. DEA is designed to take slacks into account.

10.2.3 Multiple outputs and inputs

DEA allows multiple outputs and multiple inputs to be taken into account. For example, a shirt company uses machines, workers and cotton (three inputs) in order to produce T-shirts, trousers and underwear (three outputs). DEA can account for all of these variables and even more. As a result, DEA goes far beyond the calculation of single productivity ratios such as the number of T-shirts produced per worker (one output divided by one input).

However, the total number of outputs and inputs being considered is not limitless from a practical point of view. It depends on the number of DMUs in the data set. If the number of DMUs is smaller than, roughly speaking, three times the sum of the total number of inputs and outputs, it is highly probable that several DMUs, if not all, will obtain a 100% efficiency score.[6] For example, a data set containing 21 shirt companies should be evaluated on a maximum number of seven outputs and inputs (21 divided by 3). As Cooper et al. (2006, p. 106) point out, 'if the number of DMUs (n) is less than the combined number of inputs and outputs ($m + s$), a large portion of the DMUs will be identified as efficient and efficiency discrimination among DMU is questionable due to an inadequate number of degrees of freedom. (...) Hence, it is desirable that n exceeds $m + s$ by several times. A rough rule of thumb in the envelopment model is to chose n (...) equal to or greater than $\max\{m \times s, 3 \times (m + s)\}$.'

DEA measures DMU efficiency based on multiple outputs and multiple inputs. If shirt company A produces a lot of T-shirts but only a few trousers and underwear, DEA will automatically attribute a high weighting to the T-shirts variable in order to emphasize this strength. As a result, DEA automatically optimizes the weighting of each variable in order to present each DMU in the best possible light.

Unfortunately, DEA does not work with negative or zero values for inputs and outputs. However, zero values can be substituted with very low values such as 0.01.

A distinction has to be made between variables which are under the control of management (discretionary variables) and variables which are not (non-discretionary or environmental variables). Ideally, a DEA model will exclusively include discretionary variables, although some DEA models can also accommodate non-discretionary ones. In a second step, efficiency scores can be adjusted to account for environmental variables (i.e. such variables influence the efficiency of a DMU but are not a traditional input and are not under the control of the manager).

Moreover, variables should reflect both quantitative and qualitative characteristics of DMUs' resources and services. Although it may not be easy to identify and to convert qualitative characteristics into numbers, it is desirable to include such variables in the model in order to appropriately benchmark DMUs.

10.2.4 Types of efficiency

The notion of efficiency refers to an optimal situation; the maximum output for a given level of input or the minimum input for a given level of output. Subject to data availability, several types of efficiency can be measured:

- Technical efficiency, in which both outputs and inputs are measured in physical terms.[7]

---

[6] The higher the number of inputs and outputs that are taken into consideration for a given number of firms, the more probable it is that each firm will be the best producer of at least one of the outputs. Therefore, all firms could obtain a 100% efficiency score.

[7] This chapter focuses on the measurement of technical efficiency for two main reasons. First, firms in the public sector are often not responsible for the age pyramid of their employees; therefore taking into account the wages of the employees (which often increase with seniority) would unfairly alter the efficiency of a firm with a greater proportion of senior employees. Second, firms in the public sector do not often produce commercial goods or services with a set price.

- Cost efficiency: identical to technical efficiency, except that cost (or price) information about inputs is added to the model.
- Revenue efficiency: identical to technical efficiency, except that price information about outputs is added to the model.
- Profit efficiency: identical to technical efficiency, except that cost information about inputs and price information about outputs are added to the model.

Technical efficiency is a global measure of DMU performance. However, it does not indicate the source of inefficiency. This source could be twofold:

- the DMU could be poorly managed and operated;
- it could be penalized for not operating at the right scale.

Technical efficiency can be decomposed into a 'pure' technical efficiency measure and a scale efficiency measure to reflect these two sources of inefficiency.[8]

10.2.5 Managerial implications

DEA is a benchmarking technique. The efficiency scores provide information about a DMU's capacity to improve output or input. In this sense, DEA offers strong support to decision making. Conducting an efficiency analysis and interpreting the results often raises practical questions. The following list of frequently asked questions offers some advice.

- *Is it advisable to involve the managers of the DMUs to be benchmarked in the efficiency analysis from the beginning of the process?*
  Yes, it is, and for two main reasons. First, managers know the processes of their DMUs and the data available. Therefore they are the right persons to pertinently identify which inputs and outputs to integrate into the analysis. Second, managers involved from the beginning of the process are more likely to accept the results of the analysis (rather than to reject them).
- *How should one respond to managers who claim that their DMUs are different from others, and therefore cannot be compared to them?*
  Sometimes inefficiencies can be explained by indisputable environmental variables. But sometimes not. Managers often justify the low efficiency scores of their DMUs by arguing that their situations are different compared to the situations of the other DMUs. They claim to be a 'special case' (and therefore

[8] The firm's management team will definitely be held responsible for the 'pure' technical efficiency score. In a situation where it does not have the discretionary power to modify the DMU's size, it will likely not be accountable for the scale efficiency score. However, especially in the private sector, one has the choice of the scale at which it operates: the management team can easily downsize the DMU and, with some effort, upsize it also.

it is acceptable to be inefficient). Actually, the majority of DMUs could possibly claim to be different as most possess something unique. However, it is likely that the difference of one DMU will be compensated by the difference of another. More generally, it is up to the managers to prove that they really face a hostile environment. If they cannot prove it, management measures should be taken to improve efficiency.

- *Assume that a DMU obtains an efficiency score of 86.3%. Should this number be strictly applied?*
  Not really, it should be interpreted more as an order of magnitude. This order of magnitude informs managers that they have to increase their outputs or to decrease their inputs in order to become more efficient. But one should not focus too strictly on the capacity for 13.7% improvement. Such a number could be interpreted by practitioners as too 'accurate'. Therefore it is better to consider efficiency scores as more of an objective basis to hold an open discussion about the way to improve DMU efficiency rather than a number to be strictly applied.

- *A DMU faces increasing returns to scale. It has economies of scale. What does that concretely mean from a managerial point of view?*
  Such a DMU has not yet reached its optimal size. In order to reduce its average total cost (or its average input consumption), it has to increase its size. In practice, this could be done either by internal growth (i.e. producing more output) or by merging with another DMU which is also facing increasing returns to scale. If, for some reason, managers cannot influence the scale of a DMU, they should not be held accountable for this source of inefficiency.

- *A DMU faces decreasing returns to scale. It has diseconomies of scale. What does that concretely mean from a managerial point of view?*
  Such a DMU is already oversized, having exceeded its optimal size. In order to reduce its average total cost (or its average inputs consumption), it has to decrease its size. In practice, this could be done either by internal decay (i.e. producing less output) or by splitting the DMU into two separate businesses. Note that some of the production could be transferred to a DMU facing increasing returns to scale. If, for some reason, managers cannot influence the scale of a DMU, they should not be held accountable for this source of inefficiency.

- *Is efficiency the only criterion to assess a DMU's performance?*
  Not necessarily. Basically, the assessment of a DMU's performance will depend on the management objectives. Other criteria such as effectiveness or equity are often considered alongside efficiency. If this is the case, the overall performance should be balanced with the various criteria.

- *One DMU obtains a score of 100% but all the others in the data set obtain much lower scores (e.g. starting at 40% or lower). Is this realistic?*
  It could be realistic, but the gap appears to be important. In such a case, data should be carefully checked, and especially data of the efficient DMU. If a data

problem is not identified, such results mean that the efficient DMU is likely to have completely different processes than the other DMUs. It should therefore be absolutely presented as a best practice model. However, even if they are realistic, such results are likely to be rejected by managers whose DMUs have low efficiency scores. These managers are likely to be discouraged because it is obviously unrealistic for them to improve their DMU's efficiency by 60% (or more) in the short run. Therefore it is better to exclude the efficient DMU from the sample and to run a new model.

- *Almost all the DMUs obtain an efficiency score of 100%. Does that mean that all of them are really efficient?*
  Yes, it could mean that all the DMUs are efficient. Such results would be great! But they are unlikely. Here, the total number of inputs and outputs is probably too high compared to the number of DMUs in the data set. In this case, one variable should be excluded and a new model run. If the number of DMUs obtaining a 100% score decreases, this indicates that the number of variables was too high compared to the number of DMUs. If not, all the DMUs are just efficient and should be congratulated!

- *The model does not show any results. What does that mean?*
  The data should be checked. This could happen when data with a value of zero are included. Zeros should to be substituted by a very small number (0.01).

Exercise 10.1

The following multiple-choice questions test one's knowledge on the basics of DEA. Only one answer is correct. Answers can be found on the companion website.

1. What is the main purpose of DEA?
   a) DEA measures DMUs' effectiveness
   b) DEA measures DMUs' efficiency
   c) DEA measures DMUs' profit
   d) DEA measures DMUs' productivity

2. A data set includes information about input quantity, input cost and output quantity. Which type of efficiency cannot be measured?
   a) Technical efficiency
   b) Cost efficiency
   c) Revenue efficiency
   d) Scale efficiency

3. 'Pure' technical efficiency reflects:

    a) A global measure of DMU performance

    b) The efficiency of a DMU operating at an incorrect scale

    c) A measure of profit efficiency

    d) The efficiency of a poorly managed DMU

4. DMU A is inefficient. Who is (are) its peer(s)?

    a) One or several DMUs whose efficiency scores are worse than A's efficiency

    b) One or several DMUs whose efficiency scores are better than A's efficiency, but which are not located on the efficiency frontier

    c) Any DMU located on the efficiency frontier

    d) One or several specific DMUs (i.e. a subgroup of efficient DMUs) located on the efficiency frontier

5. A DMU faces diseconomies of scale. How can the management team improve its efficiency?

    a) By merging with another DMU

    b) By producing more output

    c) By producing less output

    d) By producing the same amount of output

6. A manager plans to measure efficiency using three inputs and two outputs. What is the minimum number of DMUs that should be included in the data set?

    a) 10

    b) 6

    c) 15

    d) It does not matter

10.3 The DEA software

The user-friendly software packages for DEA incorporate intuitive graphical user interfaces and automatic calculation of efficiency scores. Some of them are compatible with *Microsoft Excel*. For a survey of DEA software packages, see Barr (2004). Several software packages have been developed:

- Free packages include *DEAP* (Timothy Coelli, Coelli Economic Consulting Services) and *Win4DEAP* (Michel Deslierres, University of Moncton), *Benchmarking package in R* (Peter Bogetoft, Copenhagen Business School, and Lars Otto, University of Copenhagen), *Efficiency Measurement System* (Holger Scheel, University of Dortmund) and *DEA Solver Online* (Andreas Kleine and Günter Winterholer, University of Hohenheim).
- Commercial packages include *DEAFrontier* (Joe Zhu, Worcester Polytechnic Institute), *DEA-Solver PRO* (Saitech, Inc.), *PIM-DEA* (Ali Emrouznejad, Aston Business School) or *Frontier Analyst* (Banxia Software Ltd). Zhu (2003) includes an earlier version of *DEAFrontier, DEA Microsoft Excel Solver*, on a CD-ROM. This software works only under *Microsoft Excel* 97, 2000 and 2003. It allows an unlimited number of DMUs and is available at little cost. Cooper et al. (2006) include a CD-ROM with a *DEA-Solver* version limited at 50 DMUs. It is also available at little cost.

This section focuses on the 'twin' DEA software packages *DEAP/Win4DEAP*.[9] These packages centre on the basics of DEA, are simple to use and are stable over time. They are freely available[10] and come with data files as examples. As *Win4DEAP* is the Windows-based interface of *DEAP* (which is a DOS program), the current section refers only to *Win4DEAP*. All screenshots and icons presented in this section from *DEAP* or *Win4DEAP* are reproduced by permission of Timothy Coelli and Michel Deslierres.

The use of *Win4DEAP* is illustrated by a case study including a sample of 15 primary schools (see Table 10.2).

Case Study 10.2

The data used in this case study are fictitious (but are very similar to real ones). Fifteen schools produce one output (number of pupils) with three inputs (number of full-time equivalent (FTE) teachers, number of full-time administrative staff and number of computers – used as a proxy for technology investment). For example, school no. 8 is teaching 512 pupils with 28.6 teachers, 1.3 administrative staff and 26 computers.

---

[9] As *DEAP* is a DOS program, a user-friendly Windows interface has been developed for it (*Win4DEAP*). These 'twin' software packages must both be downloaded and extracted to the same folder. *Win4DEAP* cannot work without *DEAP*.

[10] DEAP Version 2.1 is available from http://www.uq.edu.au/economics/cepa/deap.htm, and Win4DEAP Version 1.1.3 from http://www8.umoncton.ca/umcm-deslierres_michel/dea/install.html.

Table 10.2 Data for Case Study 10.2.

	Input			Output
School	FTE teachers	FTE admin. staff	Computers	Pupils
1	40.2	2.0	37	602
2	18.1	1.1	17	269
3	42.5	2.1	41	648
4	11.0	0.8	10	188
5	24.8	1.3	22	420
6	21.1	1.3	19	374
7	13.5	1.0	13	247
8	28.6	1.3	26	512
9	23.5	1.3	22	411
10	15.9	1.0	15	285
11	23.2	1.3	22	397
12	26.0	1.4	25	466
13	11.1	0.8	11	198
14	28.8	1.6	26	530
15	19.7	1.3	18	357

10.3.1 Building a spreadsheet in *Win4DEAP*

Win4DEAP is launched by clicking on the MD icon (**MD**). DMUs are listed in the rows, and variables (outputs and inputs) in the columns (see Figure 10.9). The opening spreadsheet contains one decision-making unit (DMU1), one output (OUT1) and one input (IN1) by default.

To edit and name DMUs, outputs and inputs, the user must click on the row and column labels DMU1 (DMU1), OUT1 (OUT1) and IN1 (IN1), respectively. The dialogue box shown in Figure 10.10 allows the user to (1) assign a long name and a label (maximum of eight characters) to any variable and (2) select the nature of

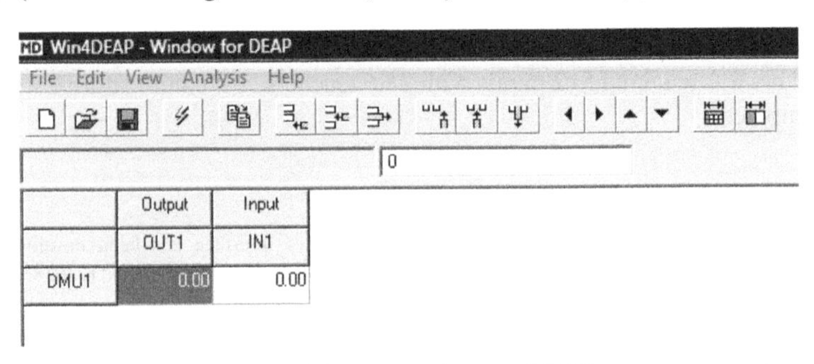

Figure 10.9 The opening spreadsheet contains one DMU, one output and one input. Reproduced by permission of Timothy Coelli and Michel Deslierres.

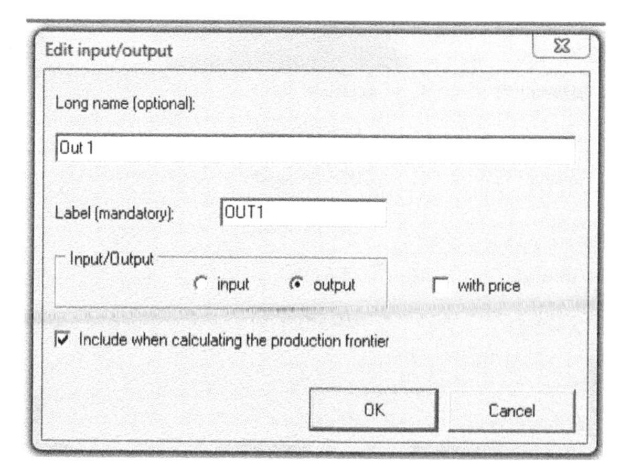

Figure 10.10 Input and output editing. Reproduced by permission of Timothy Coelli and Michel Deslierres.

the variables (either 'input' or 'output'). Finally, the user should select the *with price* option if he intends to measure cost efficiency (i.e. a *price* column will be added in the spreadsheet to the variable selected).

The icons enable the user to add DMUs. The icons enable the user to add variables (inputs or outputs). The and icons are used to delete any existing DMUs or variables. Finally, the icons allow the user to reverse the order of appearance of DMUs (rows) or variables (columns).

Data can be imported from a *Microsoft Excel* file into *Win4DEAP* by following these steps:

1. Save the *Microsoft Excel* data (only numbers, no names of DMUs or variables should be included) in CSV format (comma-delimited).

2. In *Win4DEAP*, first select the *File* menu, then the *Import* option and finally the *New data set* application.

3. Select the CSV file and open it.

The data will now appear in the *Win4DEAP* spreadsheet, though still need to be configured (DMUs and variables need to be named and variables defined as inputs or outputs).

10.3.2 Running a DEA model

To run a DEA model, the user can click on the *lightning* icon ( ). The window shown in Figure 10.11 appears. This window allows a calibration of the model by following steps 1–4 below:

1. Select an input or an output orientation (Orientation box).

2. Select the assumption about returns to scale (Returns to scale box). By checking *constant*, one assumes CRS; by checking *variable*, one assumes VRS. If one

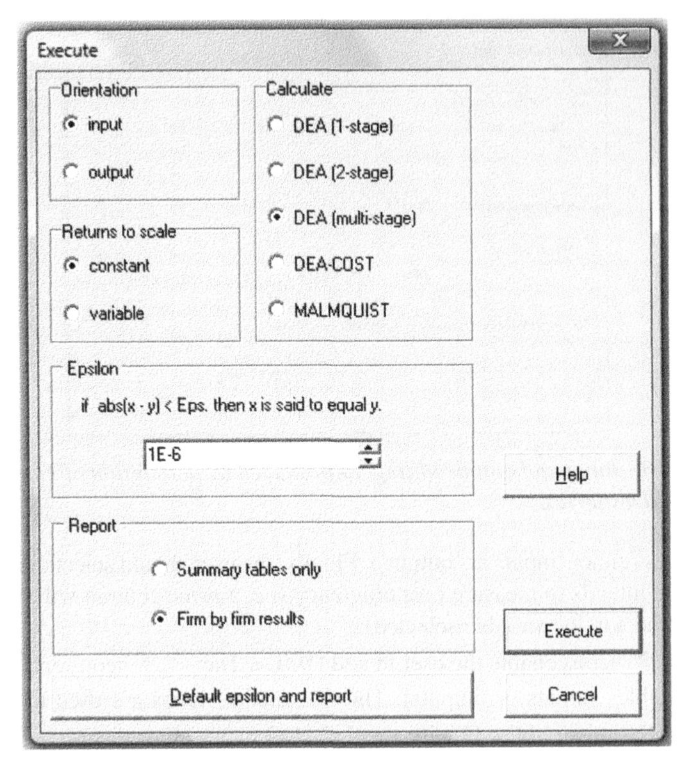

Figure 10.11 Win4DEAP *cockpit. Reproduced by permission of Timothy Coelli and Michel Deslierres.*

cannot be certain that DMUs are operating at an optimal scale, running a VRS model is recommended.

3. Select a model (Calculate box). Three main options are available:

   3.1. To calculate technical efficiency (TE) or technical (CRS), 'pure' (VRS) and scale efficiency (SE), check 'DEA (multi-stage)'. Options 'DEA (1-stage)', 'DEA (2-stage)' and 'DEA (multi-stage)' correspond to different treatments of slacks. Following Coelli (1998), the multi-stage treatment is recommended.

   3.2. To calculate cost efficiency, check DEA-COST. For this option, cost information about variables must be available and added to the spreadsheet.

   3.3. To calculate technical and scale efficiency when panel data are available, check MALMQUIST. See Section 10.5.4 to learn more about this.

4. Decide how to display the results (Report box): choose between summary tables only and firm-by-firm results.

5. Click on *Execute* to run the model.

Exercise 10.2

The objective of this exercise is to correctly calibrate a spreadsheet in *Win4DEAP* and to run a DEA model. Answers can be found on the companion website.

a) Prepare a spreadsheet in *Win4DEAP* including 15 DMUs, three inputs and one output. Name the DMUs 'School 1' to 'School 15'. The first input is 'FTE teachers', the second 'FTE administrative staff' and the third 'Number of computers'. The output corresponds to the number of pupils.

b) Feed the data appearing in Table 10.2 into the spreadsheet.

c) Save the file, preferably into the same folder containing *DEAP/Win4DEAP* (menu *File/option/ Save as*).

To run a DEA model in *Win4DEAP*, the following information is available:

- Schools are confronted with budget restrictions.
- The school system is heavily regulated.
- An obligatory school by school report is expected.

d) Calibrate the model.

e) Execute the model.

10.3.3 Interpreting results

After executing the selected model, a short notice appears with information about Timothy Coelli, the developer of *DEAP*. Results are displayed after closing this window. It is recommended that first-time users spend some time navigating through the results file in order to become familiar with it. Some results tables are commented on in this section. Table 10.3 contains a list of abbreviations with the main acronyms used in the results file.

Figure 10.12 shows an extract from the results file and features an efficiency summary. The first column contains the 15 schools. The second one displays the constant returns to scale technical efficiency (CRSTE) scores.[11] This 'total' efficiency score is decomposed into a 'pure' technical efficiency measure (VRSTE, third column) and

[11] Note that if you had run a CRS model instead of a VRS one, you would have obtained only one type of efficiency score in your results file (technical efficiency). Technical efficiency scores are strictly equal to the CRSTE scores obtained in the CRSTE column of the VRS model.

Table 10.3 A table of abbreviations to help with reading the results file.

Acronym	Full name
DEA	data envelopment analysis
CRS	constant returns to scale
VRS	variable returns to scale
TE	technical efficiency
CRSTE	constant returns to scale technical efficiency
VRSTE	variable returns to scale technical efficiency
SE	scale efficiency
IRS	increasing returns to scale
DRS	decreasing returns to scale

a scale efficiency measure (SE, fourth column). The last column indicates the nature of returns to scale (IRS, DRS or a dash):

- IRS means the DMU faces increasing returns to scale (economies of scale).
- DRS means the DMU faces decreasing returns to scale (diseconomies of scale).
- A dash means the DMU faces constant returns to scale; it is operating at an optimal scale.

```
EFFICIENCY SUMMARY:

 firm  crste  vrste  scale

   1  0.827  0.951  0.869 drs
   2  0.808  0.838  0.964 irs
   3  0.842  1.000  0.842 drs
   4  0.929  1.000  0.929 irs
   5  0.943  0.962  0.981 irs
   6  0.966  0.984  0.981 irs
   7  0.994  1.000  0.994 irs
   8  1.000  1.000  1.000  -
   9  0.951  0.963  0.987 irs
  10  0.974  0.995  0.978 irs
  11  0.930  0.943  0.986 irs
  12  0.978  0.984  0.994 irs
  13  0.969  1.000  0.969 irs
  14  1.000  1.000  1.000  -
  15  0.985  0.998  0.987 irs

mean  0.940  0.975  0.964

Note: crste = technical efficiency from CRS DEA
      vrste = technical efficiency from VRS DEA
      scale = scale efficiency = crste/vrste

Note also that all subsequent tables refer to VRS results
```

Figure 10.12 Technical efficiency (CRSTE) is decomposed into 'pure' technical efficiency (VRSTE) and scale efficiency (SE). Reproduced by permission of Timothy Coelli and Michel Deslierres.

On average, schools efficiency scores are:

- 94% for CRSTE; overall, schools could reduce their inputs by 6% whilst educating the same number of pupils.
- 97.5% for VRSTE; a better school organization would be able to reduce input consumption by 2.5%.
- 96.4% for SE; by adjusting their scale, schools could reduce their inputs by 3.6%.

All subsequent tables displayed in the results file refer to the VRSTE scores. These tables contain the following information:

- The number of the DMU under review ('Results for firm').
- The technical efficiency score ('Technical efficiency'), corresponding to the VRSTE when a VRS model has been run or to the CRSTE when a CRS model has been run.
- The scale efficiency score ('Scale efficiency'); note that the SE is mentioned only when a VRS model has been run.
- The rows of the matrix represent the outputs and the inputs of the model ('output 1', 'output 2', etc., 'input 1', 'input 2', etc.)
- The first column of the matrix recalls the original values of the variables' outputs and inputs ('original values').
- The second column of the matrix represents the movement an inefficient DMU needs in order to be located on the frontier ('radial movement').
- The third column of the matrix is the additional movement a DMU located on a segment of the frontier running parallel to the axis needs in order to become efficient ('slack movement').
- The fourth column of the matrix lists the values of the variables which enable the DMU to be efficient ('projected value'); these projected values take into account both the radial and the slack movements.
- Finally, the listing of peers is given. Each peer is identified by a number and has an associated weight ('lambda weight') representing the relative importance of the peer.

By way of illustration, three individual school tables are specifically commented on. School 1 (Figure 10.13) has a 'pure' efficiency score of 95.1% and a scale efficiency score of 86.9%. It is facing decreasing returns to scale (DRS). By improving the operation of the school, 4.9% (100 – 95.1) of inputs could be saved. By adjusting the school to its optimal size, 13.1% (100 – 86.9) of inputs could be saved.

The 'original value' column contains the original values of the school's variables: school 1 teaches 602 pupils with 40.2 teachers, 2 administrative staff and 37 computers. However, school 1 could 'produce' the same quantity of output with

```
Results for firm:       1
Technical efficiency = 0.951
Scale efficiency     = 0.869  (drs)
 PROJECTION SUMMARY:
  variable            original       radial        slack       projected
                         value     movement     movement           value
 output       1        602.000        0.000        0.000         602.000
 input        1         40.200       -1.972       -1.042          37.186
 input        2          2.000       -0.098        0.000           1.902
 input        3         37.000       -1.815        0.000          35.185
 LISTING OF PEERS:
  peer   lambda weight
    3       0.612
   14       0.373
    8       0.014
```

Figure 10.13 Results table for school 1. Reproduced by permission of Timothy Coelli and Michel Deslierres.

fewer inputs: 37.186 teachers instead of 40.2; 1.902 administrative staff instead of 2; 35.185 computers instead of 37 (see the 'projected value' column). The decreases in inputs 2 and 3 are equal to 4.9% of the original values: (−0.098/2)100 for input 2 and (−1.815/37)100 for input 3.[12] The case of input 1 is slightly different: to become efficient, it has to decrease not only by 4.9% (minus 1.972 from the 'radial movement' column) but also by an additional 1.042 (from the 'slack movement' column). Overall school 1 has to decrease its first input by minus 3.014 (−1.972 + −1.042) to become efficient. This represents 7.5% [(−3.014/40.2)100].

To improve its efficiency, school 1 has to analyze the practice of schools 3, 14 and 8, which are identified as its peers. To be a peer (or a benchmark), a DMU must have a 'pure' efficiency score of 100%. The lambda weight associated with each peer corresponds to its relative importance among the peer group. Ideally, school 1 should analyze best practice from a composite school formed by schools 3 (61.2%), 14 (37.3%) and 8 (1.4%). As such a 'virtual' school does not exist, school 1 should concentrate its best practice analysis on the peer associated with the highest lambda value (i.e. school 3).

School 2 (Figure 10.14) has a 'pure' efficiency score of 83.8% and a scale efficiency score of 96.4%. It is facing increasing returns to scale (IRS). By improving the operation of the school, 16.2% (100 – 83.8) of inputs could be saved. By adjusting the school to its optimal size, 3.6% (100 – 96.4) of inputs could be saved.

The 'original value' column contains the original values of the school's variables: school 2 teaches 269 pupils with 18.1 teachers, 1.1 administrative staff and 17 computers. However, school 2 could 'produce' the same quantity of output with fewer inputs: 15.163 teachers instead of 18.1; 0.922 administrative staff instead of 1.1; 14.242 computers instead of 17 (see the 'projected value' column). The decreases in inputs 1, 2 and 3 are equal to 16.2% of the original values ('radial movement'

[12] In a VRS model, the improvement in variables (decrease in inputs or increase in outputs) is calculated according to the VRSTE score (only). In a CRS model, it is calculated according to the CRSTE score, or TE score, including not only the pure efficiency but also the scale efficiency.

```
Results for firm:        2
Technical efficiency = 0.838
Scale efficiency     = 0.964  (irs)
 PROJECTION SUMMARY:
  variable             original        radial        slack      projected
                          value       movement     movement         value
 output       1         269.000         0.000        0.000       269.000
 input        1          18.100        -2.937        0.000        15.163
 input        2           1.100        -0.178        0.000         0.922
 input        3          17.000        -2.758        0.000        14.242
 LISTING OF PEERS:
  peer    lambda weight
   13       0.506
    4       0.261
   14       0.016
    8       0.218
```

Figure 10.14 Results table for school 2. Reproduced by permission of Timothy Coelli and Michel Deslierres.

```
Results for firm:        3
Technical efficiency = 1.000
Scale efficiency     = 0.842  (drs)
 PROJECTION SUMMARY:
  variable             original        radial        slack      projected
                          value       movement     movement         value
 output       1         648.000         0.000        0.000       648.000
 input        1          42.500         0.000        0.000        42.500
 input        2           2.100         0.000        0.000         2.100
 input        3          41.000         0.000        0.000        41.000
 LISTING OF PEERS:
  peer    lambda weight
    3       1.000
```

Figure 10.15 Results table for school 3. Reproduced by permission of Timothy Coelli and Michel Deslierres.

column). No slack movement is identified. To improve its efficiency, school should refer to schools 13, 4, 14 and 8, which are identified as its peers.

School 3 (Figure 10.15) has a 'pure' efficiency score of 100% and a scale efficiency score of 84.2%. It is facing decreasing returns to scale (DRS). This school is well managed. It cannot improve its 'pure' efficiency. The only capacity for improvement lies in a scale adjustment: 15.8% (100 – 84.2) of inputs could be saved.

The 'original value' column contains the original values of the school's variables: school 3 teaches 648 pupils with 42.5 teachers, 2.1 administrative staff and 41 computers. These values are equal to the projected ones ('pure' efficiency = 100%). As school 3 is purely efficient, it acts as its own peer.

Exercise 10.3

The objective of this exercise is the interpretation of DEA results. Figure 10.16 displays results for one of the 15 schools. It has been truncated in order to hide the VRS technical efficiency score. Answers can be found on the companion website.

```
Scale efficiency      = 0.987  (irs)
 PROJECTION SUMMARY:
  variable             original        radial          slack        projected
                          value       movement        movement        value
 output      1          411.000         0.000           0.000        411.000
 input       1           23.500        -0.864           0.000         22.636
 input       2            1.300        -0.048           0.000          1.252
 input       3           22.000        -0.809          -0.375         20.817
 LISTING OF PEERS:
  peer   lambda weight
   13       0.346
   14       0.417
    8       0.238
```

Figure 10.16 An efficiency table helps a DMU to make decisions based on objective information. Reproduced by permission of Timothy Coelli and Michel Deslierres.

Tasks

Answer the following questions:

a) The variable returns to scale technical efficiency score has been removed from the table. Find a way to calculate it.

b) Assume that the 'pure' efficiency score is equal to 96.3%. What is the main feature in need of improvement: the school's management or the school's scale?

c) Assume that the school only has time to analyze best practice in one of its peers. Which one should it select?

d) By how much must the school reduce input 3 in order to be located on the efficiency frontier?

10.4 In the black box of DEA

This section describes the two principal DEA models: the constant returns to scale model (Charnes et al. 1978) and the variable returns to scale model (Banker et al. 1984). DEA is based on the earlier work of Dantzig (1951) and Farrell (1957), whose approach adopted an input orientation. Zhu and Cook (2008), Cooper et al. (2007) and Coelli et al. (2005) provide a comprehensive description of the methodology. By 2007, Emrouznejad et al. (2008) identified more than 4000 research articles about DEA published in scientific journals or books.

DEA is a non-parametric method. Unlike parametric methods (such as ordinary least squares, maximum likelihood estimation or stochastic frontier analysis), inputs and outputs are used to compute, using linear programming methods, a hull representing the efficiency frontier. As a result, a non-parametric method does not require specification of a functional form.

10.4.1 Constant returns to scale

Charnes et al. (1978) propose a model assuming constant returns to scale (CRS model).[13] This is appropriate when all DMUs operate at the optimal scale. Efficiency is defined by Charnes et al. (1978, p. 430) as 'the maximum of a ratio of weighted outputs to weighted inputs subject to the condition that the similar ratios for every DMU be less than or equal to unity'. Following the notation adopted by Johnes (2004), let s be the number of output, m be the number of inputs, and TE_k be the technical efficiency of DMU k, $k = 1, \ldots, n$, using the m inputs to produce the s outputs. Then

$$\mathrm{TE}_k = \frac{\sum_{r=1}^{s} u_r y_{rk}}{\sum_{i=1}^{m} v_i x_{ik}} \tag{10.1}$$

where y_{rk} is the quantity of output r produced by DMU k, x_{ik} is the quantity of input i consumed by DMU k, u_r is the weight of output r, and v_i is the weight of input i.

The technical efficiency of DMU k is maximized under two constraints:

$$\text{maximize} \quad \mathrm{TE}_k = \frac{\sum_{r=1}^{s} u_r y_{rk}}{\sum_{i=1}^{m} v_i x_{ik}} \tag{10.2}$$

subject to

$$\frac{\sum_{r=1}^{s} u_r y_{rj}}{\sum_{i=1}^{m} v_i x_{ij}} \leq 1, \quad j = 1, \ldots, n, \tag{10.3}$$

$$u_r, v_i > 0 \quad \forall r = 1, \ldots, s; i = 1, \ldots, m. \tag{10.4}$$

Inequality (10.3) says that the weights applied to outputs and inputs of DMU k cannot generate an efficiency score greater than 1 when applied to each DMU in the data set. Furthermore, the weights on the outputs and on the inputs are strictly positive (10.4).

This linear programming problem can be dealt with by two different approaches. In the first, the weighted sums of outputs are maximized holding inputs constant (output-oriented model). In the second, the weighted sums of inputs are minimized holding outputs constant (input-oriented model). Note that the output and input

[13] This model is also known as the Charnes, Cooper and Rhodes model (CCR model).

orientations refer to the dual equations of each model, which are not presented in this chapter. The primal equations for each model, known as the multiplier form, are as follows. For the CRS output-oriented model we solve

$$\text{minimize} \sum_{i=1}^{m} v_i x_{ik} \tag{10.5}$$

subject to

$$\sum_{i=1}^{m} v_i x_{ij} - \sum_{r=1}^{s} u_r y_{rj} \geq 0, \; j = 1, \ldots, n, \tag{10.6}$$

$$\sum_{r=1}^{s} u_r y_{rk} = 1, \tag{10.7}$$

$$u_r, v_i > 0 \quad \forall r = 1, \ldots, s; i = 1, \ldots, m. \tag{10.8}$$

For the CRS input-oriented model we solve

$$\text{maximize} \sum_{r=1}^{s} u_r y_{rk} \tag{10.9}$$

subject to

$$\sum_{i=1}^{m} v_i x_{ij} - \sum_{r=1}^{s} u_r y_{rj} \geq 0 \quad j = 1, \ldots, n \tag{10.10}$$

$$\sum_{i=1}^{m} v_i x_{ik} = 1 \tag{10.11}$$

$$u_r, v_i > 0 \quad \forall r = 1, \ldots, s; i = 1, \ldots, m \tag{10.12}$$

Using the duality in linear programming, an equivalent form, known as the envelopment form, can be derived from this problem. It is often preferable to solve the computation using the envelopment form because it contains only $s+m$ constraints rather than $n+1$ constraints in the multiplier form. Dual equations can be found in Johnes (2004, p. 631).

Each DMU located on the sections of the efficiency frontier running parallel to the axes must be adjusted for output and input slacks. A formulation of the dual equations which integrates slacks can be found in Johnes (2004, p. 632). For an in-depth analysis on the treatment of slacks, and especially the multi-stage methodology, see Coelli (1998).

Exercise 10.4

The objective of this exercise is to programm a CRS input-oriented model using *Microsoft Excel Solver*. Answers can be found on the companion website.

	A	B	C	D	E	F	G
1		**Output 1 Birth**	**Output 2 Marriage**	**Input 1 Public servant**	**Weighted output**	**Weighted input**	**Efficiency (%)**
2	**Register office**	4	3	1	7.00	1.00	700.00
3	**Weight**	1.00	1.00	1.00			
4							
5		**Output 1 Birth**	**Output 2 Marriage**	**Input 1 Public servant**	**Weighted output**	**Weighted input**	**Weighted input minus Weighted output**
6	**Register office A**	1	6	1	7.00	1.00	-6.00
7	**Register office B**	3	8	1	11.00	1.00	-10.00
8	**Register office C**	4	3	1	7.00	1.00	-6.00
9	**Register office D**	5	6	1	11.00	1.00	-10.00
10	**Register office E**	6	2	1	8.00	1.00	-7.00

Figure 10.17 A Microsoft Excel *spreadsheet ready to use with the* Microsoft Excel Solver.

Instructions

The companion website contains a file called 'CRS with Solver'. It is recommended that users spend some time navigating around the file in order to become familiar with the formulas associated with some of the cells.

The spreadsheet is divided into two parts (Figure 10.17). The first part comprises rows 2 and 3. This section enables users to successively calculate the efficiency of the five Register Offices (one at a time), producing two outputs (birth and marriage certificates) with one input (public servants). The second part comprises rows 6 to 10. It contains the data for Register Offices A to E (output 1 = column B, output 2 = column C, input 1 = column D, weighted sum of outputs = column E, weighted sum of the input = column F). An additional column, G, is included in the spreadsheet. It is a working column used by *Microsoft Excel Solver*. Column G contains the weighted sum of the input minus the weighted sum of outputs.

In the first part, data from each Register Office must be entered successively in the dark grey cells (row 2, columns B, C and D). The spreadsheet already contains data on Register Office C. The two outputs and one input of Register Office C are assigned weights in cells B3 to D3 (light grey cells). A value of 1 has been assigned to all of them in the spreadsheet (by default). The efficiency score is calculated in cell G2 (light grey cell). The value shown in this cell (700%) exceeds 100% because the *Solver* has not yet been run and thus the constraints of the CRS model do not yet apply.

The *Solver* must first be loaded in the user's version of *Microsoft Excel*.[14]

Tasks

Answer the following questions about the *Solver*:

a) Which cell is to be optimized?

b) In this case, do you maximize output or minimize input?

c) Which equation of the CRS model is optimized by the *Solver*?

[14] Instructions on loading the *Solver* can be found in the *Microsoft Excel* help files.

d) Which variables can be changed in the optimization process?

e) The *Solver* contains two constraints. Which one corresponds to equation (10.10) in the CRS model?

f) What does the other constraint show?

g) Solve the CRS model. What is the efficiency of register office C?

h) To get a 73.08% efficiency score, what are the weights associated with output 1, output 2 and input 1?

i) Replace the values in cells B2, C2 and D2 with the values of another Register Office (A, B, D or E). What is the efficiency score?

10.4.2 Variable returns to scale

Banker et al. (1984) propose a model assuming variable returns to scale (VRS model).[15] This is appropriate when DMUs do not operate at optimal scale. As Coelli et al. (2005, p. 172) point out, 'the use of the CRS specification when not all DMUs are operating at the optimal scale, results in measures of TE that are confounded by scale efficiencies (SE). The use of the VRS specification permits the calculation of TE devoid of these SE effects.' The CRS model can be modified by relaxing the constant returns to scale assumption. A measure of return to scale for DMU k is added in the primal equation.

The linear programming problem to be solved under VRS includes a measure of returns to scale on the variables axis, c_k, for the DMU k. The primal equations are as follows. For the VRS output-oriented model we solve

$$\text{minimize} \sum_{i=1}^{m} v_i x_{ik} - c_k \tag{10.13}$$

Subject to

$$\sum_{i=1}^{m} v_i x_{ij} - \sum_{r=1}^{s} u_r y_{rj} - c_k \geq 0, \quad j = 1, \ldots, n, \tag{10.14}$$

$$\sum_{r=1}^{s} u_r y_{rk} = 1, \tag{10.15}$$

$$u_r, v_i > 0 \quad \forall r = 1, \ldots, s; i = 1, \ldots, m. \tag{10.16}$$

---

[15] This model is also known as the Banker, Charnes and Cooper (BCC) model.

For the VRS input-oriented model we solve

$$\text{maximize} \sum_{r=1}^{s} u_r y_{rk} + c_k \tag{10.17}$$

subject to

$$\sum_{i=1}^{m} v_i x_{ij} - \sum_{r=1}^{s} u_r y_{rj} - c_k \geq 0, \qquad j = 1, \ldots, n, \tag{10.18}$$

$$\sum_{i=1}^{m} v_i x_{ik} = 1, \tag{10.19}$$

$$u_r, v_i > 0 \quad \forall r = 1, \ldots, s; i = 1, \ldots, m. \tag{10.20}$$

The reader may refer to Johnes (2004) for the dual equations (p. 634) and the dual equations with slacks (pp. 634–635).

A further step needs to be taken in order to identify the nature of the returns to scale. This relates to another model, the non-increasing returns to scale (NIRS) model, derived from the VRS model (Coelli et al. 2005). In Figure 10.18 the NIRS efficiency frontier has been added (the dotted line). This corresponds to the CRS frontier from the origin to point B followed by the VRS frontier from point B. The nature of the scale inefficiencies for each DMU can be determined by comparing technical efficiency scores under NIRS and VRS. If NIRS TE $\neq$ VRS TE (as for DMUs A and D), increasing returns to scale apply. If NIRS TE $=$ VRS TE (but $\neq$ CRS TE) (as for DMUs E and C), decreasing returns to scale apply. Finally, if NIRS TE $=$ VRS TE $=$ CRS TE, as for DMU B, constant returns to scale apply.

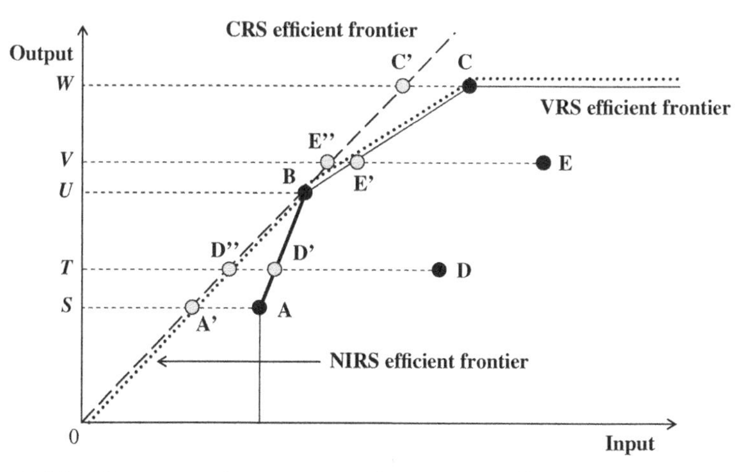

Figure 10.18 The nature of returns to scale is identified by comparing an NIRS and a VRS model.

10.5 Extensions of DEA

In this section, a selection of four extensions of DEA is briefly introduced: adjusting for the environment, preferences (weight restrictions), sensitivity analysis and time series data. For a broader overview of the major developments in DEA, see Cook and Seiford (2008). For an up-to-date review of DEA, see Cooper et al. (2011).

10.5.1 Adjusting for the environment

Environmental variables influence the efficiency of DMUs but are not under the control of the management team. In DEA, several methods are used to accommodate such variables. These include the Charnes et al. (1981) approach, the categorical model (Banker and Morey 1986a) and the non-discretionary variable model derived by Banker and Morey (1986b) (which includes the environmental variable directly in the DEA model).

The most convincing of these methods, however, is the two-stage method, the advantages of which are described in Coelli et al. (2005, pp. 194–195) and in Pastor (2002, p. 899). The two-stage method combines a DEA model and a regression analysis. In the first stage, a traditional DEA model is conducted. This model includes only discretionary inputs and outputs. In the second stage, the efficiency scores are regressed against the environmental (i.e. non-discretionary or exogenous) variables. Tobit regression is often used in the second stage. However, recent studies have shown that ordinary least squares regression is sufficient to model the efficiency scores (Hoff 2007) or even more appropriate than tobit (McDonald 2009).

The coefficients of the environmental variables, estimated by regression, are used to model the efficiency scores to correspond to an identical condition of environment (e.g. usually the average condition). Simar and Wilson (2007, p. 32) provide a selection of studies using the two-stage method. Among these are applications in education (Chakraborty et al. 2001; McMillan and Datta 1998; McCarty and Yaisawarng 1993), hospitals (Burgess and Wilson 1998), defence (Barros 2004), police (Carrington et al. 1997), farming (Binam et al. 2003) and banking (O'Donnell and van der Westhuizen 2002). Sueyoshi et al. (2010) and Sibiano and Agasisti (2012) provide more recent applications in the manufacturing sector and education.

10.5.2 Preferences

For different reasons (e.g. the weights assigned to variables by DEA are considered as unrealistic for some DMUs; the management team may wish to give priority to specific variables), preferences about the relative importance of individual inputs and outputs can be set by the decision maker. This is done by placing weight restrictions on outputs and inputs (also called multiplier restrictions). Cooper et al. (2011) and Thanassoulis et al. (2004) provide a review of models regarding the use of weights restrictions. An earlier review can be found in Allen et al. (1997). Generally, the imposition of weight restrictions worsens efficiency scores. Three main approaches are identified to accommodate preferences:

- Dyson and Thanassoulis (1988) propose an approach which imposes absolute upper and lower bounds on input and output weights. This technique is applied in Roll et al. (1991) to highway maintenance units and in Liu (2009) to garbage clearance units.
- Charnes et al. (1990) develop the cone-ratio method. This approach imposes a set of linear restrictions that define a convex cone, corresponding to an 'admissible' region of realistic weight restrictions. See Brockett et al. (1997) for an application to banks.
- Thompson et al. (1986, 1990) propose the assurance region method. This approach is actually a special case of the cone ratio. It imposes constraints on the relative magnitude of the weights. For example, a constraint on the ratio of weights for inputs 1 and 2 can be included, such as $L_{1,2} \leq \nu_2/\nu_1 \leq U_{1,2}$, where $L_{1,2}$ and $U_{1,2}$ are lower and upper bounds for the ratio of the weight of input 2 (ν_2) to the weight of input 1 (ν_1). As a result, the assurance region method limits the 'region' of weights to a restricted area by prohibiting large differences in the value of those weights. An application of this model is provided by Sarica and Or (2007) in the assessment of power plants.

10.5.3 Sensitivity analysis

Cooper et al. (2006, p. 271) mention that the term 'sensitivity' corresponds to stability or robustness. For Zhu (2003, p. 217), 'the calculated frontiers of DEA models are stable if the frontier DMUs that determine the DEA frontier remain on the frontier after particular data perturbations are made'. Sensitivity analysis aims to identify the impact on DMU efficiency when certain parameters are modified in the model.

The first way to test the sensitivity of DEA results is to add/remove DMUs to/from DEA models. Dusansky and Wilson (1994, 1995) and Wilson (1993, 1995) provide different approaches to deal with this concern. The approach of Pastor et al. (1999) allows users to identify the observations which considerably affect the efficiency of the remaining DMUs. It also determines the statistical significance of efficiency variations which are due to the inclusion of a given DMU in the sample.

Another way to test the sensitivity of DEA results is to modify the values of outputs and inputs. They focus on the maximum data variations a given DMU can endure whilst maintaining its efficiency status. Approaches include:

- perturbation of a single variable of an efficient DMU (Charnes et al. 1985), data of other DMUs remaining fixed;
- simultaneous proportional data perturbation of all outputs and inputs of an efficient DMU (Charnes and Neralic 1992), data of other DMUs remaining fixed;

- simultaneous data perturbation of an efficient DMU in a situation where outputs and inputs can be modified individually (Seiford and Zhu 1998a; Neralic and Wendell 2004), data of other DMUs remaining fixed;
- simultaneous proportional data perturbation of all outputs and inputs of all DMUs (Seiford and Zhu 1998b).

For more on sensitivity analysis, see Zhu (2001).

10.5.4 Time series data

In DEA, panel data are considered using two methods: window analysis and the Malmquist index.

Window analysis, introduced by Charnes et al. (1985), examines the changes in the efficiency scores of a set of DMUs over time. A 'window' of time periods is chosen for each DMU. The same DMU is treated as if it represented a different DMU in every time period. In this sense, window analysis can also be considered as a sensitivity analysis method. For instance, a model including n DMUs with annual data and a chosen 'window' of t years will result in $n \times t$ units to be evaluated. For each DMU, t different efficiency scores will be measured. The 'window' is then shifted by one period (one year in our example) and the efficiency analysis is repeated. Yue (1992) provides a didactical application of window analysis. Other applications include Yang and Chang (2009), Avkiran (2004) and Webb (2003).

The Malmquist total factor productivity index was first introduced by Malmquist (1953) before being further developed within the framework of DEA. It is used to measure the change in productivity over time. The Malmquist index decomposes this productivity change into two components The first is called the 'catch-up'; this captures the change in technical efficiency over time. The second is called the 'frontier-shift'; this captures the change in technology that occurs over time (i.e. the movement of efficiency frontiers over time). Readers may refer to Färe et al. (2011) and Tone (2004) for actual reviews. Applications of the Malmquist index can be found in Coelli and Prasada Rao (2005) and Behera et al. (2011).

References

Allen, R., Athanassopoulos, A., Dyson, R. G., and Thanassoulis, E. (1997). Weights restrictions and value judgements in data envelopment analysis: Evolution, development and future directions. *Annals of Operations Research*, *73*, 13–34.

Avkiran, N. K. (2004). Decomposing technical efficiency and window analysis. *Studies in Economics and Finance*, *22*(1), 61–91.

Banker, R. D., and Morey, R. C. (1986a). The use of categorical variables in data envelopment analysis. *Management Science*, *32*, 1613–1627.

Banker, R. D., and Morey, R. C. (1986b). Efficiency analysis for exogenously fixed inputs and outputs. *Operations Research*, *34*, 513–521.

Banker, R. D., Charnes, A., and Cooper, W. W. (1984). Some models for estimating technical and scale inefficiencies in data envelopment analysis. *Management Science*, *30*(9), 1078–1092.

Barr, R. (2004). DEA software tools and technology: A state-of-the-art survey. In W. W. Cooper, L. M. Seiford, and J. Zhu, *Handbook on Data Envelopment Analysis* (pp. 539–566). Boston: Kluwer Academic Publishers.

Barros, C. P. (2004). Measuring performance in defence-sector companies in a small NATO member country. *Journal of Economic Studies, 31*, 112–128.

Behera, S. K., Faroquie, J. A., and Dash, A. P. (2011). Productivity change of coal-fired thermal power plants in India: A Malmquist index approach. *IMA Journal of Management Mathematics, 22*, 387–400.

Binam, J. N., Sylla, K., Diarra, I., and Nyambi, G. (2003). Factors affecting technical efficiency among coffee farmers in Côte d'Ivoire: Evidence from the centre west region. *R&D Management, 15*, 66–76.

Brockett, P. L., Charnes, A., Cooper, W. W., Huang, Z. M., and Sun, D. B. (1997). Data transformations in DEA cone ratio envelopment approaches for monitoring bank performance. *Journal of Operational Research, 98*, 250–268.

Burgess, J. F., and Wilson, P. W. (1998). Variation in inefficiency among US hospitals. *Canadian Journal of Operational Research and Information Processing (INFOR), 36*, 84–102.

Carrington, R., Puthucheary, N., Rose, D., and Yaisawarng, S. (1997). Performance measurement in government service provision: The case of police services in New South Wales. *Journal of Productivity Analysis, 8*, 415–430.

Chakraborty, K., Biswas, B., and Lewis, W. C. (2001). Measurement of technical efficiency in public education: A stochastic and nonstochastic production approach. *Southern Economic Journal, 67*, 889–905.

Charnes, A., and Neralic, L. (1992). Sensitivity analysis in data envelopment analysis. *Glasnik Matematicki, 27*, 191–201.

Charnes, A., Clarke, C., Cooper, W. W., and Golany, B. (1985). A development study of DEA in measuring the effect of maintenance units in the U.S. Air Force. *Annals of Operations Research, 2*, 95–112.

Charnes, A., Cooper, W. W., and Rhodes, E. L. (1978). Measuring the efficiency of decision making units. *European Journal of Operational Research, 2*, 429–444.

Charnes, A., Cooper, W. W., and Rhodes, E. L. (1981). Evaluating program and managerial efficiency: An application of DEA to program follow through. *Management Science, 27*(6), 668–697.

Charnes, A., Cooper, W. W., Huang, Z. M., and Sun, D. B. (1990). Polyhedral cone-ratio DEA models with an illustrative application to large commercial banks. *Journal of Econometrics, 46*, 73–91.

Coelli, T. J. (1998). A multi-stage methodology for the solution of oriented DEA models. *Operations Research Letters, 23*, 143–149.

Coelli, T. J., and Perelman, S. (1996). Efficiency measurement, multiple-output technologies and distance functions: With application to European railways. Center of Research in Public Economics and Population Economics Discussion Paper 96/05.

Coelli, T. J., and Perelman, S. (1999). A comparison of parametric and non-parametric distance functions: With application to European railways. *European Journal of Operational Research, 117*(2), 326–339.

Coelli, T. J., and Prasada Rao, D. S. (2005). Total factor productivity growth in agriculture: A Malmquist index analysis of 93 countries, 1980–2000. *Agricultural Economics, 32*, 115–134.

Coelli, T. J., Prasada Rao, D. S., O'Donnell, C. J., and Battese, G. E. (2005). *An Introduction to Efficiency and Productivity Analysis.* New York: Springer-Verlag.

Cook, W. D., and Seiford, L. M. (2008). Data envelopment analysis (DEA) – Thirty years on. *European Journal of Operational Research, 192*, 1–17.

Cooper, W. W., Ruiz, J. L., and Sirvent, I. (2011). Choices and uses of DEA weights. In W. W. Cooper, L. M. Seiford, and J. Zhu, *Handbook on Data Envelopment Analysis* (pp. 93–126). New York: Springer-Verlag.

Cooper, W. W., Seiford, L. M., and Tone, K. (2006). *Introduction to Data Envelopment Analysis and its Uses.* New York: Springer-Verlag.

Cooper, W. W., Seiford, L. M., and Tone, K. (2007). *Data Envelopment Analysis: A comprehensive Text with Models, Applications, References and DEA-Software.* New York: Springer-Verlag.

Dantzig, G. B. (1951). Maximization of a linear function of variables subject to linear inequalities. In T. C. Koopmans, *Activity Analysis of Production and Allocation* (pp. 339–347). New York: John Wiley & Sons, Inc.

Dusansky, R., and Wilson, P. W. (1994). Measuring efficieny in the care of developmentally disabled. *Review of Economics and Statistics, 76*, 340–345.

Dusansky, R., and Wilson, P. W. (1995). On the relative efficiency of alternative modes of producing public sector output: The case of the developmentally disabled. *European Journal of Operational Research, 80*(3), 608–618.

Dyson, R. G., and Thanassoulis, E. (1988). Reducing weight flexibility in DEA. *Journal of the Operational Research Society, 39*(6), 563–576.

Emrouznejad, A., Parker, B. R., and Tavares, G. (2008). Evaluation of research in efficiency and productivity: A survey and analysis of the first 30 years of scholarly literature in DEA. *Socio-Economic Planning Sciences, 42*, 151–157.

Färe, R., Grosskopf, S., and Margaritis, D. (2011). Malmquist productivity indexes and DEA. In W. W. Cooper, L. M. Seiford, and J. Zhu, *Handbook on Data Envelopment Analysis* (pp. 127–150). New York: Springer-Verlag.

Farrell, M. J. (1957). The measurement of productive efficiency. *Journal of Royal Statistical Society, Series A, 120*(3), 253–281.

Giannoulis, C., and Ishizaka, A. (2010). A web-based decision support system with ELECTRE III for a personalized ranking of British universities. *Decision Support Systems, 48*, 488–497.

Hoff, A. (2007). Second stage DEA: Comparison of approaches for modeling the DEA score. *European Journal of Operational Research, 181*, 425–435.

Johnes, J. (2004). Efficiency measurement. In G. Johnes, and J. Johnes, *International Handbook on the Economics of Education* (pp. 613–742). Cheltenham: Edward Elgar.

Liu, C.-C. (2009). A study of optimal weights restriction in data envelopment analysis. *Applied Economics, 41*(14), 1785–1790.

Malmquist, S. (1953). Index numbers and indifferences surfaces. *Trabajos de Estatística, 4*, 209–242.

McCarty, T. A., and Yaisawarng, S. (1993). Technical efficiency in New Jersey school districts. In H. O. Fried, C. A. Lovell, and S. S. Schmidt, *The Measurement of Productive Efficiency: Techniques and Applications* (pp. 271–287). New York: Oxford University Press.

McDonald, J. (2009). Using least squares and tobit in second stage DEA efficiency analyses. *European Journal of Operational Research, 197*, 792–798.

McMillan, M. L., and Datta, D. (1998). The relative efficiencies of Canadian universities: A DEA perspective. *Canadian Public Policy – Analyse de Politiques*, *24*, 485–511.

Neralic, L. (2004). Preservation of efficiency and inefficiency classification in data envelopment analysis. *Mathematical Communications*, *9*, 51–62.

O'Donnell, C. J., and van der Westhuizen, G. (2002). Regional comparisons of banking performance in South Africa. *South African Journal of Economics*, *70*, 485–518.

Pastor, J. M. (2002). Credit risk and efficiency in the European banking system: A three-stage analysis. *Applied Financial Economics*, *12*, 895–911.

Pastor, J. M., Ruiz, J. L., and Sirvent, I. (1999). A statistical test for detecting influential observations in DEA. *European Journal of Operational Research*, *115*, 542–554.

Roll, Y., Cook, W. D., and Golany, B. (1991). Controlling factor weights in data envelopment analysis. *IEE Transactions*, *23*(1), 2–9.

Sarica, K., and Or, I. (2007). Efficiency assessment of Turkish power plants using data envelopment analysis. *Energy*, *32*, 1484–1499.

Seiford, L. M., and Zhu, J. (1998a). Stability regions for maintaining efficiency in data envelopment analysis. *European Journal of Operational Research*, *108*(1), 127–139.

Seiford, L. M., and Zhu, J. (1998b). Sensitivity analysis of DEA models for simultaneous changes in all the data. *Journal of the Operational Research Society*, *49*, 1060–1071.

Sibiano, P., and Agasisti, T. (2012). Efficiency and heterogeneity of public spending in education among Italian regions. *Journal of Public Affairs*, to appear.

Simar, L., and Wilson, P. W. (2007). Estimation and inference in two-stage, semi-parametric models of production processes. *Journal of Econometrics*, *136*, 31–64.

Sueyoshi, T., Goto, M., and Omi, Y. (2010). Corporate governance and firm performance: Evidence from Japanese manufacturing industries after the lost decade. *European Journal of Operational Research*, *203*, 724–736.

Thanassoulis, E., Portela, M. C. S., and Allen, R. (2004). Incorporating value judgments in DEA. In W. W. Cooper, L. W. Seiford and J. Zhu (Eds.), *Handbook on Data Envelopment Analysis* (pp. 99–138). Boston: Kluwer Academic Publishers.

Thanassoulis, E., Portela, M. C. S., and Despic, O. (2008). Data Envelopment Analysis: The Mathematical Programming Approach to Efficiency Analysis. In H. O. Fried, C. A. Lovell, and S. S. Schmidt, *The Measurement of Productive Efficiency and Productivity Growth* (pp. 251–420). New York: Oxford University Press.

Thomson, R. G., Langemeier, L. N., Lee, C., Lee, E., Thrall, R. M., and Smith, B. A. (1990). The role of multiplier bounds in efficiency analysis with application to Kansas farming. *Journal of Econometrics*, *46*, 93–108.

Thomson, R. G., Singleton Jr., F. D., Thrall, R. M., and Smith, B. A. (1986). Comparative site evaluations for locating a high-energy physics lab in Texas. *Interfaces*, *16*(6), 35–49.

Tone, K. (2004). Malmquist productivity index – Efficiency change over time. In W. W. Cooper, L. M. Seiford, and J. Zhu, *Handbook on Data Envelopment Analysis* (pp. 203–227). Boston: Kluwer Academic Publishers.

Webb, R. (2003). Level of efficiency in UK retail banks: a DEA window analysis. *International Journal of the Economics of Business*, *10*(3), 305–322.

Wilson, P. W. (1995). Detecting influential observations in data envelopment analysis. *Journal of Productivity Analysis*, *6*, 27–45.

Wilson, P. W. (1993). Detecting outliers in deterministic non-parametric frontier models with multiple outputs. *Journal of Business and Economic Statistics*, *11*, 319–323.

Yang, H.-H., and Chang, C.-Y. (2009). Using DEA window analysis to measure efficiencies of Taiwan's integrated telecommunication firms. *Telecommunications Policy*, *33*, 98–108.

Yue, P. (1992). Data envelopment analysis and commercial bank performance: A primer with applications to Missouri banks. *Federal Reserve Bank of St Louis Review*, *74*, 31–45.

Zhu, J. (2003). *Quantitative Models for Performane Evaluation and Benchmarking*. New York: Springer-Verlag.

Zhu, J. (2001). Super-efficiency and DEA sensitivity analysis. *European Journal of Operational Research*, *129*(2), 443–455.

Zhu, J., and Cook, W. D. (2008). *Data Envelopment Analysis: Modeling Operational Processes and Measuring Productivity*. Seattle: CreateSpace.

Part IV

INTEGRATED SYSTEMS

11

Multi-method platforms

11.1 Introduction

The multi-criteria decision aid literature is overloaded with various different methods and algorithms, with the aim of helping a decision maker faced with a complex, real-world decision problem. These methods differ in the type of problem they tackle, the underlying philosophy on which they are based, the assumptions they rely on and the input information they require, etc. Nevertheless the methods share some commonalities. Most of the well-known and commonly used methods have been implemented in different software packages, as the contents of this book attest.

It might be difficult for a decision maker to choose the most appropriate software for solving the decision problem. Furthermore, the decision problems change and evolve over time and might require the method to be changed for a more appropriate one. From a practical point of view, using different methods for the same problem (if the input conditions allow and assumptions are valid) may strengthen confidence in the results.

Two main projects have arisen over the last decade in order to combine several methods and to strengthen interoperability: Decision Deck and DECERNS. Although these two projects differ considerably in their approach, they both enable the user to test different methods in one package.

In this chapter Decision Deck is briefly described and a user-oriented description of DECERNS is put forward. As our aim is to describe methods and software packages that require no *a priori* technical knowledge and can be used by beginners and experienced practitioners alike, we describe the DECERNS project in more detail. The DECERNS project includes several methods that have been described in the previous chapters of this book. It also contains an integrated Geographical Information System that can be useful if the decision maker is dealing with with spatial alternatives.

Multi-Criteria Decision Analysis: Methods and Software, First Edition. Alessio Ishizaka and Philippe Nemery.

11.2 Decision Deck

The aim of Decision Deck (http://www.decision-deck.org/) is 'collaboratively developing Open Source software tools implementing MultiCriteria Decision Aid (MCDA) techniques which are meant to support complex decision aid processes'. Decision Deck is a collaborative project involving 18 researchers (from several European countries such as France, Spain, the Netherlands, Luxembourg and Belgium) and is led by an executive committee.

Decision Desktop, also called D2, which is a rich open source Java client, includes the following MCDA methods:

- the IRIS sorting method (Dias and Mousseau 2003);
- the RUBIS and VIP choice method (Bisdorff et al. 2008; Dias and Climaco, 2000);
- the UTA-GMS/GRIP ranking method (Greco et al. 2008; Figueira et al. 2009); see Chapter 4.

Furthermore, Decision Deck proposes a standardized XML schema to represent MCDA objects and data structures. This XML schema permits better interoperability between software packages. Based on this schema, the user has access to XMDA web services, enabling the distribution of the RUBIS Python solver and the Kappalab R-script (Grabisch et al. 2012).

The Decision Deck project is also active in the development of the so-called Diviz open source Java client and server (Veneziano et al. 2009) which is for designing, executing and sharing several MCDA methods by composition of web services, workflow management and deployment.

It must be stressed that the main feature of these software solutions is that they are interoperable in order to create a coherent ecosystem.

Decision Deck is targeted at practitioners, teachers and researchers who want to compare and test different methods. However, to use it effectively requires a strong technical and theoretical background.

11.3 DECERNS

The aim of DECERNS (which stands for Decision Evaluation for Complex Environmental Risk Network Systems) is 'the development of an integrated, user-intuitive software platform which can use diverse data sources including spatial and temporal data, value and judgment criteria and quantitative environmental models output, to provide a comprehensive risk management tool' (Grebenkov et al. 2007). The DECERNS project is thus orientated 'to the creation of web spatial decision support systems, including web-based and desktop software tools for multicriteria decision analysis on land use planning, risk management, and, in general, on a wide range of problems on (multicriteria) analysis of alternatives' (Yatsalo et al. 2010a).

The DECERNS platform, illustrated in this section, provides the user with all the functionalities and tools for data input, scoring, weighting and sensitivity analysis, etc. for several MCDA methods. However, it does not include all method-specific aspects, for example the Gaia map, preference matrix or uni-criterion flows of the PROMETHEE method. DECERNS, to the best of our knowledge, is the only platform that easily permits the use and comparison of different methods on the same data set. The user can define the weights for each method by choosing from various weight elicitation methods; for example, Ishizaka et al. (2013) have compared the use of three multi-criteria methods, namely the weighted sum, PROMETHEE and TOPSIS, for site selection for the construction of a casino in the greater London area.

DECERNS was an internationally collaborative project carried out by Obninsk State Technical University of Nuclear Power Engineering (Russia), Joint Institute for Power and Nuclear Research (Belarus), Brookhaven National Laboratory (New York) and Cambridge Environmental Inc. (Massachusetts). It was funded by the United States Department of Energy (DoE) Initiatives for Proliferation Prevention (IPP) programme, where its development was based on the previous web version of PRANA Decision Support System (Yatsalo 2007).

Now a family of DECERNS systems/software tools have been developed and are supported by the Decision Evaluation & Software Development company (DeE&Soft); see http://deesoft.ru/lang/en.

The DECERNS Spatial Decision Support System (SDSS) has been developed on the basis of Java technologies (Java EE5) as a distributed application constructed on the information–application–client layer architecture. DECERNS SDSS is thus a client–server application. The client tier that constitutes the user interaction point with the system is presented by a set of html pages, Java applets and JavaScript modules (Yatsalo et al. 2010a). This means that DECERNS SDSS is a web application (provided with an advanced security access control) relying on different components that can be (re)used in different applications as explained below. For applications and case studies, see Yatsalo et al. (2010a, 2010b, 2011b), Sullivan et al. (2009) and Gritsyuk et al. (2011).

The DECERNS platform contains three core modules:

- Geographical Information System (GIS) module;
- Multi-Criteria Decision Aid (MCDA) module;
- Group Decision Support System (GDSS) module.

These modules are fully integrated but can work independently and are briefly described in the following sections.

11.3.1 The GIS module

The GIS module provides 'map functionalities' that allow a user-friendly, two-dimensional visualization of spatial data and spatial data analysis (Yatsalo et al. 2010a):

- multi-layered map visualization and colouring;
- measurements (distance along polylines, polygon area);
- zooming and panning (dragging and dropping of the map) to navigate in the interactive map;
- feature selection (single and multi-selection of features) and searching;
- overlay operations such as intersection, union and subtraction;
- buffering (i.e. defining zones around specific map objects);
- vector format and rasterization (enables the conversion of vector formats to raster images);
- spatial interpolation and geostatistical tools;
- defining spatial alternatives and criteria, using GIS layers, attributive data, queries and selections.

The module is complemented with specific map data manipulation tools, used for digital layer uploading and map attribute editing. There is a ‘map set manager’ for the creation, editing and loading of map layers to a geo-database and a ‘map attributes editor’, which easily updates the attributes of a map.

Figure 11.1 is a screenshot of the GIS module where the user has chosen to display four layers: land use, railroads, main roads and rivers. The user has selected a specific area to investigate.

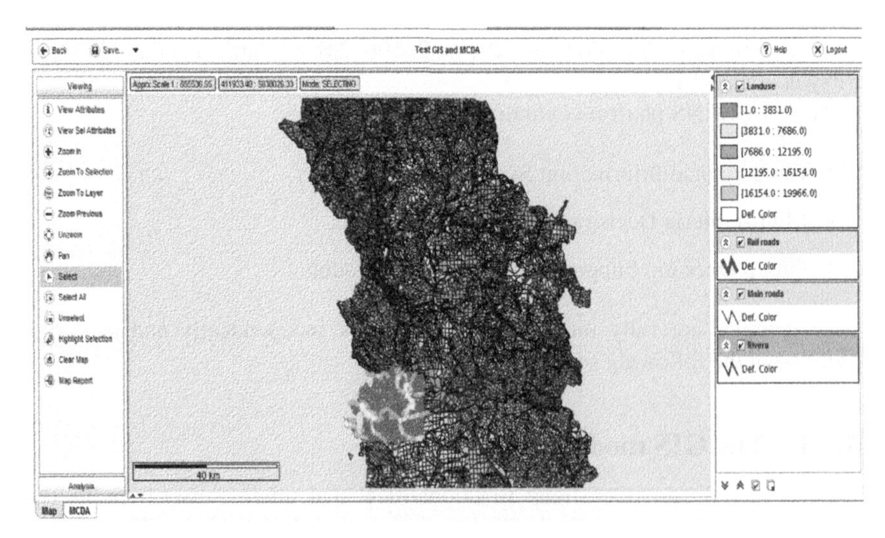

Figure 11.1 GIS module of DECERNS. Reproduced by permission of Boris Yatsalo.

11.3.2 The MCDA module

The MCDA module aims to structure a decision problem (i.e. define the criteria and alternatives), evaluate the alternatives and weight the criteria, thereby solving the problem. First, the decision maker has to choose from the following decision aid methods:

- ranking methods:
  - MAUT (see Chapter 4)
  - MAVT (multi-attribute value theory)
  - AHP (see Chapter 2)
  - PROMETHEE (see Chapter 6)
  - TOPSIS (see Chapter 8)
- sorting method:
  - FlowSort (see Chapter 6.5.2).

It is not necessary for the input data to be precisely defined. Probabilistic distributions and 'fuzzy' numbers can be used in the case of uncertainty. Different graphical and tabular tools are implemented to introduce probabilistic input/output data (i.e. density distributions). The weights and the evaluations can be roughly defined by means of distributions such as the normal, uniform, log-normal and delta distributions. This is illustrated in Figure 11.2. Different graphical and tabular tools are also

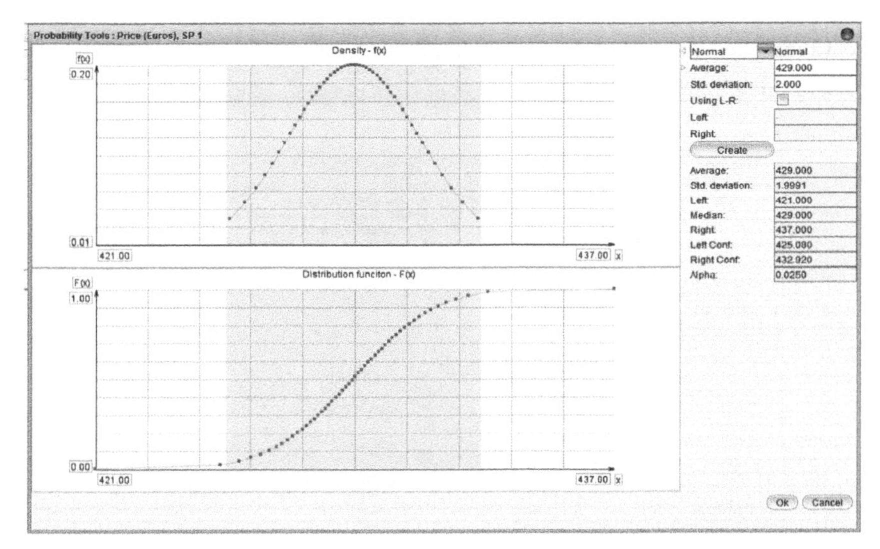

Figure 11.2 The probability menu in DECERNS. Reproduced by permission of Boris Yatsalo.

Criteria		
	Price	Customer Review
Name	Price	Customer Review
Description		
Scale	local \ - \ minimize \ vf: linear	local \ - \ maximize \ vf: linear
Weight		

Performance Table		
Alternatives / Criteria	Price	Customer Review
SP 1	429.000	4.000
SP 2	629.000	4.000

Figure 11.3 Performance table. Reproduced by permission of Boris Yatsalo.

available to present fuzzy input/output data. Fuzzy numbers, for example, triangular, trapezoidal, piecewise linear or singleton, can be used to represent the evaluations and weights. The following fuzzy and probabilistic multi-criteria methods have been implemented in DECERNS (Yatsalo et al. 2011a):

- PROMAA (Probabilistic Multi-criteria Acceptability Analysis);
- FMAA (Fuzzy Multi-criteria Acceptability Analysis);
- Fuzzy MAVT (Fuzzy Multi Attribute Value Utility).

In these methods, the user can enter the data via a performance table (Figure 11.3) or a value tree (Figure 11.4). Changing the value of the weight or preference functions performs a sensitivity analysis.

As an illustrative example, refer to the Case Study 4.1, where the choice of five smartphones is evaluated based on four criteria. In this section, we have chosen, by way of illustration, to regroup the screen size and storage size criteria into the 'technical parameters'subgroup.

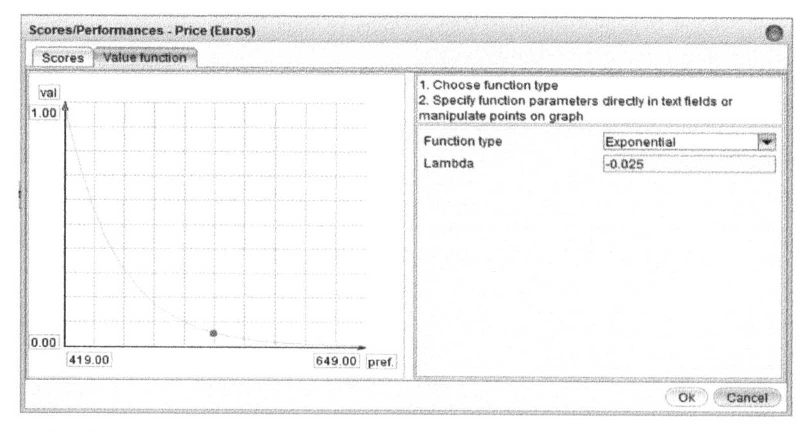

Figure 11.4 Determination of the value function. Reproduced by permission of Boris Yatsalo.

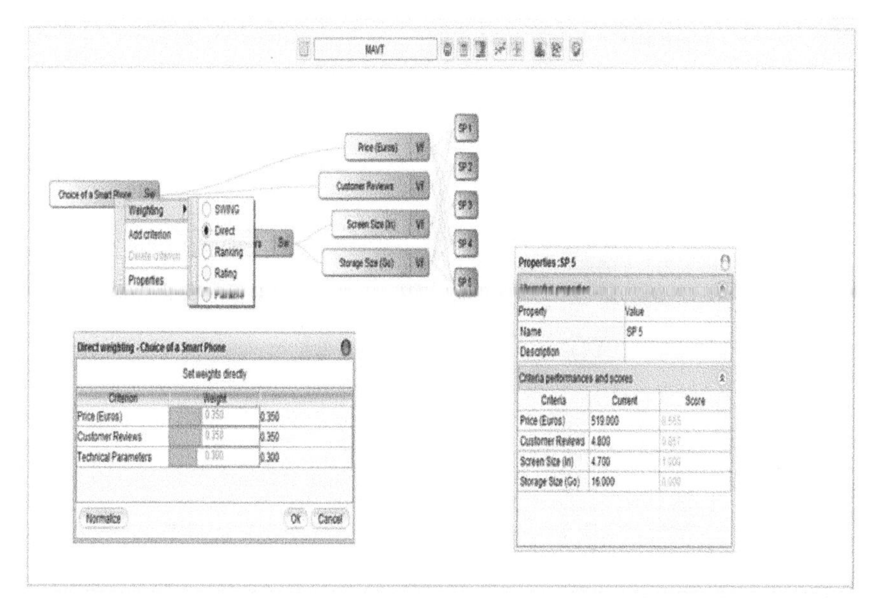

Figure 11.5 Tree structure of a decision problem in DECERNS, showing the Direct weighting dialogue as well as the performances of SP5. Reproduced by permission of Boris Yatsalo.

As illustrated in Figure 11.5, the tree structure gives a user-friendly representation of the decision problem. The task (choice of a smartphone), criteria (price, customer reviews, etc.) and alternatives (SP1, SP2, etc.) can be viewed in this tree.

The dialogue boxes have been added to allow the introduction of direct weights and performance of alternatives. There is another view of the decision problem, which displays the performance table as in Figure 11.3.

Figure 11.5 shows that the weight determination can be achieved in various ways. The user can choose from:

- direct weight determination;
- the SWING method;
- ranking of the weights;
- rating of the weights;
- pairwise comparison of the weights as with the AHP method (see Chapter 2).

The preference value functions of the criteria can be 'drawn' easily according to the decision maker's preference (see Figure 11.4). The user has a choice between piecewise linear functions and exponential functions. The final results are illustrated in Figure 11.6.

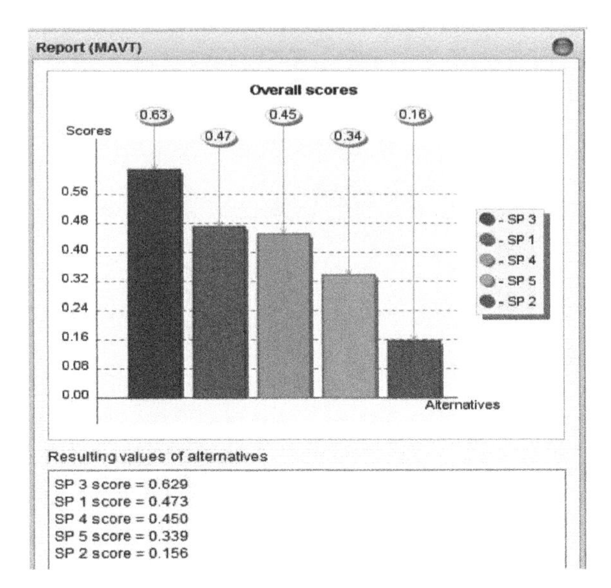

Figure 11.6 Results according to the MAUT method. Reproduced by permission of Boris Yatsalo.

The user can perform two different types of sensitivity analysis by:

- modifying the weights of the criteria or displaying the 'Line weights' representation which illustrates which weight value in a specifically chosen criterion changes the ranking (Figure 11.7);
- modifying the shape of the utility function of a criterion and analysing the corresponding change in the ranking (Figure 11.8).

The user can easily change the decision aid method, for example, when changing the decision aid in the TOPSIS method, the user needs to redefine the model-specific parameters, such as the preference functions in PROMETHEE. Figure 11.9 represents the results obtained with the TOPSIS method while defining identical weight values (the user needs to consider the meaning of the parameters for each method as they are often different).

Figure 11.10 shows the sensitivity analysis of the result obtained with the TOPSIS method when changing the weight values. This differs significantly from the analysis obtained with the MAUT method.

11.3.3 The GDSS module

The DECERNS project has a specific GDSS module which permits the creation and administration of online surveys. The results of these surveys are automatically collected and analysed. This feature is essential when alternatives have to be assessed by various decision makers.

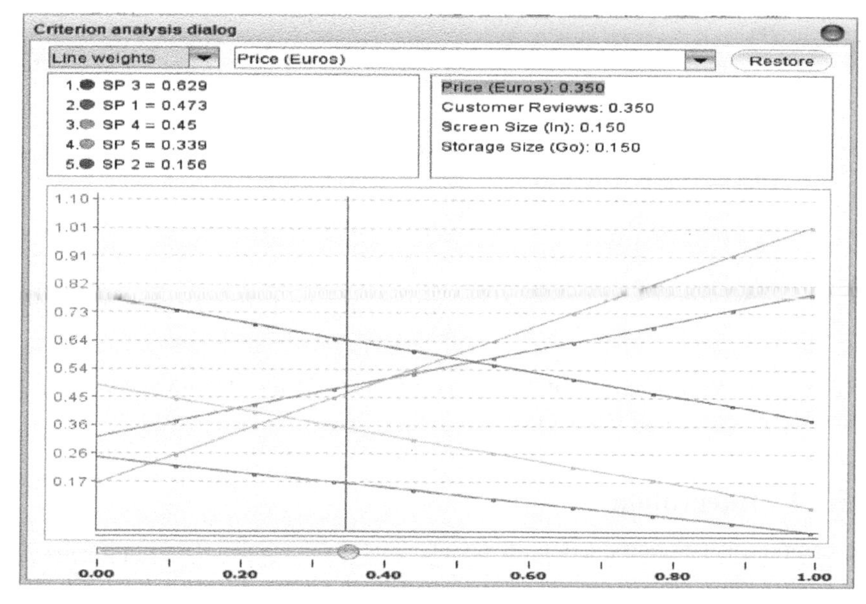

Figure 11.7 Criterion analysis window. Reproduced by permission of Boris Yatsalo.

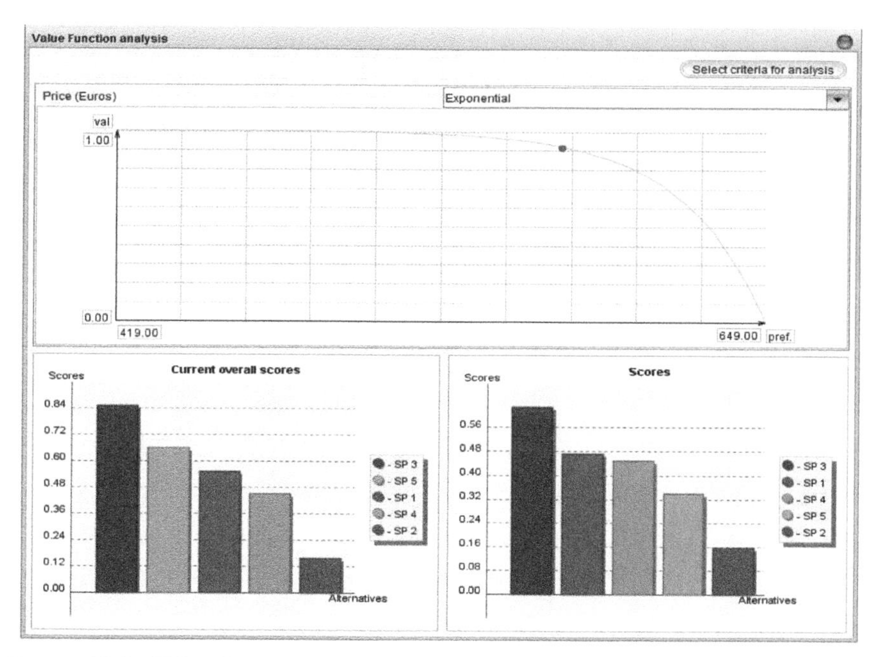

Figure 11.8 Value function analysis window. Reproduced by permission of Boris Yatsalo.

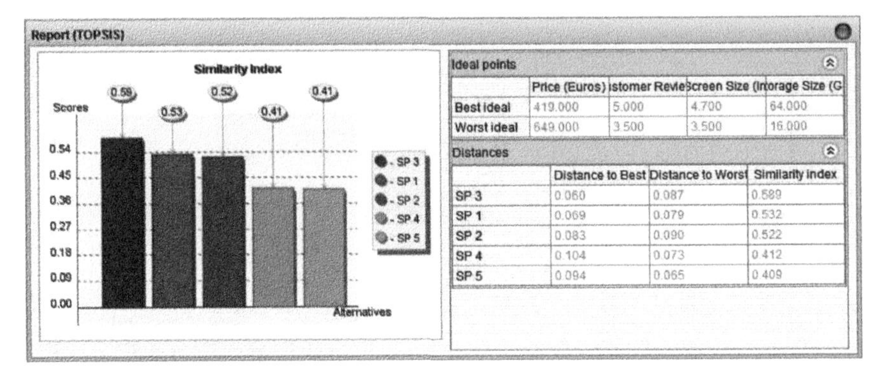

	Price (Euros)	istomer Revie	Screen Size (In	torage Size (G
Best ideal	419.000	5.000	4.700	64.000
Worst ideal	649.000	3.500	3.500	16.000

	Distance to Best	Distance to Worst	Similarity index
SP 3	0.060	0.087	0.589
SP 1	0.069	0.079	0.532
SP 2	0.083	0.090	0.522
SP 4	0.104	0.073	0.412
SP 5	0.094	0.065	0.409

Figure 11.9 Representation of the results obtained with the TOPSIS method. Reproduced by permission of Boris Yatsalo.

11.3.4 Integration

As mentioned in the introduction, all modules are fully integrated. This implies that the user can define areas in a map and define them as alternatives of the decision problem. The alternatives from a map can automatically be transferred into the performance table and vice versa. This integration is a strong advantage of the decision support system.

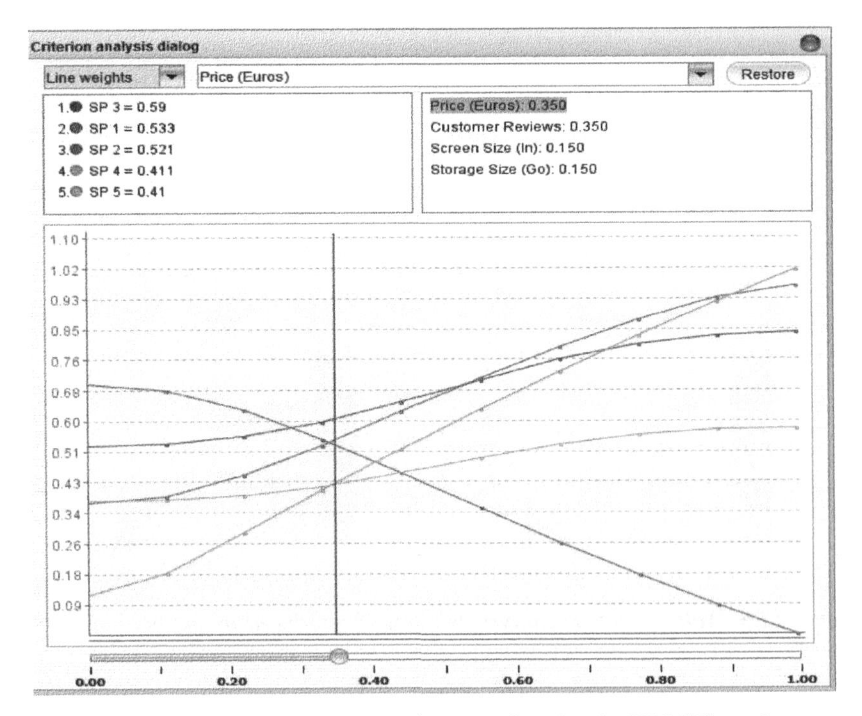

Figure 11.10 The criterion analysis window for the TOPSIS results.

References

Bisdorff, R., Meyer, P., and Roubens, M. (2008). Rubis: A bipolar-valued outranking method for the choice problem. *4OR*, *6*(2), 143–165.

Dias, L., and Climaco, J. (2000). Additive aggregation with variable interdependent parameters: the VIP analysis software. *Journal of the Operational Research Society*, *51*, 1070–1082.

Dias, L., and Mousseau, V. (2003). IRIS: A DSS for multiple criteria sorting problems. *Journal of Multi-Criteria Decision Analysis*, *12*(4–5), 285–298.

Figueira, J., Greco, S., and Slowinski, R. (2009). Building a set of additive value functions representing a reference preorder and intensities of preference: GRIP method. *European Journal of Operational Research*, *195*(2), 460–486.

Grabisch, M., Meyer, P., and Kojadinovic, I. (2012). *Kappalab R-Script*. Retrieved June 2012, from Kappalab: http://cran.r-project.org/web/packages/kappalab/kappalab.pdf

Grebenkov, A., Yatsalo, B., Sullivan, T., and Linkov, I. (2007). *DECERNS: Methodology and Software for Risk-Based Land Use Planning and Decision Support (in EnviroInfo)*. Aachen: Shaker Verlag.

Greco, S., Mousseau, V., and Slowinski, R. (2008). Ordinal regression revisted: Multiple criteria ranking using a set of additive value functions. *European Journal of Operational Research*, *191*, 416–436.

Gritsyuk, S., Yatsalo, B., Babutski, A., Mirzeabasov, O., and Didenko, V. (2011). Multicriteria decision analysis with the use of DECERNS DSS. *MCDM Conference 2011*.

Ishizaka, A., Nemery, P., and Lidouh, K. (2013). Location selection for the construction of a casino in the greater London region: A triple multi-criteria approach. *Tourism Management*, *34*(1), 211–220.

Sullivan, T., Yatsalo, B., Grebenkov, A., and Linkov, I. (2009). Decision Evaluation for Complex Risk Network Systems (Decerns) software tool. In A. Marcomini, G. W. Suter II, and A Critto (eds), *Decision Support Systems for Risk-Based Management of Contaminated Sites*. New York: Springer Science + Business Media.

Veneziano, T., Bigaret, S., and Meyer, P. (2009). Diviz: An MCDA components workflow execution engine. *Euro XXIII: 23td Euroepan Conference on Operational Research*. Bonn, Germany.

Yatsalo, B. (2007). Decision support system for risk based land management and rehabilitation of radioactively contaminated territories: PRANA approach. *International Journal of Emergency Management*, *4*(3), 504–523.

Yatsalo, B., Didenko, V., Tkachuk, A., and Gritsyuk, S. (2010a). Application to land use planning. *International Journal of Information Systems and Social Change*, *1*(1), 11–30.

Yatsalo, B., Gritsyuk, S., Didenko, V., Vasilevskaya, M., Mirzeabasov, O., and Babutski, A. (2010b). Land-use planning and risk management with the use of web-based multi-criteria spatial decision support system (DECERNS). *25th Mini Euro-Conference: Uncertainty and Robustness in Planning and Decision Making*. Coimbra, Portugal.

Yatsalo, B., Gritsyuk, S., Mirzeabasov, O., and Vasilvskaya, M. (2011a). Uncertainty treatment within multicriteria decision analysis with the use of acceptability concept. *Control of Big Systems*, *32*, 5–30.

Yatsalo, B., Sullivan, T., Didenko, V., and Linkov, I. (2011b). Environmental risk management for radiological accidents: Integrating risk assessment and decision analysis for remediation at different spatial scales. *Integrated Environmental Assessment and Management*, *7*(3), 393–395.

Appendix

Linear optimization

A.1 Problem modelling

Linear optimization (or linear programming) is a mathematical method for determining the value of decision variables in order to obtain the best outcome (e.g. the highest profit or lowest cost) under given constraints. We will illustrate the method with a transportation problem described in Case Study A.1.

Case study A.1

A company has a transportation problem. It needs to transfer two products, nails and screws, to a warehouse. The company owns a small van, which can transport a maximum of 5 tonnes in weight and 10 m^3 in volume of goods. The transport needs to take into account the following data: 1 tonne of nails has a volume of 1 m^3 and brings in a revenue of £200; 1 tonne of screws has a volume of 5 m^3 and brings in a revenue of £300. What quantity of each product should the company transport in order to maximize the benefit?

The solution to a linear programming problem is achieved in four steps:

1. *Identify the objective of the problem.*
   The objective of the problem is to either maximize (e.g. profit) or minimize (e.g. costs) a function. In Case Study A.1, the objective is to maximize the benefit.

2. *Indentify the decision variables*
   Decision variables are independent variables that are changed until the desired benefit is obtained (i.e. maximum or minimum). In Case Study A.1, the weight

Multi-Criteria Decision Analysis: Methods and Software, First Edition. Alessio Ishizaka and Philippe Nemery.

of the nails and screws should be adjusted to maximize the benefit under the given constraints.

3. *Identify the objective function*
   The objective function describes a linear relation between the decision variables and the objective of the problem. In Case Study A.1, we need to maximize the weight of nails (x) and screws (y) sold in order to maximize the profit:

$$\max 200x + 300y.$$

4. *Identify the constraints*
   The constraints define the limit of the decision variables. They are also linear. In Case Study A.1, the constraints are set by the capacity of the van:

$$\begin{array}{ll} x + y \leq 5 & \text{(constraint on weight)} \\ x + 5y \leq 10 & \text{(constraint on volume)} \\ x, y \geq 0 & \text{(non-negative constraint)} \end{array}$$

A.2 Graphical solution

If the problem only contains two decision variables, a solution can be found graphically. Each axis represents a decision variable and the straight lines of the constraints, obtained by replacing the inequality with equality, are sketched:

$$\begin{array}{ll} y = -x + 5 & \text{(constraint on weight)} \\ y = (-x + 10)/5 & \text{(constraint on volume)} \end{array}$$

These lines determine the feasible region, which is the collection of all the points that satisfy all constraints. The direction of the arrows (left or right of the lines) is decided by testing one point, generally the origin. If this point satisfies the constraint, then the arrow points in that direction. For example, if we introduce the origin (0,0) in the constraint weight ($0 \leq 5$) and volume ($0 \leq 10$), both constraints are satisfied.

Finally, the coordinates of each corner point should be substituted into the objective function to determine the optimal value because the solution is necessarily on one of the corner point. In Case Study A.1, the optimal value is 3.75 tonnes of nails and 1.25 tonnes of screws (Figure A.1).

For an analytic solution, where there are more than two decision variables, the simplex algorithm is used.

A.3 Solution with Microsoft Excel

In Figure A.2 the problem is modelled in *Microsoft Excel*. The first three lines are the given data. The variable parameters (Figure A.2) have to be entered in *Solver* (Figure A.3):

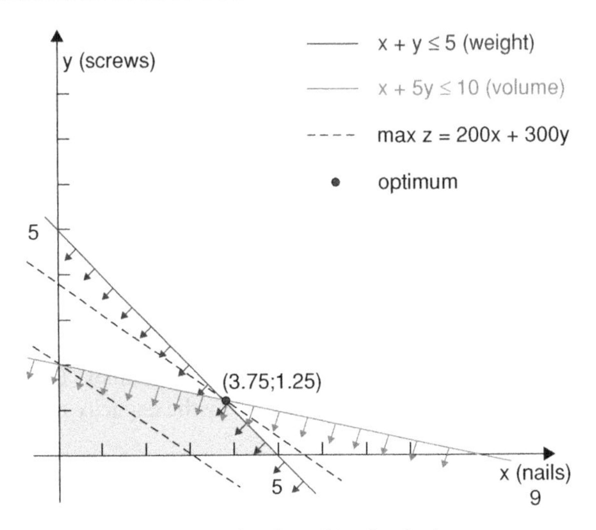

Figure A.1 Graphical solution.

- The objective of the problem to be maximized is given by the benefit in cell D12.
- The decision variables are set in cells B6 and B7.
- The constraint on weight is given by cell B12 (which must be less than or equal to B14).
- The constraint on volume is given by cell C12 (which must be less than or equal to C14).

The *Solver* changes the initial data in B6 and B7 until the maximum in D12 is obtained.

	A	B	C	D
1	**Transport of (x,y)**	Weight (t)	Volume (m^3)	Revenue (£)
2	Nails (x)	1	1	200
3	Screws (y)	1	5	300
4				
5		decision variables		
6	x = Weight nails	3.75		
7	y = Weight screws	1.25		
8				
9		Weight (t)	Volume (m^3)	Revenue (£)
10	x (nails)	3.75	3.75	750
11	y (screws)	1.25	6.25	375
12	Total :	5	10	1125
13				
14	Maximum authorised :	5	10	<= Constraints
15				
16				

Figure A.2 Modelling of the problem in Microsoft Excel.

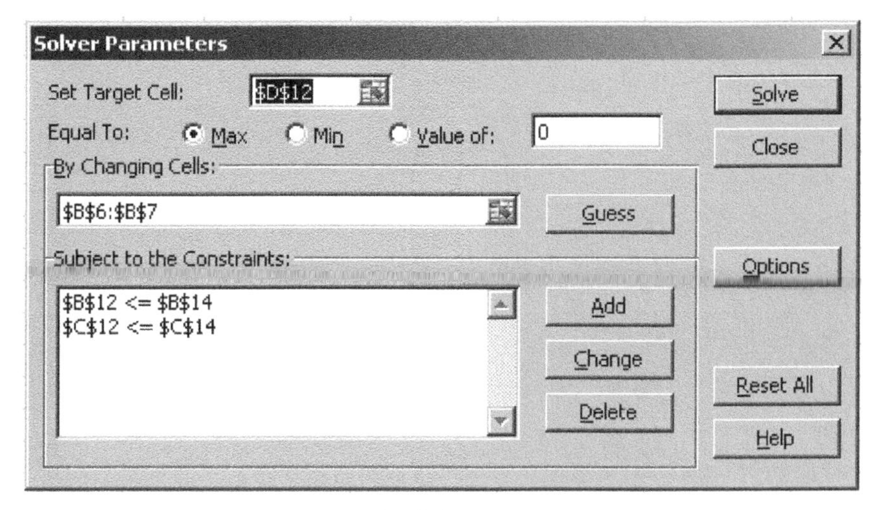

Figure A.3 Solver parameters.

Exercise A.1

Here you will learn to use the *Microsoft Excel Solver.*

Learning Outcomes

- Understand the modelling of a linear optimization problem
- Understand the configuration of *Microsoft Excel Solver*

Tasks

Open the file Transport.xls. It contains a spreadsheet with the modelling of the problem of the Case Study A.1.

Answer the following questions:

a) In the spreadsheet, find the objective of the problem, the decision variables and the constraints. (Read the comments in the red square in case of difficulty.)

b) Open the *Solver*. What is entered in the set target cell? What is entered in the 'By Changing Cells' box? What is entered in the 'Subject to Constraints' box?

Index

Multi-Criteria Decision Analysis: Methods and Software, First Edition. Alessio Ishizaka and Philippe Nemery.

Printed and bound by CPI Group (UK) Ltd, Croydon, CR0 4YY
12/01/2025
14624491-0002